CU01261386

Arbitration in Context Series

VOLUME 1

Series Editors:
Lucy Reed
Alexander Yanos

There are by now many excellent sources on international commercial arbitration, addressing procedures and practice in general and in relation to specific arbitral institutions and rules. There are also many excellent sources on the business sectors in which commercial disputes often go to international arbitration: to name only some, construction, sports, intellectual property, energy, insurance, securities.

The purpose of Kluwer Law International's sector-by-sector *Arbitration in Context Series* is to merge these two types of sources into stand-alone guides for the busy in-house lawyer and transactional practitioner facing the prospect of specialized arbitration. We are pleased to launch the first of this series with *International Construction Arbitration Law* by (our colleagues in London) Jane Jenkins and Simon Stebbings.

Lucy Reed and Alexander Yanos
Freshfields Bruckhaus Deringer
New York, April 2006

International Construction Arbitration Law

Jane Jenkins
and
Simon Stebbings

KLUWER LAW
INTERNATIONAL

A C.I.P. catalogue record for this book is available from the Library of Congress.

ISBN 90 411 23415

Published by:
Kluwer Law International
P.O. Box: 316
2400 AH, Alphen aan den Rijn
The Netherlands

Sold and distributed in North, Central and South America by:
Aspen Publishers Inc.
7201 McKinney Circle
Frederick, MD 21704
USA

Sold and distributed in all other countries by:
Turpin Distribution Services Ltd
Stratton Business Park
Pegasus Drive
Biggleswade
Bedfordshire SG18 8TQ
United Kingdom

© 2006 Kluwer Law International B T, The Netherlands

All rights reserved. No part of this publication may be reproduced, stored in a retrieval system, or transmitted in any form or by any means, mechanical, photocopying, recording or otherwise, without prior written permission of the publishers.

Permission to use this content must be obtained from the copyright owner. Please apply to: Permissions Department, Wolters Kluwer Law & Business, 111 Eight Avenue, 7th Floor, New York, NY10011-5201, United States of America. E-mail: permissions@kluwerlaw.com. Website: www.kluwerlaw.com

Table of Contents

Preface   1

Acknowledgement   5

Chapter 1
Introduction   7

I.   What is Special about International Construction Disputes?   7

II.   The Aims of this Book and its Scope   9

III   The Intended Audience   12

IV   A Practical Guide   13
       A.   FIDIC Conditions of Contract   13
       B.   ENAA Model Forms   15
       C.   ICE Standard Forms   15
            1.   The ICE Conditions of Contract   15
            2.   The New Engineering Contract   16
       D.   AIA Standard Forms   16
            1.   A191 DB-1996 Standard Form of Agreement Between Owner and Design / Builder   17
            2.   A201-1997 General Conditions of the Contract for Construction   17

Chapter 2
Key Features of Construction Contracts 19

I.		Introduction	19
II.		Key Players	19
III.		Key Documents in a Construction Contract	20
IV.		Forms of Contractual Structure	21
	A.	'Design-Bid-Build' or 'Build Only'	21
	B.	Design and Build or 'Turnkey'	23
	C.	Construction Management	24
	D.	Management Contracting	25
	E.	Partnering and Alliancing	26
V.		Pricing Methodologies	27
	A.	Cost Plus	27
	B.	Target Cost	27
	C.	Lump Sum	28
	D.	Provisional Lump Sum	28
	E.	Guaranteed Maximum Price	28
	F.	Unit Price or Measured Works	29
	G.	Mix and Match?	30
VI.		Payment Arrangements	31
	A.	Progress Payments	31
	B.	Milestone Payments	31
VII.		Administration of the Contract	32
	A.	Background	32
	B.	The Independent Engineer	32
	C.	Reality – The Employer's Representative	33
	D.	The Role Of Certificates	33
	E.	Challenging Certificates	34
VIII.		Variation Orders	35
	A.	Why are Variation Orders Required?	35
	B.	Controlling Variations and Their Costs	36
	C.	Variation Orders and Disputes	36

Table of Contents

IX.		Completion, Extensions of Time and Liquidated Damages	37
	A.	Role of Completion	37
	B.	Defects Liability Period	38
	C.	Date(s) for Completion	38
	D.	Completion Certificate(s)	38
	E.	Completion Tests	39
	F.	Liquidated Damages for Delay	39
	G.	Extensions of Time	40
	H.	The Role of the Programme	41
	I.	Critical Path Analysis	41
	J.	Concurrent Delays	42
	K.	Work-Around Measures	42
	L.	Acceleration / Constructive Acceleration	43
	M.	Extension of Time Claims and Disputes	43
	N.	Liquidated Damages for Performance Failures	43
X.		Liability	44
	A.	Limitations of Liability	44
	B.	Delay and Performance	45
	C.	Defects	45
	D.	Entitlement to Loss and Expense Due to Delay and Disruption	46
	E.	A Conclusive Final Certificate?	46
XI.		Project Security	47
	A.	Types of Security	47
	B.	Advance Payment Bonds	47
	C.	Retention Funds	48
	D.	Performance Bonds and Guarantees	48
	E.	Parent Company Guarantees	49
	F.	Payment Guarantees	49
	G.	Other Forms of Security	50
XII.		Project Financed Construction Projects	50
	A.	Introduction	50
	B.	The Lenders' Engineer	51
	C.	Lenders' Step-In Rights	51
	D.	Variation Orders	52
	E.	Liquidated Damages	52
	F.	Project Security	52

Chapter 3
Dispute Avoidance and Resolution 55

I. Introduction 55

II. Dispute Avoidance 56

III. Options for Tiered Dispute Resolution Procedures 58
    A. Mandatory Discussions 58
    B. ADR – Alternative Dispute Resolution 60
        1. Mediation and Conciliation 60
        2. Mini- Trial / Executive Tribunal 62
        3. Early Neutral Evaluation 62
    C. Factors to Consider in Drafting ADR Provisions 63
    D. Adjudication and The Use of Dispute Review Boards 64
    E. Advantages and Disadvantages of Adjudication 68
    F. Factors to Consider in Drafting Adjudication Provisions 70
    G. Statutory Adjudication 72
        1. The United Kingdom Experience 72
        2. Excluded Activities and Contracts 75
    H. The Final Tier: Litigation or Arbitration? 76
    I. Factors to Consider in Drafting Arbitration Provisions 81
        1. Institutional Versus Ad Hoc Arbitration 81
        2. The Seat of the Arbitration 83
        3. The Language of the Arbitration 83
        4. Number of Arbitrators 83
        5. Right to Appeal 83
        6. Continuing Performance 84
        7. Interim and Injunctive Relief 85
        8. Multiple Parties or Contracts 85
        9. Limitation Periods 86
        10. Drafter Beware 87

IV. BITs and MITs 88

V. Related Provisions 89
    A. Governing Law 90
    B. State Immunity 90
    C. Jurisdiction and Service Abroad 91

VI. Drafting Checklist 92

Chapter 4
Claims Administration   **93**

I.   Introduction   93

II.   Early Warning Provisions   94

III.   Requirements for Provision of Particulars   94

IV.   Conditions Precedent to Claims   95

V.   Rights to Access to Information   96

VI.   Audit Rights   97

VII.   'Keep Working' Provisions   98

VIII.   Provision for Communication/Reporting Lines   98

IX.   Partnering and Alliancing   99

X.   Effective Contract Management   102

Chapter 5
Adjudication and Dispute Review Boards   **105**

I.   Introduction   105

II.   Commencing the Reference   105

III.   The Powers of the Dispute Review Board   106

IV.   Do the Rules of Natural Justice Apply to Dispute Review Boards?   107

V.   Institutional Rules   109
       A.   World Bank Dispute Review Boards   109
       B.   FIDIC Dispute Adjudication Boards   111

	C.	American Arbitration Association (AAA) Dispute Review Boards	112
	D.	International Chamber of Commerce (ICC) Dispute Boards	113
VI.		Comparison of Rules and Procedures for Dispute Adjudication Boards/Dispute Review Boards	114
VII.		Finding the Best Approach	122
	A.	Preparation of Submissions	122
	B.	Disclosure of Documents	122
	C.	Hearings	122
VIII.		The Dispute Review Board's Decision	123
	A.	The Status of a DRB Decision	123
	B.	Enforcing a DRB Decision	124
		1. No Jurisdiction Over an Issue	124
		2. Lack of Fairness	125

Chapter 6
Forms of ADR   127

I.		Introduction	127
II.		Types of ADR	129
	A.	Mediation and Conciliation	131
	B.	Mini-Trial	136
	C.	Early Neutral Evaluation	137
III.		Which ADR Technique Should be Used?	138
IV.		ADR's Interface with Arbitration	139
	A.	ADR and Arbitration – A Question of Timing	139
	B.	Arbitrators as Mediators: Med-Arb	141

Chapter 7
Commencement of an Arbitration   143

I.	Introduction	143

II.	Selection of the Tribunal	144
	A. Number of Arbitrators	145
	B. Considerations when Selecting a Tribunal	146
	C. Methods of Appointment	148
	D. Duties of Arbitrators	149
III.	The Parties: Joinder/Consolidation	152
	A. Consolidation by Consent of the Parties	153
	B. Consolidation by Court Order	154
	C. Advantages and Disadvantages of Joinder and Consolidation	155
	D. No Joinder: Alternative Solutions	155
IV.	The Request for Arbitration	157
V.	Selection of the Parties' Representatives	158
VI.	Conclusion	159

Chapter 8
Control of the Arbitration — **161**

I.	Introduction	161
II.	Source of the Arbitral Tribunal's Powers	161
III.	General Principles	163
	A. Party Autonomy	163
	B. Restrictions on Party Autonomy	165
	C. Preliminary Steps	168
	D. Written Submissions	168
	E. Evidence	169
	F. Experts	173
	G. Interim Measures	174
	H. Hearings	175
	I. Post-Hearing Matters	176
IV.	The Importance of Effective Case Management	176
	A. Procedural Fairness	176
	B. The Arbitral Tribunal's Duty to Act Expeditiously	180

Chapter 9
Preparation and Collection of Evidence 183

I. Introduction 183

II. The Value of a Chronology and Other Aids 184

III. Document Management 186
    A. Document Review and Disclosure 186
        1. The Scope of the Disclosure 187
        2. Electronic Document Management 187
        3. Market Trends 189
        4. EDMS Software and Web-based Packages 189
    B. Managing Documents During the Hearing 190

IV. Scott Schedules 191

V. Evidence Required for Common Construction Claims 193
    A. Introduction 193
    B. Contractors' Claims for Delay or Disruption 194
        1. Claims for Overheads 195
            a. Site Overheads and Preliminaries 195
            b. Head Office Overheads 195
        2. Net Lost Profits 196
        3. Increased Costs in Executing Remaining Works Following Delay 197
    C. Disruption and Acceleration 197
    D. Contractors' Claims for Repudiation 198
        1. Repudiation Before the Date for Possession of the Site 198
        2. Repudiation Subsequent to the Date for Possession of the Site 199
    E. Global Claims 199
        1. Preparing a Global Claim 201
    F. Employers' Delay Claims 201
        1. Loss of Profits 201
        2. Wasted Expenditure 202
        3. Repudiation by the Contractor 202
        4. Defective Works 203

VI.		Expert Evidence	204
	A.	Introduction	204
	B.	The Role of Experts in Arbitration	205
		1. Appointing Experts	205
		a. Party-Appointed Experts	205
		b. Tribunal-Appointed Experts	205
		2. Tribunal and Party-Appointed Experts	206
		3. Selecting an Expert	207
		a. Finding Candidates	207
		b. Qualifications and Expertise	207
		c. Credibility	208
		d. Timing of Appointment	209
	C.	Briefing an Expert	209
		1. Briefing Party-Appointed Experts	209
		a. Scope	210
		b. What to Give the Expert	210
		c. Instructions	211
		2. Directing Tribunal-Appointed Experts	212
	D.	What Form Does Expert Evidence Take?	212
		1. Reports	212
		2. Oral Evidence	213

Chapter 10
Programme Analysis   215

I.	Introduction	215
II.	Date for Completion and the Role of the Programme	216
III.	Critical Path and Delay Analysis	222
	A. The Critical Path	223
	B. As-Planned vs. As-Built	224
	C. As-Planned, Impacted	225
	D. As-Built, But For	226
	E. Time Impact	227
	F. Choosing an Appropriate Method of Analysis	228
IV.	Concurrent Delay	230
V.	Ownership of Float	233

Chapter 11
Procedural Issues   **237**

I.   Introduction   237

II.   Objective of Procedural Rules   237

III.   Derivation of Procedural Rules   238

IV.   Contrasting Civil and Common Law Procedural Approaches   240
     A.   Introduction   240
     B.   Extent of Disclosure   241
     C.   Written Submissions   242
     D.   Preparation of Evidence   242

V.   Typical Procedural Directions – A Fusion Between Common and Civil Law   243
     A.   Exchange of Written Statements of Case   244
     B.   Disclosure of Documents   244
     C.   Exchange of Witness Statements   246
     D.   Exchange of Expert Reports   247
     E.   Exchange of Submissions   248
     F.   Hearings   248
     G.   Closing Submissions   248

VI.   Additional Issues   249
     A.   Preliminary Issues   249
     B.   Security for Costs   249
     C.   Interim Measures   250
     D.   Site Visits   255

VII.   Administrative Issues   256

Chapter 12
The Conduct of the Hearing   **257**

I.   Introduction   257

II.   Documents   259
     A.   Bundles   259

		B.	Authenticity of Documents	260
		C.	Presentation of Documentary Evidence	261
III.	Submissions			261
	A.		Who May Appear?	261
	B.		How Long Should Oral Submissions Last?	262
	C.		What Approach Should be Adopted?	263
	D.		Written Submissions	263
	E.		Visual Aids and Other ' Bells and Whistles'	264
IV.	Presentation of Witness Evidence			264
	A.		Witness Statements	264
	B.		Witness Preparation	265
	C.		Witness Examination	267
		1.	Order of Presentation	267
		2.	Oath or Affirmation of Truthful Evidence	268
		3.	The Process of Examination	269
			a. Direct Examination	269
			b. Cross-Examination	269
			c. The Role of the Tribunal	269
			d. The Scope of Examination	270
	D.		Witness Conferencing	270
	E.		Expert Witnesses	271
	F.		Video Conferencing	272
V.	Practicalities and Other Issues			272
	A.		Attendance	272
	B.		Scheduling of the Hearing	273
		1.	Duration	273
		2.	Cancellation Penalties	273
	C.		Logistics of the Hearing	274
		1.	Who Organizes the Hearing?	274
		2.	Venue for Hearing	274
		3.	Arrangement of Hearing Room	275
		4.	Stenographers or Transcribers	275
		5.	Translators	275
	D.		*Ex Parte* Hearings	276
		1.	What is an *Ex Parte* Hearing?	276
		2.	What is the Procedure in an *Ex Parte* Hearing?	276

Chapter 13
Effect of the Award 279

I.	Introduction		279
	A.	Types of Award	280
	B.	Res Judicata Effect of the Award	281
	C.	Notification of the Award	283
	D.	Options for the Losing Party	283
II.	Challenging the Award		284
	A.	The Meaning of 'Challenge'	284
	B.	The Purpose of a Challenge	284
	C.	Prior Exhaustion of Other Available Options	285
	D.	Challenging an Award in Court	286
		1. In Which Court Should an Application be Made?	287
		2. When Should an Application be Made?	287
	E.	The Grounds for Challenge Under the UNCITRAL Model Law	288
		1. Application for Setting Aside Under the UNCITRAL Model Law	288
		2. Other Grounds for Challenge	292
	F.	Remedies Available as a Result of a Successful Challenge	293
	G.	Conclusion	294
III.	Enforcement of the Award		295
	A.	Methods of Enforcing an Award	295
	B.	Recognition and Enforcement	296
	C.	Domestic and Foreign Awards	296
		1. Domestic Awards	297
		2. Foreign Awards	297
IV.	The New York Convention		298
	A.	Scope of Application of the New York Convention	298
	B.	Recognition and Enforcement	299
	C.	Formalities for Enforcement	300
	D.	Resisting Enforcement Under the New York Convention	300
	E.	The New York Convention Grounds for Refusal of Recognition and Enforcement	301
V.	Other Bilateral and Multilateral Conventions		306
	A.	Bilateral Conventions	306

	B.	Multilateral Conventions	307
		1. The 1965 Washington Convention	307
		2. The 1975 Panama Convention	307
		3. The Amman Convention	308
VI.	Conclusion		308

Chapter 14
Closing Thoughts — 309

I.	Introduction	309
II.	The Response to ADR	309
III.	Introduction of New Technology	311
IV.	Cultural Contributions	312

Annex 1
Advantages and Disadvantages of Arbitration and Litigation — 315

Annex 2
Sample Model Arbitration Clauses — 317

LCIA Rules	317
Rules of Arbitration of the International Chamber of Commerce	317
American Arbitration Association's International Dispute Resolution Procedures	318
American Arbitration Association's Construction Industry Arbitration Rules	318

Annex 3
Important Drafting Considerations for Dispute Resolution Clauses — 319

Preliminary Issues	319
Negotiation	319

ADR	319
Adjudication	321
Arbitration	321
Litigation	322
Other Considerations	322

Annex 4
World Bank Standard Bidding Documents for the Procurement of Works — 323

Clause 20 – Claims, Disputes And Arbitration	323
Appendix A – General Conditions of Dispute Board Agreement	328
Procedural Rules	334

Annex 5
American Arbitration Association Dispute Resolution Board Guide Specifications — 337

1.01	General	337
1.02	Membership	339
1.03	Operation	343
1.04	Review of Disputes	344
1.05	Alternative Dispute Resolution	347
1.06	Board Member Fees and Expenses	347
1.07	Administrative Assistance of AAA	348

Table of Contents   xix

Annex 6
ICC Dispute Board Rules, Dispute Board Clauses and Model Dispute Board Member Agreement   349

    ICC Dispute Board Rules   349

        Standard ICC Dispute Board Clauses   349
        ICC Dispute Review Board Followed By ICC Arbitration If Required   349
        ICC Dispute Adjudication Board Followed By ICC Arbitration If Required   350
        ICC Combined Dispute Board Followed By ICC Arbitration If Required   350

    Dispute Board Rules of the International Chamber of Commerce   351

    Model Dispute Board Member Agreement   367

Annex 7
Convention on the Recognition and Enforcement of Foreign Arbitral Awards 1958 (The New York Convention)   371

    Convention on the Recognition and Enforcement of Foreign Arbitral Awards   371

Annex 8
UNCITRAL Model Law on International Commercial Arbitration   377

    UNCITRAL Model Law on International Commercial Arbitration   377

    Explanatory Note by the UNCITRAL Secretariat on the Model Law on International Commercial Arbitration   390

Annex 9
IBA Rules on the Taking of Evidence in International Commercial Arbitration 405

List of Cases **415**

Index **421**

Preface

Despite its title, this book is not simply about Construction Arbitration. It covers a much wider field than this and does so with admirable clarity and assurance.

There are of course many different types of construction projects. They may be as relatively straight-forward as the building of an extension to a private house; or as complex as the construction of tunnels and viaducts to carry a busy motorway across a range of mountains. But the 'holy grail' of any construction project is the same: to complete the work to the required standard and to do so on time and on budget.

There are many ways of achieving, or at least of trying to achieve, this objective. There is first, what the authors describe as 'the most traditional model of construction contract for major projects' – namely 'design-bid-build' or 'build only', in which the employer has a direct contractual relationship with the design team, the main contractor and the sub-contractors. This puts the employer firmly in charge of the project, but carries with it the risk of delays between design and construction, together with the risk that if things go wrong – as they often do – it may be difficult to assign responsibility as between the different participants. An alternative model is that of the 'design and build' or 'turnkey' project. This was at one time highly popular, since it aimed to deliver a ready-to use-facility to the employer, with responsibility for completion and performance to specification placed firmly on the contractor. However, since such projects were – and are – generally put out to tender on the basis of a preliminary design only, there is plenty of scope for dispute if the detailed design differs, as the employer's requirements become more clearly defined.

Attempts to avoid such problems have led to other models of construction contracts, including management contracting, partnering and alliancing; and attempts to stop costs spiralling out of control have led to different pricing mechanisms, including 'cost plus', 'target cost', 'lump sum' and so forth. These different ways of proceeding are discussed clearly and helpfully in the early chapters of this book. 'Alliancing', for example, which it seems is becoming increasingly popular in many projects, aims to create a community of interest between employer and contractor, with joint teams and shared procedures specifically designed to avoid what the authors describe as 'the traditional claims-orientated adversarial approach adopted on traditional construction projects'.

The focus of the book then shifts to the effective management of construction contracts and the avoidance of disputes. Various methods of alternative dispute resolution are described. Here the emphasis is on moving away from conflict and trying to find an 'interest based' solution to any problems which arise. Such an approach may start (for instance) with express contractual provisions for senior management to meet and try to reach a settlement; and if this fails, for a solution to be found – or at least to be sought – by the use of other techniques, such as mediation, expert determination, adjudication by a single individual (or, more commonly, by a Panel of Experts or a Disputes Review Board) and so forth. In this way, different tiers of dispute resolution are built up, like the tiers of a wedding-cake; and only if these do not work, and divorce becomes more and more likely, does it become necessary to have recourse to a truly authoritative tribunal.

The choice at this stage becomes a choice between arbitration and litigation – a private tribunal, chosen by or on behalf of the parties or a court of law, chosen by the state. As between the two, it is arbitration which is the clear favourite, particularly in international construction contracts. The advantages and disadvantages of arbitration are well-rehearsed; but the key point, in an international context, is that the parties are assured of a hearing in a neutral place and before a neutral tribunal, working almost invariably in the language of the contract. A common complaint of clients compelled to litigate in a foreign court (and in a foreign language) is that they had no idea of what their Counsel was saying to the judge and whether it was right or wrong!

Having made the point that in major construction projects, 'parties typically select arbitration rather than litigation, as the final tier of dispute resolution', the authors give a general description of the arbitral process – choice of tribunal, place of arbitration and so forth. There are, of course, other books which provide information of this kind. What is particularly useful about this book is that it focusses specifically on issues which are relevant to construction arbitrations.

Such arbitrations are notoriously heavy on documentation, with the walls of the hearing room typically being lined with lever-arch files of site diaries, progress meetings, management reports, labour returns, correspondence, internal memoranda and so forth. The authors' discussion of document review and disclosure, and of the growing importance of electronic document management systems (EDMSs), is only one example of the practical help and guidance to be found in this book. Other examples would necessarily include the authors' discussion of claims for overheads and for delay and disruption; acceleration; critical path analysis; and such arcane and much disputed matters as ownership of the float! And finally mention should be made of the sections on programme analysis, on the effective presentation of evidence, on oral and written submisssions and other matters of practical significance.

This book should be of great assistance to those engaged in drawing up or advising upon construction contracts, and on any disputes that may arise or threaten to arise from such contracts. This is because the views which are expressed, and the guidance which is given, are based on the day-to-day knowledge and the hard-won experience of two highly skilled lawyers who know what they are talking about.

Alan Redfern, London, March 2006.

Acknowledgement

The authors would like to thank (in no particular order) each of Michelle MacPhee, Nicola Taylor, Gisele Stephens, Leilah Bruton, Nuala Shaw, Elizabeth Snodgrass, Sarah Keller, and Akin Akinbode, all London members of the Engineering, Procurement and Construction and International Arbitration groups at Freshfields Bruckhaus Deringer, Patrick Schroeder of the firm's Frankfurt office and Suzi Mills (now in pastures new) for their significant contributions in writing this book. Warmest thanks also go to Briana Young for project managing the delivery of the manuscript to Kluwer Law International and liaising with them to finalize the publication for print. Thanks also go to Sally Charin for arranging the copyright consents with great efficiency. The authors are also indebted to Alan Redfern for kindly reviewing the publication and preparing the preface. More relevantly, Alan must be credited with introducing both Simon and Jane to construction and engineering disputes and providing them with invaluable guidance and experience.

Thanks also go to Lucy Reed, the series editor, for her constant encouragement and enthusiasm, helpful comments and guidance on the content of the book. We do hope that it will provide a valuable and practical guide to in-house counsel and lawyers in private practice when approaching the management and resolution of disputes arising on complex construction projects.

As always the authors remain, alone, responsible for any and all shortcomings this work may have.

Jane Jenkins and Simon Stebbbings

Chapter 1
Introduction

I.   WHAT IS SPECIAL ABOUT INTERNATIONAL CONSTRUCTION DISPUTES?

Why is a book devoted to construction arbitration required? What differentiates construction contracts from any other commercial agreement? In essence, why are international construction disputes special? We suggest that a number of issues differentiate construction disputes from others and have given rise to particular practices being adopted to assist in the resolution of such disputes.

First, construction disputes are frequently technically complex, requiring fact intensive investigations which necessitate efficient management of the claims process. Guidance has been developed by arbitral institutions focused specifically on suitable procedures for construction disputes. For example, the ICC has published a Report on Construction Industry Arbitrations, recognizing that particular practices and procedures assist in the effective case management of such disputes. In the authors' view, it is helpful to focus right back at the stage of contract drafting on the means to facilitate the early resolution of disputes and efficient contract administration. In the event that the parties are unable to settle a dispute in advance of arbitration, efficient contract administration can only help in the presentation of evidence before the arbitral tribunal.

A second relevant feature is that construction disputes invariably necessitate a rapid, if temporary, decision in order to permit work to progress on the project. It is in the interests of all parties that completion should be achieved on time or at least at the earliest possible opportunity and recourse to a lengthy arbitration to decide questions necessary for completion of the works, such as the correct technical solution to overcome design problems, is clearly not practical. This has given rise to the increasingly important role of dispute review boards in resolving disputes at the first instance.

A further complicating factor is that construction projects invariably involve many parties. The method of procurement adopted may introduce a fragmentation of responsibilities, for example, for design and fabrication of particular equipment. Even where a turnkey arrangement is adopted with one contractor accepting single point responsibility for design and construction of the project, there will be subcontractors and suppliers involved in chain contracts below the turnkey contract. The methods for resolving disputes will ideally contemplate joinder or consolidation of related proceedings and the management of the interfaces between the contracts is an important role for a contract administrator. The use of project finance on international projects introduces an added layer of complexity, as the interests of lenders and other credit providers have to be taken into account. Lenders may, for example, have the opportunity to step into a project in the event of threatened termination of the principal project agreements. There will no doubt be reporting requirements to lenders and also insurers as to the status of disputes and they may be entitled to dictate or influence the manner in which claims are presented and pursued.

A related issue is the involvement on major international construction projects of parties from different jurisdictions with different cultural approaches, and indeed with different legal systems. They will come to a project with different expectations perhaps both as to the effect of their substantive rights under the contract and also as to the treatment of disputes as and when they arise. These are aspects that anyone dealing with disputes needs to be aware of, and that will be of particular relevance in any formal dispute resolution procedure where, for example, the contrasting approaches of civil and common law jurisdictions to procedural issues may give rise to difficulties if not effectively managed.

Another factor to bear in mind when approaching construction disputes is the influence of the widespread use of standard forms in the industry. Parties may choose to use a standard form, typically making amendments to tailor the form to the project in question, or they may use the standard form as the starting point for drafting a bespoke contract.

Some standard forms have been in use for many years and have been litigated from time to time. Certain provisions may therefore have a well-settled meaning under the governing law of the contract in question. For instance, English courts have ruled on the meaning of 'contemporary records' in clause 53 of the FIDIC Conditions of Contract for Works of Civil Engineering Construction (4th Edition),[1] on the employer's power of forfeiture under clause 63 of the FIDIC Conditions of Contract for Works of Civil Engineering Construction (2nd Edition),[2] on the liability of the employer's engineer to the contractor under clauses 60 and 67 of the FIDIC Conditions of Contract for Works of Civil Engineering Construction (2nd Edition),[3] on valuations for extra work under clause 52(1)(b) of the ICE Conditions of Contract

---

1 *Attorney-General for the Falkland Islands v Gordon Forbes Construction (Falklands) Ltd* (No. 2) [2003] BLR 280.
2 *Mvita Construction Co Ltd v Tanzania Harbours Authority* [1989] 46 BLR 19.
3 *Pacific Associates Inc and anor. v Baxter and Ors* [1988] 44 BLR 33.

(6th Edition),[4] and on the scope of unforeseen physical conditions falling within clause 12 of the ICE Conditions of Contract (5th Edition).[5]

In addition, arbitral decisions, although not binding precedent, may also provide a useful guide to interpretation. Summaries or extracts from ICC awards are published on a no-names basis in specialist journals including the *ICC International Court of Arbitration Bulletin, Yearbook of Commercial Arbitration, Journal du droit international* and *Revista de Derecho Internacional y del Mercosur*.[6]

Another feature differentiating construction contracts from other commercial agreements is the impact of sector-specific legislation, which addresses particular practices relevant to the construction industry. An example is the UK Housing Grants Construction and Regeneration Act 1996 (the UK Construction Act 1996) which outlaws pay-when-paid provisions, introduces a right to stop work for non-payment and a right for disputes under construction contracts to be referred to fast-track adjudication. Similar legislation has recently been adopted in Singapore – the Building and Construction Industry Security of Payment Act 2004.

The factors addressed above appear and reappear throughout this book, providing the rationale for practices adopted for effective dispute resolution or, perhaps, the reason disputes arise in the first place. They demonstrate why construction disputes justify, and indeed require, a particular treatment that differentiates them from other commercial and contractual disputes.

A hostile local environment may give rise to legal as well as technical challenges. The absence of a sophisticated legal system may undermine legal devices adopted readily in other jurisdictions, such as the means of providing security to lenders, presenting the need for novel techniques to be adopted. These may provide a fertile source of disputes when projects run into difficulties.

II. THE AIMS OF THIS BOOK AND ITS SCOPE

This book is designed to provide a practical guide to dispute resolution under construction contracts. The majority of its focus is on arbitration as a means of resolving construction disputes. As a starting point, however, Chapter 2 provides a basic introduction to the key features of a construction contract, with a view to setting the scene and identifying the issues which commonly give rise to disputes. Examples of standard form provisions are given by way of illustration.

In Chapter 3 the issue of dispute avoidance and dispute resolution is addressed, looking at the different techniques available for resolving disputes. In every project the client will want to see the work delivered on time and within budget. The client's

---

4   *Henry Boot Construction Ltd v Alstom Combined Cycles Ltd* [2000] BLR 247.
5   *Humber Oil Terminal Trustee Ltd v Harbour and General Works (Stevin) Ltd* [1991] 59 BLR 1.
6   Comprehensive annual lists of ICC awards published, the names of journals and various languages of publication are available at:
    http://www.iccwbo.org/court/english/awards/awards.asp (accessed 22 July 2005).

financiers, whether lending to the client or to the project (in the case of public private partnerships and project-financed projects) have a similar interest, often with the additional concern of ensuring that the project generates the anticipated revenues within the period over which the finance is to be repaid. Contractors, particularly where they are involved only as contractors and not also as equity participants (as is commonly the case in project-financed projects), are looking to maximize cash flow and the profit to be earned in the construction of the project. All have a keen interest in seeing disputes resolved swiftly so as not to disrupt or delay the construction process, to improve cash flow and maximize returns.

However, it may be that swift solutions to disputes are at the expense of a comprehensive and rigorous review of all relevant circumstances and the consequences flowing from the events in question. This lack of rigour may be acceptable for low value disputes but will not usually be suitable for larger and more important disputes. It is common, therefore, to see a tiered approach to dispute resolution, with the parties agreeing to a series of fast-track procedures which are either non-binding or which have only interim effect. Arbitration or litigation is then the last step in the process leading to a final and binding decision if the parties fail to achieve an acceptable resolution to their problem.

Increasingly, the approach taken by standard form contracts in use internationally is to incorporate multi-tiered dispute resolution provisions providing for intermediate stages with arbitration as the final tier.

Negotiations at a senior level may be expressly provided for as a condition precedent to service of any formal notice of dispute. Thereafter dispute review boards, being essentially a panel of experts, have taken the place in many cases of the engineer in resolving disputes at first instance. Typically, the decision of a dispute review board is binding unless and until challenged in arbitration or, rarely in international projects, litigation. Multi-tiered dispute resolution provisions are considered in Chapter 3, along with the advantages and disadvantages of arbitration and litigation (with the balance falling fairly decisively in favour of arbitration as the final tier of dispute resolution on international projects).

Chapter 3 also touches on the increasingly important role played by investment treaties in international construction projects. Often overlooked, such treaties (be they multilateral or bilateral) provide parties with recourse to arbitration against a host nation state absent a contractual arbitration agreement. They therefore provide a valuable means of investor protection and are a useful adjunct to contractual remedies.

In the authors' view, it seemed important also to cover the effective administration of construction contracts and claims management. Tools for effective contract management and claims administration are therefore addressed in Chapter 4, including suggested provisions to include in contracts in order to facilitate claims handling. This chapter also considers partnering and alliancing techniques, which are proving increasingly popular as a means of establishing joint working between clients, contractors and suppliers with a view to solving disputes at the project level without the need for recourse to formal dispute resolution.

Chapter 5 focuses on proceedings before dispute review boards or expert panels. As noted above, these are now commonly seen in standard form construction contracts as the first tier of binding dispute resolution techniques, being binding only on an interim basis unless and until revised by arbitration. Typical procedures before dispute review boards are considered and the relevant procedural rules compared. The chapter also comments on how to approach preparation of references tactically.

The decisions of dispute review boards or expert panels will not be enforceable in the same way as an arbitral award. Instead, they are binding only on the basis of the parties' contractual agreement. Accordingly, in the event of default in compliance by a losing party a claim for breach of contract would have to be pursued – a more cumbersome and costly process than enforcement of arbitral awards. In the context of attempts to enforce the decisions of dispute review boards, we look at grounds for attack of such decisions, an issue of great significance in the event that the parties have agreed in advance that certain decisions shall be final.

Chapter 6 considers mediation, conciliation and other forms of alternative dispute resolution. Such techniques for alternative dispute resolution are becoming increasingly popular, certainly in common law jurisdictions. Cost orders may be made by courts in the United States or the United Kingdom for example, against parties who have refused to participate in such proceedings. The impetus behind this development is to avoid the costs and delay associated with the hearing of a complex construction dispute. Perhaps it is because procedures in civil law systems are more streamlined that mediation is not seen so often in such jurisdictions. In international projects, however, in the authors' experience, it is increasingly being used and with reasonable success.

Chapter 7 then turns to arbitration and the steps for commencement of arbitration proceedings. First, we look at the selection of the tribunal and the procedure for agreeing on a sole arbitrator or the appointment of two nominated arbitrators and a third to act as chairman. Where institutional arbitration has been selected by the parties, Chapter 7 looks at the role that the institutions play in the appointment of the tribunal. It also considers the duties of arbitrators and grounds for challenge on the basis of impartiality or conflict of interest. The need for joinder of related disputes (involving third parties) is considered, together with consolidation (where the same parties are involved in related disputes which may potentially be heard before different tribunals). The aim, obviously, is to try to avoid multiplicity of proceedings and to bring the same issues before the same tribunal to avoid the risk of increased costs and inconsistent decisions. While the domestic courts in most sophisticated legal systems will have powers to order joinder of third parties to proceedings and consolidation of existing proceedings, this is not so in arbitration. As arbitration is a consensual process based on the agreement of the parties, any such arrangements need to be by consent, either by including appropriate provisions in the contract from the outset or by achieving consent as and when disputes arise (which may be more difficult if joinder is perceived not to be in the interests of one or more of the relevant parties at such time).

Chapter 7 then addresses the form of a typical request for arbitration. It considers the basic information to be included in such a document having regard, where relevant, to the rules of the widely used institutions. The chapter then considers the early selection of the parties' lawyers.

In Chapter 8 we consider the importance of effective case management in the arbitration, addressing the powers of the tribunal to control the proceedings in default of, or sometimes as an adjunct to, the agreement of the parties. Reference is made to the UNCITRAL, ICC, LCIA and AAA arbitration rules and also to the UNCITRAL Model Law. Examples are also given from selected jurisdictions in relation to the role of the domestic courts in supporting the arbitral process in default of agreement by the parties.

Chapters 9 and 10 look at the preparation and collection of evidence. Chapter 9 provides a general summary, looking at the value of a chronology, document management, technical evidence, the use of so-called 'Scott schedules' (which essentially set out all relevant particulars of a claim in tabular form) and the evidence required for common construction claims, including possible approaches to the quantification of claims. Chapter 10 then focuses specifically on programme analysis, looking at the role of the programme, critical path analysis, questions of concurrent delay and the issue of ownership of 'float' in the programme.

Procedural issues are then addressed in Chapter 11. A topic of great interest to arbitrators over recent years is the 'internationalization' of arbitration: the coming together of the most useful elements of civil and common law procedural approaches. These issues are addressed in the context of the preparation of evidence, cross examination of witnesses, expert reports and the extent of document disclosure. Typical procedural directions are then addressed, with suggestions as to additional issues that may be relevant, including identification of any preliminary issues suitable for early determination, security for costs, interim measures and site visits.

In Chapter 12 we consider the conduct of the hearing, including the likely timing, the division of issues, the presentation of submissions and the taking of evidence, as well as essential logistical considerations.

Chapter 13 addresses the effect of the award, potential grounds for challenge and enforcement of the award.

Finally we sum up and present some conclusions, with comments on possible future developments in construction contracting and dispute resolution.

III   THE INTENDED AUDIENCE

It will perhaps be apparent from the above introduction that this work is designed to provide guidance to those unfamiliar or inexperienced in the workings of construction contracts and means of dispute resolution. It is aimed at in-house lawyers who may have many years experience as commercial contract negotiators but have not had to live through a construction dispute nor manage a construction contract during the life of a project. We hope that lawyers in private practice embarking on a construction dispute for the first time will also find this book of value.

IV   A PRACTICAL GUIDE

The aim in drafting this book is to provide a practical guide. Where relevant, checklists and executive summaries of key points are given. Annexes provide relevant supporting material.

This work is intended for an international audience. Its authors are both English trained lawyers but who work on international construction projects. Increasingly in our experience techniques for dispute resolution are becoming 'internationalized' although, of course, parties must always beware the idiosyncrasies of the governing law and local law. Local law, for example, may restrict the parties' freedom to agree dispute resolution procedures of their choice. It may instead impose compulsory adjudication as the first tier of dispute resolution for certain types of construction contracts. It comes as a surprise to many international participants in construction projects taking place in England, Wales and Scotland, that there is legislation imposing mandatory fast-track adjudication for disputes arising under construction contracts (see further Chapter 5). This work does not and cannot attempt to address all such regimes that may apply on international projects: examples of English law procedures and rules are provided as examples only, to flag up the sometimes surprising results of choice of governing law, seat of arbitration or location of the project in question.

Standard form contracts used on international construction contracts are cited, again to provide illustrations and examples. Those referred to most frequently are the standard forms published by the Fédération Internationale des Ingénieurs-Conseils (FIDIC), the Engineering Advancement Association of Japan (ENAA), the Institution of Civil Engineers ICE and the American Institute of Architects (AIA). Each of these is briefly addressed below.

A.   FIDIC CONDITIONS OF CONTRACT

FIDIC was created in 1913 by Belgium, France and Switzerland and represents national associations of consulting engineers. The United Kingdom became in a member in 1949 and the United States followed in 1958. FIDIC's membership is currently drawn from 68 countries across five continents. In addition, various organizations representing non-engineering/construction professionals, such as lawyers and insurers, who nevertheless have an interest in the field, have affiliated member status. FIDIC maintains its secretariat in Geneva, Switzerland.

FIDIC published its first standard form in 1957, the conditions of contract for works of civil engineering construction (old Red Book). The conditions of contract for electrical and mechanical works (old Yellow Book) followed in 1963. In 1995, the conditions of contract for design-build and turnkey (Orange Book) was released.

From the start, FIDIC's standard forms were intended for use on projects with an international dimension. For example, the old Red Book was based on United Kingdom standard conditions of contract and was for use on the various development projects being undertaken by British and other engineers in the Commonwealth as well as other jurisdictions.[7]

In the late 1990s, FIDIC carried out a major overhaul of all of its standard forms and released a suite of four contracts in 1999 to replace the previous contracts. The new suite of contracts consisted of three standard forms for major works and one for minor works:

– Conditions of Contract for Building and Engineering Works Designed by the Employer (Red Book);
– Conditions of Contract for Plant and Design-Build for Electrical and Mechanical Plant, and for Building and Engineering Works, Designed by the Contractor (Yellow Book);
– Conditions of Contract for EPC/Turnkey Projects (Silver Book); and
– Short Form of Contract (Green Book)

The Red Book replaces the old Red Book and is for use on all types of projects where the main responsibility for design lies with the employer. Today, FIDIC's Red Book, albeit an amended version,[8] remains the first choice of all major international development banks and agencies for use as general conditions of contract on a construction project financed by them. These institutions include the Asian Development Bank (ADB), African Development Bank (AfDB), Caribbean Development Bank (CDB), Commission of the European Communities (CEC), European Bank for Reconstruction and Development (EBRD), European Investment Bank (EIB), Inter-American Development Bank (IDB), International Bank for Reconstruction and Development (IBRD) and the United Nations Development Programme (UNDP).

The Yellow Book is based on the old Yellow Book and on the Orange Book and is intended to replace both.

The Silver Book is an entirely new standard form added to the FIDIC contract stable and is meant to cater for the growing trend in the international market for projects where the contractor bears more risk (at extra cost to the employer) and a more definite final price and completion time is agreed from the outset.

The Green Book is another brand new standard form from FIDIC and is designed for use on non-complex international projects such as rural roads, water

---

7 See Chris Wade, *An Overview of FIDIC Contracts*, notes from presentation at an ICC-FIDIC conference in Paris (April 2004).
8 For example see the World Bank's Standard Bidding Document for Procurement of Works (May 2005) available at: http://siteresources.worldbank.org/INTPROCUREMENT/Resources/Works-5-05-ev1.pdf (accessed 27 July 2005).

supply, sewage disposal, and electricity transmission in developing countries. The Green Book can be used irrespective of whether the employer or contractor is carrying out design or whether payment is on a lump sum or other basis.[9]

B.   ENAA MODEL FORMS

ENAA is a non-profit making organization, established in August 1978, supported by the Ministry of International Trade and Industry (MITI) of Japan and various other national and local government agencies, universities and research organizations. The Association has the support of over 1,500 specialists from member companies, as well as various experts in their respective fields.

In 1986, the ENAA developed a turnkey style contract designed specifically for process plant construction called the Model Form International Contract for Process Plant Construction. The model form was subsequently revised and a second edition was published in 1992 (ENAA Process Plant Model Form 1992[10]). The form consists of five volumes: Agreement and General Conditions (Volume 1), Samples of Appendices (Volume 2), Guide Notes (Volume 3), Work Procedures (Volume 4) and Alternative Form Without Process Licence (Volume 5).

In 1996 the ENAA published the Model Form International Contract for Power Plant Construction (ENAA Power Plant Model Form 1996), a turnkey contract for the construction of power plants. The 1996 model form consists of three volumes: Agreement and General Conditions (Volume 1), Samples of Appendices (Volume 2) and Guide Notes (Volume 3).

C.   ICE STANDARD FORMS

The ICE was established in 1818 and is based in the United Kingdom. The ICE draws its membership from over 70,000 professionally qualified civil engineers in the United Kingdom, China, Russia, India and 140 other countries. The ICE publishes two families of standard forms:

– the ICE Conditions of Contract; and
– the New Engineering Contract (NEC)

1.   The ICE Conditions of Contract

The ICE Conditions of Contract are standard forms drafted by the Conditions of Contract Standing Joint Committee (CCSJC) for use on civil engineering works.

---

    9   See Wade, *Overview of FIDIC Contracts.*
   10   Engineering Advancement Association of Japan, *Model Form International Contract for Process Plant Construction*, 2nd edition 5 vols. (ENAA, 1992) vol. 3, p. 1.

The CCSJC comprises representatives from the ICE, the Civil Engineering Contractors Association (CECA) and the Association of Consulting Engineers (ACE). The suite consists of:

- Measurement Version 7th Edition;
- Design & Construct 2nd Edition;
- Term Version 1st Edition;
- Minor Works 3rd Edition;
- Partnering Addendum;
- Tendering for Civil Engineering Contracts;
- Agreement for Consultancy Work in Respect of Domestic or Small Works.

The ICE Design & Construction Conditions of Contract (ICE D&C) is the most relevant to international construction projects.

2.   The New Engineering Contract

The New Engineering Contract (NEC) was first published by the ICE in 1993. A second edition was published in 1995 and renamed the standard form as the Engineering and Construction Contract (ECC). The third edition of the NEC (NEC3) was published in June 2005. The term 'NEC' is now used to refer to the whole suite of contracts which includes the ECC as well as a subcontract, a professional services contract, an adjudicator's contract, a short contract, a short subcontract and the NEC partnering option. The ECC is the relevant contract for international construction projects. The ECC is designed for use in:

- engineering and construction work whether civil, electrical, mechanical or building work;
- projects whether the contractor has full, partial or no design responsibility; and
- United Kingdom and international projects.

The ECC also allows a choice of six contract strategies: Option A – Priced contract with activity schedule; Option – B Priced contract with bill of quantities; Option C – Target contract with activity schedule; Option D – Target contract with bill of quantities; Option E – Cost reimbursable contract; and Option F – Management contract.

D.   AIA STANDARD FORMS

The AIA is a national association of architects founded in 1857. The AIA has over 115 years of experience creating standard forms, publishing its first uniform contract for use between an owner and a contractor in 1888. In 1911, the AIA published its first standard general conditions for construction, the precursor of the A201-1997 General Conditions of the Contract for Construction discussed in more detail below. The AIA contracts are now a standard in the construction industry of the United States. The A191 DB-1996 Standard Form of Agreement Between Owner and Design/Builder and A201-1997 General Conditions of the Contract for Construction are the two main standard forms used on construction projects.

1.   A191 DB-1996 Standard Form of Agreement Between Owner and Design/Builder

The A191 DB-1996 Standard Form of Agreement Between Owner and Design/Builder (A191) is intended for use where the employer contracts directly with one entity for both design and construction services. It is a two-part agreement, evolving the project and the design through two separate contracts, one for preliminary design and budgeting and one for final design and construction. The first is used to map out the design and pricing of the project, while the second provides construction arrangements where the employer decides to move forward with the project. The A191 will be retired on 31 August 2006 and replaced by A141-2004 Agreement Between Owner and Design-Builder (A141). The A141 consists of the agreement and three exhibits, Exhibit A (Terms and Conditions), Exhibit B (Determination of the Cost of the Work) and Exhibit C (Insurance and Bonds).

2.   A201-1997 General Conditions of the Contract for Construction

The A201 1997 General Conditions of the Contract for Construction (A201) is for use where design responsibility belongs to the architect, as distinct from the contractor. The A201 sets out the rights and obligations of the employer, contractor and architect. Though the architect is not a party to the construction contract, he nevertheless participates in the preparation of the contract documents and performs certain construction phase duties and responsibilities described in detail in the A201. Supplementary conditions are usually added to amend or supplement portions of the A201 as required by the exigencies of a particular project.

Chapter 2
Key Features of Construction Contracts

I.   INTRODUCTION

The aim of this chapter is to provide an introduction to the shape and structure of a construction contract, with particular attention paid to those features that commonly give rise to disputes. It focuses on the key players and key provisions that construction contracts usually contain. It also considers briefly the role of ancillary documents such as bonds and guarantees which, if they do not play a central role, are often an important part of the background facts or motivation for a dispute.

II.   KEY PLAYERS

The obvious starting point is the client, who is sometimes referred to in construction contracts as the 'employer' (or the 'principal'). The employer may be, for example, a developer, a government or state entity, an investor or an end-user.

Next there is the 'contractor' who builds and possibly designs, or manages, the construction process. These various roles are discussed later in this chapter.

The main or prime contractor typically subcontracts some or all of its work to 'subcontractors' (either chosen by the contractor or nominated by the employer). It may also enter into contracts with 'suppliers' for the supply of equipment or materials.

There will also be a design team, the membership of which will depend upon the type of project.

III. KEY DOCUMENTS IN A CONSTRUCTION CONTRACT

Construction contracts are often voluminous and consist of many documents of different origin, authorship and development. The standard documents forming a construction contract are:

- *Articles of agreement* – These are usually only a few pages long and set out the names of the parties with a brief description of the project. The articles incorporate the conditions of contract upon which the parties are to carry out the project. Sometimes the articles of agreement and the conditions of contract may form a single document. It is the articles of agreement that are executed by the parties.
- *Conditions of contract* – These contain the bulk of the substantive clauses of the contract. Their content depends upon the particular form of contractual structure chosen for the project. These forms are addressed in the next section of this chapter.
- *Specification* – The specification sets out the project requirements. Specifications fall into two main categories, although a single document may contain elements of both. A technical specification sets out the actual technical criteria of the project (methods of construction etc). This type of specification often accompanies the contract drawings. It can usually be found in a 'build only' contract[1] where the design of the project is largely complete before the contract is entered into. The other form of specification is a performance specification, which usually forms part of a 'turnkey' contract[2] as it sets out the standard of outputs that the constructed project must achieve.
- *Drawings* – In an ideal world the drawings incorporated in a 'build only' contract should be the construction stage or 'as for construction' drawings which means that the contractor can build the project straight from the drawings. Quite often, however, the contract drawings even in a build only contract are not at such an advanced stage and therefore subsequent revisions are issued. Each time a drawing is issued there is usually something to indicate that it is a revision and the relevant amendment is noted either in a box on the drawing or on the drawing itself (for example, by a circle around the affected area). In turnkey and construction management[3] contracts, any contract drawings will show only an outline design as the detailed design is carried out pursuant to the contract.
- *Bill of Quantities* – This document is more common in small projects, however it may also appear in large international projects, for example, as a tool to value variations.[4] It identifies the quantity of each item of material that is

---

1 As to which see further below.
2 As to which see further below.
3 See further below.
 As to which see further below.

required in order to construct the project. Sometimes a contract may refer to an approximate bill of quantities if the design is far from finalized at the time the contract is entered into. A bill of quantities usually contains a section called the preliminaries where the site conditions and restrictions under which the works are to be carried out will be set out. These may not appear anywhere else in the contract documentation.

Given the different origin, authorship and development of these documents, it is not uncommon to find that a construction contract is internally inconsistent (or ambiguous) and, in the absence of a well-drafted precedence clause, disputes frequently involve resolution of competing clauses. For example, a common problem is where a technical specification prepared early in the life of a project by a consultant engineer contains substantive clauses that contradict the conditions of contract prepared much later, based on a standard form or by a lawyer who is never shown the technical specification. Another is where the specification envisages procedures for design development that are inconsistent with the conditions of contract.

IV.   FORMS OF CONTRACTUAL STRUCTURE

Construction contracts come in various different forms depending on the particular needs of the project. The most commonly encountered forms on major projects are:

A.   'DESIGN-BID-BUILD' OR 'BUILD ONLY'

This is the most traditional model of construction contract. The design stage involves the employer appointing its own design team to design the proposed project. Usually, when the outline design is completed, this is followed by tendering for construction of the works only, by a single contractor or in work 'packages' by a number of contractors. Finalization of the detailed design for construction can be carried out in tandem with the negotiation of the construction contracts, so allowing for resolution of any conflicting design and construction issues. There is significant flexibility in the timing of the letting of the construction packages. For example, civil works may be let at an early stage of the project with a series of M&E packages divided into specialist activities, such as ventilation, power supply and communications being let at a later stage. Alternatively, specialty work packages, such as signalling, track works and station infrastructure packages, may be let early in the life of a rail project.

The employer is not directly linked to subcontractors or suppliers although the employer may create a direct contractual relationship with subcontractors/suppliers by entering into collateral warranties enabling the employer to pursue the subcontractors/suppliers for breach of contract and possibly allowing the employer to 'step into the contractor's shoes' under the subcontract in the event of impending termination of the subcontract by the subcontractor.

This model can be diagrammatically represented as follows:

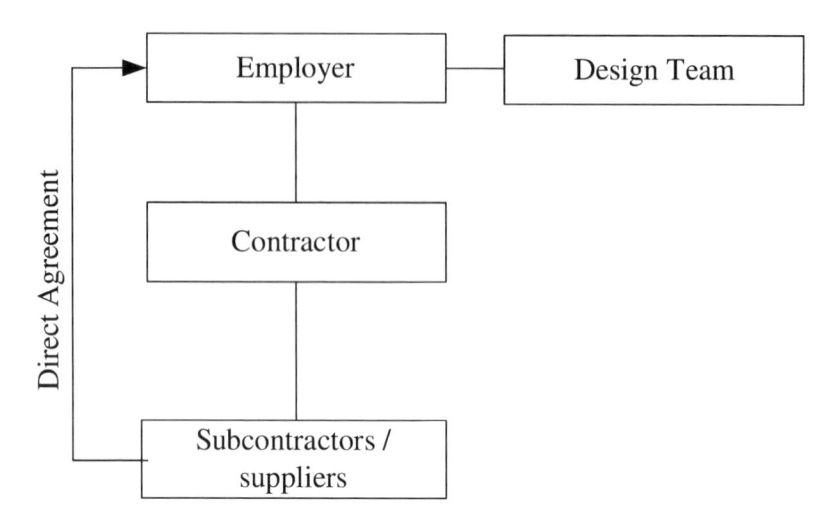

```
                    ┌──────────────┐      ┌──────────────┐
           ┌───────▶│   Employer   │──────│  Design Team │
           │        └──────────────┘      └──────────────┘
           │                │
  Direct   │        ┌──────────────┐
Agreement  │        │  Contractor  │
           │        └──────────────┘
           │                │
           │        ┌──────────────────┐
           └────────│ Subcontractors / │
                    │    suppliers     │
                    └──────────────────┘
```

Of the commonly used international standard forms, the FIDIC Red Book is a build-only contract.

The advantage of this structure is that the employer has a direct contractual link with each member of its design team and the contractor(s) and therefore controls the entire design process and can be assured that the design will meet its objectives. The employer does assume the risk of the design in relation to the contractor although this risk would be passed on to the relevant design professional. The employer also bears responsibility for the co-ordination of design development with construction.

The principal disadvantage of the model is that the separation of the design and construction phases results in longer periods for project completion. Also, the employer may be exposed to the problem of fragmented liabilities and it may be difficult to determine whether responsibility lies with a member of the design team or a contractor – a common area of dispute. In the case of project financed construction,[5] there is an additional disadvantage in that there is no guarantee at the time the design work is contracted for that the works will be built either for a cost, or within a period which will enable the project to be financed.

A variation of this model is where the design team is contracted to the employer at the outset of the project and then the employer novates the design team contracts to the contractor for the construction period of the project. This variation is more akin to a hybrid between the 'build only' and 'turnkey' models.

---

5   As to which see further below.

B.	DESIGN AND BUILD OR 'TURNKEY'

This model involves all detailed design, construction and procurement obligations being assumed by a single contractor (or more frequently in large infrastructure projects, a consortium of contractors). Essentially the contractor is responsible for turning over to the client a ready-to-use facility. The contractor, in addition to building the works, assumes the duties of the design team. It manages the design process as well as the construction process and, importantly, takes the risk of the design and the risk of the interfaces between the design team, its own works and those of its subcontractors.

As in the case of the build only model, the employer does not have a direct contractual link with the subcontractors/suppliers, as they contract only with the contractor. However, again the employer commonly creates a direct contractual relationship with subcontractors/suppliers by entering into direct agreements.

This model can be diagrammatically represented as follows:

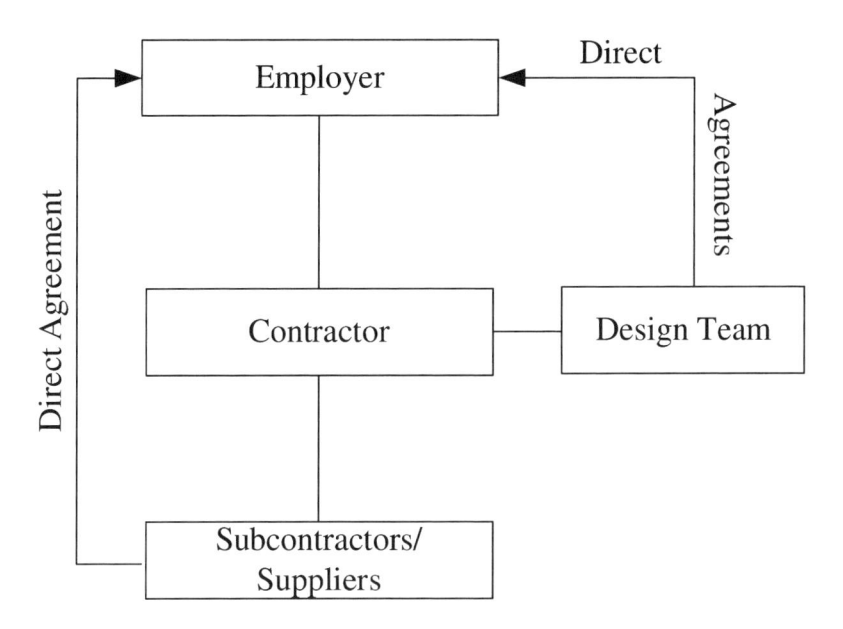

```
                              Direct
         ┌─────────────┐  ◄────────────┐
    ┌───►│  Employer   │               │
    │    └─────────────┘             Agreements
    │                                  │
  Direct Agreement                     │
    │                                  │
    │    ┌─────────────┐     ┌─────────────┐
    │    │ Contractor  │─────│ Design Team │
    │    └─────────────┘     └─────────────┘
    │           │
    │    ┌─────────────┐
    └────│Subcontractors/│
         │  Suppliers  │
         └─────────────┘
```

Of the commonly used international standard forms, the FIDIC Yellow Book, ICE D&C, ENAA Process Plant Model Form 1992 and AIA A191 DB-1996 are turnkey contracts.

The advantage of the turnkey model is that the project can progress on a 'fast-track' timetable with design, construction and procurement progressing in parallel. Another advantage for the employer is that the contractor takes the full risk of design and also of changes to the design (other than design changes requested by the employer). Also, the employer will have no difficulty in deciding who is responsible as there is a single point of responsibility. Of course, the contractor must have the requisite co-ordination capabilities to manage contractual interfaces among its own consortium members and sub-contractors and suppliers. While the contractor might be assumed to possess these skills, the practice may be different. Mismanagement and lack of co-ordination is a common complaint and catalyst for dispute between employers and contractors and/or contractors and their subcontractors especially where consortium members take responsibility for different aspects of the contractor's duties.

The associated danger with this approach is that projects generally go out to tender on a preliminary design, which will represent only between 20 per cent and 35 per cent of the total design development required to finalize the design. Detailed design by the contractor will be based on this preliminary design and parameters set by the employer's design criteria including performance criteria. The employer can then change its requirements only if it accepts responsibility for attendant delay and costs. Changes in the design will often have a significant impact in terms of delay and disruption to (and, therefore, cost of) the construction process. This risk of increases in project cost and time is particularly acute where construction commences before the design is sufficiently advanced. Responsibility for time and cost impacts of significant design changes is a common area of dispute.

An additional problem can also arise as the turnkey contractor may be tempted to assume too much of the work itself, rather than contracting out the work to the subcontractors best qualified to carry out the work, as might have happened if the works had been tendered out in separate packages. Even where this is done and particularly complex work is subcontracted out to contractors not forming part of the turnkey contractor's consortium, this will normally result in the turnkey contractor adding a percentage mark-up on the subcontract price.

C.   CONSTRUCTION MANAGEMENT

In the construction management model, the employer appoints the trade contractors directly to undertake the works. In addition, the employer appoints a construction manager to undertake the role of co-ordination of design and construction. The construction manager does not carry out any construction works. It acts in a role similar to that of a project manager where the level of responsibility is likely to be that of 'reasonable care and skill'. Generally there is no guarantee by the construction manager as to price of the works or the time for completion. Instead, the construction manager is obliged to use its reasonable skill and care to endeavour to achieve specified dates and budgets. To mitigate this, construction management contracts may include targets as incentives for the construction manager to complete the works on time and to budget – for example, the construction manager may be entitled to recover only its costs (i.e. without profit) if it fails to achieve longstop dates, or it may receive an increased profit margin if it completes on time or within budget.

If the construction manager is involved early in the project, it can also assist with 'buildability' issues and arranging the works into work packages.

This model can be diagrammatically represented as follows:

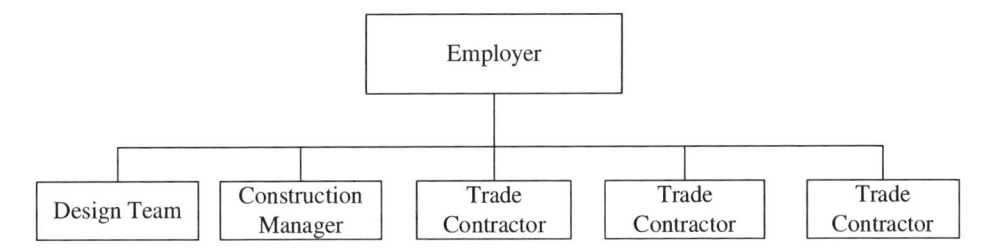

```
                          ┌──────────┐
                          │ Employer │
                          └────┬─────┘
       ┌──────────┬───────────┼───────────┬───────────┐
┌─────────────┐ ┌──────────────┐ ┌────────────┐ ┌────────────┐ ┌────────────┐
│ Design Team │ │ Construction │ │   Trade    │ │   Trade    │ │   Trade    │
│             │ │   Manager    │ │ Contractor │ │ Contractor │ │ Contractor │
└─────────────┘ └──────────────┘ └────────────┘ └────────────┘ └────────────┘
```

The advantage of this model is that the phases of design, planning and construction can also overlap easily and the employer has direct rights against the trade contractors who are carrying out the works.

The drawback of construction management is that the employer bears the interface risk between each individual trade contract package and also between the design team and construction manager. Therefore, liabilities are fragmented and the employer may be unclear where liability lies. Conversely, the employer is also at risk of being a defendant to a multiplicity of claims. Despite the management responsibilities of the construction manager, there remains a significant managerial burden on the employer who is responsible for the interface of all parties including the construction manager.

D.   Management Contracting

Management contracting is a hybrid between the traditional forms of procurement and the construction management model. The employer appoints its design team and, in addition, a management contractor. The management contractor generally does little or no direct construction work itself, but organizes and co-ordinates those that do. The management contract can be let at a far earlier stage than a traditional form contract and before any of the detailed design has been done or even before planning permission has been obtained. Commonly, the management contract will comprise two stages – pre and post construction – with an option on the part of the employer either to abandon the project altogether or to re-tender the construction works before the second stage. Management services can be in respect of the works only or both the works and the design. The works (and, in some cases, the design) are done by 'works contractors' who are subcontractors to the management contractor. The management contractor is responsible for the management of the construction process for the benefit of the employer and must pursue any defaulting works contractors on the employer's behalf. It is usually stipulated that any shortfall in recovery is borne by the employer and any other loss which is suffered directly by the management contractor (for example, in meeting claims against the management contractor by a works contractor who has been disrupted by a defaulting work contractor) may also be passed to the employer if the defaulting works contractor fails to pay. Commonly, the employer creates a direct contractual relationship with the subcontractors by entering into collateral warranties which are, in effect, direct contracts between the employer and subcontractors.

This model can be diagrammatically represented as follows:

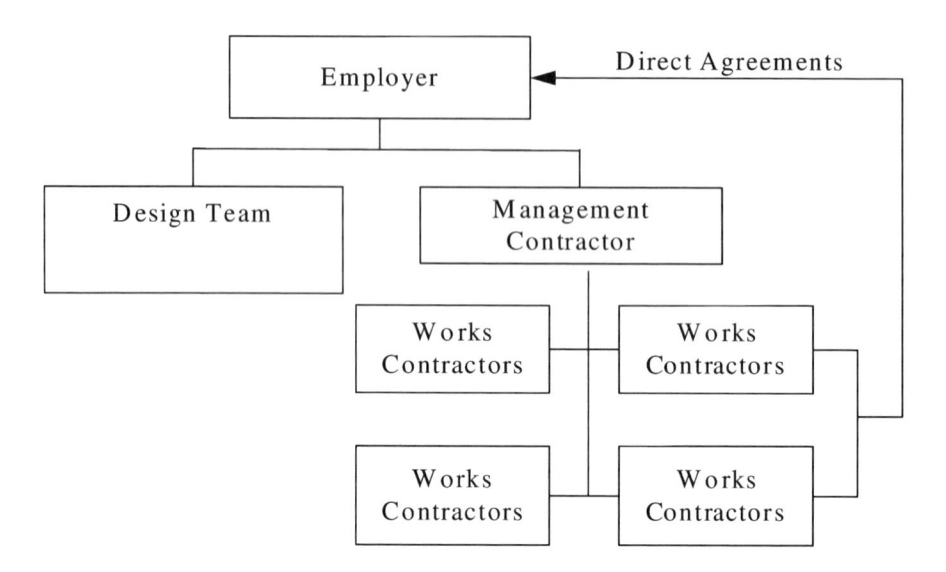

```
                    ┌──────────────┐  Direct Agreements
                    │   Employer   │◄──────────────────┐
                    └──────┬───────┘                   │
              ┌────────────┴────────────┐              │
      ┌───────────────┐         ┌───────────────┐      │
      │  Design Team  │         │  Management   │      │
      │               │         │  Contractor   │      │
      └───────────────┘         └───────┬───────┘      │
                          ┌─────────────┼─────────────┐│
                   ┌────────────┐              ┌────────────┐
                   │   Works    │              │   Works    │
                   │Contractors │              │Contractors │
                   └────────────┘              └────────────┘
                   ┌────────────┐              ┌────────────┐
                   │   Works    │              │   Works    │
                   │Contractors │              │Contractors │
                   └────────────┘              └────────────┘
```

This type of contractual structure affords a significant time advantage as the management contractor can be appointed when design is at a very early stage. Another advantage for the employer is that there is a single co-ordinating entity with whom it must deal (although this does not necessarily translate to a single point of liability).

The disadvantage of management contracting is that the employer bears the risk of the default of the works contractors, save in the case of failure by the management contractor in its management role. Further, when a defaulting works contractor causes loss and expense to an innocent works contractor, the innocent works contractor will look to the management contractor for reimbursement, who will in turn look to the employer for reimbursement. The employer also bears the risk of insolvency of the works contractors.

E.   PARTNERING AND ALLIANCING

Partnering and alliancing are alternative methods of procurement that have largely grown out of dissatisfaction with the traditional design and build procurement methods and particularly the adversarial culture they tend to promote. Their aim is to improve co-operation and joint working between various parties and to encourage

Key Features of Construction Contracts

the early resolution of problems and potential disputes with collaborative effort. Partnering and alliancing techniques are addressed in more detail in Chapter 4.

V. PRICING METHODOLOGIES

Whichever contract structure is selected, the work can still be priced in a number of different ways. The most common bases for pricing under construction contracts are as follows:

A. COST PLUS

This method involves paying the contractor its actual costs, reasonably and properly incurred, plus a profit fee. It may be appropriate where the construction presents unusual difficulties, for example, innovative design or complex engineering with high risks, in which case a lump sum price that adequately takes account of all contingencies would be prohibitively high. However, as neither the employer (nor any lenders) would be willing (save in exceptional circumstances) to accept an open-ended commitment to meet the contractor's costs without limit there is usually some mechanism to encourage the works to be built for the lowest possible cost and in the shortest possible timescale. Sometimes the risk is mitigated by incorporating a cap on the contractor's fee. Alternatively, it may be done by introducing ceilings – or targets – beyond which costs are shared (see below). In contracts of this kind the employer's rights to audit thoroughly the costs claimed to ensure that they have been reasonably and properly incurred, and to require the contractor to maintain comprehensive records for audit purposes, are key. The most common area for dispute in pure cost plus arrangements is whether the costs incurred by the contractor are both reasonable and properly incurred.

B. TARGET COST

Cost plus contracts incorporating ceilings or targets are known as target cost contracts. Essentially the employer and the contractor will agree on an estimate for the likely cost of the work. This is known as the target cost. The contractor is paid the actual costs incurred, together with a fee or other mechanism to cover profit and indirect costs (for example, head office overheads). However, if actual costs incurred exceed the target cost, the employer and the contractor will share liability for the overspend in agreed proportions. As an additional incentive to the contractor to keep costs to a minimum, a target cost contract may also include a system of bonus payments by which the contractor shares in the benefit of cost savings. As well as the question with this basis for pricing whether costs have been reasonably and properly incurred, another common area for dispute is whether, and the extent to which, the target cost should be adjusted where variations are ordered on the project.

C. Lump Sum

At the other end of the scale of pricing structures is the lump sum contract that entitles the contractor to a pre-agreed contract sum regardless of actual costs incurred, subject only to adjustment for variations to the works and in a limited number of other circumstances. The lump sum should be sufficient to cover the contractor's anticipated actual costs including overheads plus a profit component. The main advantage for the employer is (at least apparent) cost certainty. A further advantage is that the administration of the contract is usually more straightforward than under the cost plus target cost methods, where costs have to be verified. The apparent appeal of a lump sum price, however, may be illusory where the contingencies required to be included in the contractor's bid are such as to make the contractor's offer prohibitively high or where the design of the project is sufficiently uncertain to make variations (and thus additional payment) a virtual certainty. The most common area of dispute in lump sum contracts is the question of precisely what work is covered by the contract and what work amounts to variations for which the contractor is entitled to extra payment. Turnkey contracts, where the specification of the works is general or defined by reference to performance criteria, are particularly prone to these types of disputes.

D. Provisional Lump Sum

A provisional lump sum involves taking a firm price for part of the works, with provisional rates to apply to a series of contingencies. For example, a range of potential ground conditions could be separately priced and the rates would simply be applied to match the conditions actually encountered. The employer then knows the 'worst case' scenario, but will benefit from paying the lower rates if the conditions encountered prove to be straightforward. A provisional lump sum also mitigates the potential problems where the design is defined by reference to open-ended or general criteria, by allowing for alternative design solutions to be priced.

E. Guaranteed Maximum Price

Increasingly popular is the Guaranteed Maximum Price (GMP) Contract. This involves a two-stage procurement process. The first stage includes preliminary investigation, feasibility study and outline design, sometimes with enabling works. The second stage is the development of the design and project construction. The first stage of the project is paid for on a cost plus basis. The contract then 'converts' to a guaranteed maximum price once the scope and definition of the works becomes more certain and at this point the contract effectively operates as a lump sum contract. The employer retains the option not to continue with the project at all after the first stage if the GMP offered is prohibitively high or to break with the contractor and tender the job anew on the basis of the completed preliminary design.

F.   UNIT PRICE OR MEASURED WORKS

The pricing method midway between lump sum and cost plus is the unit price. The price of the project is calculated per task in accordance with a bill of quantities which comprises a list of items giving the quantities and brief descriptions of work comprised in the contract. It forms the basis upon which tenders are obtained and affords a means of comparing tenders received once priced. Terminology used in connection with bills of quantities include:

- *Daywork*: the method of valuing work on the basis of the time spent by the workmen, the materials used and the plant employed. Daywork rates are effectively 'cost plus' rates.
- *Prime Cost* (or the initials 'PC'): a sum entered in the bill of quantities at the time of tender as the notional sum provided to cover the cost of specific articles or materials to be supplied or work to be done, after deducting all trade discounts and any discount for cash. The contractor is paid the amount the work (usually carried out by others) actually costs.
- *Provisional Sum*: any sum of money included in the bill of quantities to provide for work not otherwise included therein or for unforeseen contingencies arising out of the contract. It is generally only to be expended, either wholly or in part, under the employer's direction and at its discretion. The only real difference between prime cost items and provisional sums is that the employer intends to have prime cost items included in the works (though by people it specifies and for whose costs it effectively remains responsible) whereas provisional sums are for work which may or may not be carried out.

Where work is priced on the basis of agreed rates and prices, the contract should also specify a chosen standard method of measurement. Standard methods of measurement specify how virtually all commonly-encountered construction activities are to be measured. A unit price or measured works contract without an agreed method of measurement is prone to dispute. To illustrate: the cost of excavation of a pit to accommodate a tank with vertical sides might be determined by the volume of material removed to allow the tank to be erected. However, it is generally impossible to excavate a pit of any depth with vertical sides, so a larger pit with sloping sides needs to be excavated. In the absence of an agreed method of measurement, is the work to be priced on the basis of actual volume of excavation or merely on the amount required to fit the tank in? Sometimes, however, standard forms of measurement create a dangerous priority over the drawings and specifications contained in the contract, and can be used to increase the final price by their manipulation. Standard forms of measurement may also be used as guides in the quantification of disputed construction claims generally, for example, valuation of variation claims or quantum meruit claims.

G.   MIX AND MATCH?

Larger projects may adopt a mix of payment arrangements. For example, on the Channel Tunnel project for the construction of a railway link between England and France by way of two running tunnels and a service tunnel, the underground civil works (the structure of the tunnels, the cross passages for emergency escape and the various underground ducts and machine rooms) were paid for on a target cost basis. The terminals (buildings and platforms) and the mechanical and electrical equipment (track, catenary, power supplies, signalling, ticketing systems, fire mains, pumps etc) were paid for on the basis of a single lump sum. Finally the rolling stock was procured on a cost reimbursement basis. Interestingly, the largest cost overrun (in cash terms) was for the mechanical and electrical equipment – procured on the basis of a so-called lump sum contract.

Different pricing methodologies may also appear within a contract. For example, a lump sum contract may include a provisional sum component for a part of the works that may or may not go ahead and may provide for variations (i.e. works additional to the original scope of works) to be paid on the basis of daywork rates.

Overall, mixing payment and pricing mechanisms within the same contract is not recommended and can give rise to disputes. For instance, in the Channel Tunnel project the contractor, as part of the lump sum works, decided to replace miles of expensive high current, low voltage cable in the tunnels with a system of high voltage cables, intermediate step down transformers and shorter runs of less expensive low voltage cables. This made sense to the contractor from an economic perspective (the cost of the components required for the high voltage solution was cheaper) and, it is fair, also made sense from an engineering perspective. The contractor was entitled to do this as the work fell within the lump sum element of the works and, for the same reason, the savings on the equipment costs accrued to the contractor. However, the transformers had to be put somewhere and in order to accommodate them additional space had to be excavated in the undersea machine rooms. This work formed part of the target works which was paid for in full by the employer. The overall effect, therefore, was that the contractor was able to take the benefit of the new engineering solution as a saving in the lump sum part of the works while the cost of the target works (and the overall cost to the employer) increased.

VI.   PAYMENT ARRANGEMENTS

It is very rare for contractors (other than on very small or short-term projects) to be paid by way of a single payment at the completion of works. Cash-flow is vital for the success of any construction project as contractors must be able to cover the considerable costs incurred throughout the entire construction period in order to continue working and complete the project. Therefore, some system of payment at intervals is required. In some jurisdictions, legislation has been enacted to require employers to pay contractors on a regular basis throughout the construction period. At the same time as the contractor receives interim payments, the employer needs to ensure that it is obtaining value for the payments made and that the contractor is sufficiently incentivized to remain and carry out the project in accordance with the requirements of the contract.

Interim payments may take a variety of forms (sometimes in combination). The main ones are:

A.   PROGRESS PAYMENTS

Most construction contracts will provide for staged payments to the contractor, frequently on a monthly basis during the course of the project. The amount due to the contractor at each stage will reflect its progress on the works – for example, as a percentage of the total project works – and will be recorded in an interim certificate issued by the certifier. The certificate will set out the cumulative amounts paid to the contractor to date and the additional amount due to the contractor in the relevant payment period. The certificate will then record the total amount due to the contractor in the payment period.

B.   MILESTONE PAYMENTS

Some construction contracts provide for payments to contractors upon achievement of milestones throughout the construction period. Milestones are set against the completion of certain activities or a section of work and, upon completing those works or that section, the contractor is paid for the works/section. Milestones are often incorporated in, and shown on, the project programme.[6] This type of payment method incentivizes the contractor to complete works rapidly and in accordance with the programme.

---

6   As to which see further below.

VII. ADMINISTRATION OF THE CONTRACT

A. BACKGROUND

Essential to many construction contracts is the role of the contract administrator. The administrator will basically perform one or more of the following three discrete functions:

- the employer's agent in overseeing the progress of the works and generally giving instructions as to how the work should proceed;
- certifier under the contract – construction contracts invariably provide for critical decisions which are determinative of the rights of the parties to be recorded in certificates, for example certificates of payment, certificates of completion, certificates of making good defects etc;
- (possibly) resolving disputes between the parties in the first instance.

In relation to the last function, the construction industries in France and other civil law jurisdictions do not give the same precedence to the role of the engineer as does the United Kingdom construction industry. Most notably, the engineer – or Maître d' Oeuvre – does not normally have any role in dispute resolution in the French construction industry.[7]

B. THE INDEPENDENT ENGINEER

Traditionally the role of contract administrator was given to the 'engineer' – see for example the ICE and FIDIC Red Book standard forms of contract. When acting as certifier and in resolving disputes, the engineer was required to exercise his judgment fairly and impartially between the parties. Essentially, he was supposed to act as a professional exercising his judgment in an even-handed manner. In his other duties, he was free to act as the employer's representative or agent.

The reality is – and probably always was – somewhat different. The so-called 'independent' engineer is appointed by the employer. One of his roles is to act as the employer's agent in overseeing the progress of the works. The strong perception of contractors is that the engineer will tend to favour the employer's position. The contractor's concerns are even more acute where the engineer is resolving disputes between the parties in the first instance. Almost invariably he will be

---

7 See: F Einbinder, *'The role of an intermediary between contractor and owner on international construction projects: a French contractor's viewpoint'*, ICLR 11 (1994), 175; T Kreifels, *'Construction or project management in Germany – the structure and its impact on project participants – an overview'*, ICLR 10 (1993), 326; S Nicholson, *'Effectiveness of the FIDIC contract under Argentine law'*, ICLR 9 (1992), 261; and F Nicklisch, *'The role of the engineer as contract administrator and quasi-arbitrator in international construction and civil engineering projects'*, ICLR 7 (1990), 322.

second-guessing his own decisions – for example where one of his certificates is challenged.

C.   REALITY – THE EMPLOYER'S REPRESENTATIVE

The realities of the relationship have led to the 'independent' engineer being replaced by the 'employer's representative' as contract administrator. There is no pretence here at independence. The employer's representative is not only often an employee but is acknowledged to act in the employer's best interests when administering the contract. However, English courts[8] have determined that the employer's representative must act honestly, fairly and reasonably when issuing certificates (as an implied term of the contract).

As a counter-balance to the consolidation of control of certification and contract administration in the hands of the employer, dispute resolution provisions frequently provide for disputes to be resolved in the first instance by an independent panel or expert who has no role in contract administration. This subject is addressed in Chapter 3.

D.   THE ROLE OF CERTIFICATES

Construction contracts provide for critical decisions that are determinative of the parties' rights to be recorded in certificates issued by the contract administrator. The most important certificates are:

- *interim payment certificates* (see above);
- *completion or handover certificates*, which mark the point in time at which the employer is entitled to take over the completed works and the point when the risk in the works (and responsibility for insurance) transfers from the contractor to the employer (see further below);
- *certificate(s) of non-completion*, which record the failure by the contractor to complete by the due date (sometimes a condition precedent to a deduction by the employer of liquidated damages);
- *certificate(s) of making good defects*, which are issued on inspection at the end of the defects liability period(s),[9] provided that the certifier is satisfied that all defects have been adequately remedied. It will also usually trigger release of any final tranche of retention held by the employer; and
- *the final certificate*, which is essentially a final reconciliation statement between the parties. It will take account of all necessary adjustments which have to be made to determine the final figure due to be paid to the contractor for all of

---

8   For example, in *Balfour Beatty v Docklands Light Railway Limited* [1996] 78 BLR 42.
9   As to which see further below.

the works. Deductions will be made as appropriate to reflect the cost of remedial works to correct any outstanding defects notified to the contractor. The final certificate may be given a conclusive effect, subject to any disputes referred to dispute resolution within a limited period following the date of issue of the final certificate. This question is addressed further below in the context of limitation of liability.

E. CHALLENGING CERTIFICATES

Some construction contracts confer the ability on an arbitrator to open up, review and amend certificates issued under a construction contract. This type of provision led the English Court of Appeal[10] to say at one time that such power was reserved for arbitrators alone, to the exclusion of the courts. Subsequently, the English House of Lords[11] overruled the Court of Appeal's *obiter* comments and found that the conferral of such power on an arbitrator does not limit the jurisdiction of the courts to determine the rights and obligations of the parties to the contract. In other words, a court can make findings that are contrary to those forming the basis of the administrator's certification under the contract. The court is free to investigate the facts and to interpret the parties' respective contractual rights and obligations; the administrator's decision on the issue will simply be part of the evidence that the court will weigh up.

Despite this there was some divergence between the individual judgments of the House of Lords on the question of whether the court has the power to issue a fresh certificate. Lord Lloyd and Lord Hoffman suggested that the Court does have this power while Lord Hope disagreed. If the view prevails that the court does not have this power, arbitrators in England will continue to have a broader power than the courts, although in many if not all cases the court's ability to determine the rights and liabilities under the contract will render this distinction irrelevant. However, there may be situations where the issue of a certificate is fundamental – for example if the certificate triggers drawdown of funds from banks or other credit provisions on a development supported by private funding – in which case the difference between arbitral and judicial powers remains of importance.

One proviso is where the construction contract contains a clear statement that certificates issued under the contract are conclusive. In this case, an English court would give effect to the contractual provision and would not open up the certificate. Save in exceptional circumstances such as fraud, the current position under English law, therefore, is that unless a certificate is expressly conclusive and binding it will be open to review by any tribunal called upon to determine the rights of the parties whether arbitral or judicial.

---

10 *Northern Regional Health Authority v Derek Crouch Construction Co. Ltd* [1984] 26 BLR 1.
11 *Beaufort Developments (N.I.) Ltd v Gilbert-Ash N.I. Ltd* [1998] 2 All ER 778.

VIII. VARIATION ORDERS

A. WHY ARE VARIATION ORDERS REQUIRED?

In theory, every construction contract will define the project to be built, whether by reference to a set of drawings and/or technical specifications or, where the contractor is to carry out the design, by reference to a performance specification. In practice, however, the employer's requirements will often change or, occasionally, the contractor may propose a change to the employer. It is therefore customary to incorporate a requirement in construction contracts that the contractor must execute the changes (or variations). Variations are effected by way of 'variation orders' (sometimes called 'change orders'). The only limit imposed on the employer's power to order variations as a matter of English law is that the contractor cannot be compelled to carry out something that is completely different from the original proposal.[12]

Of course, the ability to change the scope of the works is normally matched by an obligation to adjust the contractor's remuneration and, if appropriate, to grant an extension of time to the agreed completion date if required to complete any additional works. Variation orders generally require additional work, in which case the contractor is usually entitled to additional payment and time to complete the works. Negative variation orders or omissions involve deleting work from the original scope of works and, therefore, usually a reduction in the contractor's payment and programme. For this reason, there are usually limits on the circumstances in which an employer may issue a negative variation order – for example, it is one thing to delete works if those works are no longer required for the project but it is quite another for the employer to take work out of the hands of the contractor for the sole purpose of awarding it to another.

One additional constraint on the exercise of the power to order variations is not immediately obvious and arises where the contractor is carrying out the design and has agreed to meet specified performance criteria. In addition to granting the contractor additional time and money, the effect of the change on the ability of the contractor to meet the performance criteria should be taken into account. Accordingly turnkey contracts often contain a provision permitting the contractor to flag up any proposed changes which would affect its ability to meet the contract's performance requirements, giving the employer the opportunity to withdraw or modify the variation order. Of course, if the contractor does not exercise this right, the usual consequence would be for it to be required to meet the performance requirements regardless of the effect of any variation order.

---

[12] *See* Ian Duncan Wallace, *Hudson's Building and Engineering Contracts*, 11th edition, 2 vols. (Sweet & Maxwell, 1995), Vol. I, p. 929.

B.	CONTROLLING VARIATIONS AND THEIR COSTS

Variation orders may only be given by the employer (though in some cases the contractor may initiate a proposed variation). Since their effect is usually for the contractor to become entitled to additional payment, the manner in which variations may be instructed or authorized is normally closely controlled and involves a degree of formality. Variation orders are usually required to be in writing and given by specified individuals in accordance with an agreed procedure.

As well as controlling the issue of variation orders, construction contracts usually ensure that the issue of the variation orders is made on an informed basis – so that, at the time of issue, the employer is aware of the time, cost and other consequences which will flow from the variation. Contracts frequently provide for a procedure whereby the contractor is, save in urgent cases, required to provide an estimate of the likely cost of a proposed variation and the likely impact on the timetable for completion of the project. The employer can then decide, in the light of such information, whether it wishes to implement the variation.

Generally, a contract will require the contractor to give notice of a claim for additional payment promptly, for example within 28 days of issue of the variation order, together with full supporting particulars as to the sum claimed. If the variation is accepted in whole or in part, then either the whole of the additional payment will be included in the next following payment certificate to be issued to the contractor or the amount will be paid to the contractor progressively as the variation works are carried out.

Alternatively, a contract may provide that the variation order will be implemented on an agreed basis as regards time and cost, with no opportunity to claim additional impacts. If the parties cannot agree, the employer is normally entitled to order the contractor to comply with the variation order whilst the matter is referred to dispute resolution in accordance with the agreed procedure. The employer may be required to pay some costs against the variation order to the contractor in the interim.

C.	VARIATION ORDERS AND DISPUTES

Variations, together with extension of time claims, are the most common source of construction disputes. Occasionally a poorly defined specification is at fault, but just as often it is the administration of the variation clause or the financial pressures of the contract as a whole that give rise to variation disputes. On a lump sum contract, variation claims are one of the few opportunities for the contractor to recover additional money.

Disputes in relation to variations concern in the first instance whether the alleged variation is in truth within the original scope of the contract (in which case the contractor is not entitled to any additional payment). The second issue is determination of the appropriate adjustment to the price – this throws up questions of applicability of contract rates and prices and the appropriate mark-up for overheads. The third issue is the impact of the variation on the contract programme.

Less common are disputes over whether the employer is entitled to issue the variation at all – for example if the subject of the variation order is out of proportion with the scope of works for the project or if a negative variation order is made for an apparently improper purpose.

Disputes also arise in relation to the notification or other procedural requirements of the contract regarding variation orders – for example, employers commonly rely on the contractor's failure to comply strictly with the variation procedure in defence of a claim for additional payments where such requirements are said to be conditions precedent to any claim. As well as debating whether there was in fact compliance with such requirements, disputes frequently involve allegations by the contractor that the formal requirements were waived or that the employer is otherwise not entitled to rely on strict requirements for compliance with conditions precedent.

IX. COMPLETION, EXTENSIONS OF TIME AND LIQUIDATED DAMAGES

A. ROLE OF COMPLETION

Completion is a critical stage in the project. Works are normally said to be complete when they achieve 'practical completion' (also known as 'substantial completion'). 'Practical completion' is a term of art used in construction contracts to record substantial completion of the works save in respect of minor items which do not affect the use of the works for which they are intended.

Typically it has the following consequences:

- it is the point at which the completed works are handed over to the employer and risk in the works passes from contractor to employer;
- it stops the clock running for the purposes of the contractor's liability to pay liquidated damages for late completion;
- it triggers the start of the defects liability or warranty period, which is essentially a period during which the contractor has an obligation to return to the site to remedy at its own cost any defects which appear in the works and which are notified to the contractor prior to the expiry of the defects liability period; and
- it also generally triggers the release of half of any retention which has been deducted (or secured by a bond) from amounts due to the contractor under interim payment certificates by the employer.

B. DEFECTS LIABILITY PERIOD

Following 'practical completion', the works will normally be subject to a 'defects liability period' or warranty period. This is the period – often 12 months – during which the contractor must return to site and repair any defects that are discovered, or arise, in the works. It is at the end of the defects liability period when, if all defects have been adequately remedied, the certificate of making good defects may be issued. The employer will generally retain a portion – usually half – of the retention or other security to secure performance of the contractor's defects liability period obligations.

C. DATE(S) FOR COMPLETION

A construction contract may state a single date for completion by which the contractor must practically complete the whole of the project works. More commonly on large or complex projects, there will be a series of dates by which the contractor will be required to complete and hand over to the employer various sections or stages of the works. For example, a multi-unit power station may be handed over unit-by-unit or an office building may be handed over floor-by-floor. This is known as 'sectional completion' (or 'staged completion'). Sectional completion may enable the employer to commence operations, fit-out or staff training on those pats of the project completed early. Or sectional completion in a civil works contract – for example on a water filtration plant – may be required to allow structural and M&E contractors progressively to commence their works on the project. Sectional completion can also advantage the contractor who would usually be entitled on completion of a section to transfer liability for insurance and security to the employer and be relieved from liability to pay liquidated damages in respect of the completed section.

D. COMPLETION CERTIFICATE(S)

Given the criticality of completion, it is usually marked by issue of a completion certificate issued by the certifier that states the date on which practical or substantial completion of the works was achieved. Practical completion certificates are invariably issued subject to a list of outstanding items or defects that have been identified on inspection prior to issue of the certificate. This list of defects is sometimes known as 'the snagging list' or 'punch list. In a contract which provides for sectional completion there will be a practical completion certificate for each section of the works.

E.　　　　COMPLETION TESTS

On any sophisticated project there will be a regime of tests on completion. In the case of a power plant, for example, there will be a series of tests culminating in a reliability run, which involves the units running continually for a period of, say, 30 days. The tests on completion will define the standards that the plant has to achieve, for example in the case of a power plant in relation to electricity generation, fuel consumption and levels of heat production. In the case of a transportation system the tests will turn on levels of service, noise, power or fuel consumption, for example.

F.　　　　LIQUIDATED DAMAGES FOR DELAY

Construction contracts invariably provide for the payment of liquidated damages by the contractor for failure to achieve completion by the agreed completion date, or sectional completion dates. Liquidated damages are commonly payable in a specific amount on a daily or weekly basis. They should include a profit element. On larger projects the total amount of liquidated damages for which a contractor can be liable is often capped. Contracts normally make it clear that payment of liquidated damages does not relieve the contractor from its obligation to complete or perform any of its other obligations.

Liquidated damages serve a number of purposes. Most notably they avoid any need for the employer to prove loss in the event of contractor delay. For contractors, liquidated damages provide certainty as to their exposure to delay damages – and, where the damages are capped (which is common on major projects), a limit of liability for delay. For both parties, liquidated damages can also operate as a goal or incentive for project completion.

In common law jurisdictions, liquidated damages must represent a genuine pre-estimate of loss, otherwise they may be subject to attack by the contractor on the grounds that they represent a penalty and are therefore unenforceable.[13] Occasionally, liquidated damages clauses are poorly drafted so that the drafting rather than the specified rate renders the liquidated damages penal. Contracts that provide for sectional completion are particularly prone to such errors, for example, if the contract only states a single figure for liquidated damages or where the effect of cumulative delays across sections has not been appropriately addressed.[14]

13　In contrast, Dutch law, for example, does not require liquidated damages to represent compensation for loss.
14　See *Gleeson (M J) (Contractors) Ltd v London Borough of Hillingdon* [1970] 215 Estates Gazette 165; *Zornow (Bruno) (Builders) Ltd v Beechcroft Developments Ltd* [1989] 51 BLR 16; *Stanor Electric Ltd v R Mansell Ltd* [1988] CILL 399; and *Bramall & Ogden Ltd v Sheffield City Council* [1983] 29 BLR 73.

G. EXTENSIONS OF TIME

Construction contracts invariably provide that the contractor is entitled to an extension of time (or 'EOT') to the date or dates for completion if it can demonstrate that the progress of the works has been delayed by one or more specified events ('EOT events'), for example:

- changes to the works introduced by the employer;
- events of force majeure (to the extent wider than employer's retained risks); and
- unforeseen ground conditions (to the extent that this risk is not borne by the contractor under express terms of the contract).

Essentially the grounds for entitlement to an extension of time agreed between the contracting parties reflect the allocation of risk between the parties for events which are likely to cause delay to the project. Further examples of common EOT events are exceptionally inclement weather, industrial disputes (strikes, lock-outs etc) and unforeseeable shortages in availability of personnel or goods caused by epidemic or government action.

On its face, an extension of time clause is for the benefit of the contractor. However, in common law jurisdictions, it is also an essential requirement to ensure that a liquidated damages provision will be enforceable where delays are caused by the employer, or those for whom it is responsible. The employer will not be able to rely on its own wrong in attempting to recover liquidated damages. If the employer has caused the contractor's delay, and there is no adequate mechanism provided for in the contract for extending time to reflect such delay, the liquidated damages provision becomes unenforceable. The time for completion is then said to be 'at large', i.e. the contractor is required to complete within a reasonable time. The employer is not left without a remedy; it will be entitled to general damages as opposed to a pre-determined liquidated sum.[15] This, however, is a highly unsatisfactory situation – there is no fixed time for completion and, because there is no pre-agreed figure representing the employer's loss, there will inevitably be disputes as to the employer's right to set off unliquidated sums against sums certified due to the contractor in any subsequent payment certificate. Exclusive remedies clauses or caps or exclusions of certain heads of loss, such as loss of profit or business, may also operate to limit or exclude recoveries which would otherwise fall within the scope of an agreed liquidated damages provision.

---

15 *Peak Construction (Liverpool) Ltd v McKinney Foundations Ltd* [1970] 1 BLR 114; *Rapid Building Group Ltd v Ealing Family Housing Association Ltd* [1984] 29 BLR 5.

Typical features of extension of time clauses include:

- timely notice of claims as an express condition precedent (to prevent the contractor stockpiling claims and inhibiting the timely investigation of the reasons for delay);
- a requirement that the contractor should use best endeavours to avoid or reduce delays; and
- the ability for the employer to order acceleration of the works in lieu of a grant of an extension,[16] sometimes coupled with procedures for obtaining quotes of cost and likely time consequences.

H. THE ROLE OF THE PROGRAMME

Progress of works is planned and measured against an agreed programme, which may or may not be specified as a contract document. The programme maps out the contract works and shows the order in which the works will be completed and the expected duration of each activity, all in order that the works (or each section thereof) will be completed by the date for completion stated in the contract. Turnkey contract programmes will also include the design period of the project. The programme will identify generally the 'critical path' or paths through the works – the critical path is the linkage between activities which the contractor is required to complete in order to avoiding delaying the date for completion.

Programmes come in various forms and levels of detail, depending on the size and complexity of the project and the requirements of the contract. Programming is a sophisticated science and normally utilizes a standard, commercially available software package, first to generate a programme at the outset of the project and then to manipulate it during the project period. In addition to the information described above, programmes may show dates by which the employer is required to provide the contractor with required information and milestones or target dates with bonuses payable on the contractor meeting those dates. Contracts frequently oblige the contractor to update the programme, either periodically or upon the occurrence of delay events, in order to ensure that there always exists a current programme.

The role of the programme is considered in more detail in Chapter 10.

I. CRITICAL PATH ANALYSIS

The contractor will not be entitled to any extension of time unless the delay is critical to completion of the relevant works. In other words, the delay must affect works on the critical path so that delay to those works will cause direct delay to the

---

16   See further below.

date for completion of the works (or the relevant section of works). For example, a delay to landscaping works is unlikely to be critical until the very end of the construction period, as it is normally one of the very last tasks to be completed and does not usually have other activities dependent on its completion. The question whether the works are critical to completion is decided by reference to an up-to-date contract programme showing actual progress to date, the activities still to be completed and their interrelationship. This topic is discussed in detail in Chapter 10.

J. CONCURRENT DELAYS

Concurrent delay arises where two events occur in the same time period, one of which is a delay under the contract that entitles the contractor to an extension of time, and the other is not, and each event, were it to have occurred in isolation, would have caused the work to be delayed. True concurrency is rare. Analysing what appear to be concurrent delays with the aid of programming tools that allow analysis of the impact of the events on the critical path often reveals co-existing events that may have been concurrent, but are not co-effective.

There exist various approaches to resolve true concurrency. One approach is the 'dominant' cause approach – i.e. if there are two events causing a delay, the contractor will succeed in its claim for an extension of time where it can establish that the EOT event (as opposed to the event for which it is not entitled to an EOT) is the effective, or dominant, cause rather than merely the occasion for delay. The question of which is the dominant cause is a question of fact. The generally held view, though, is that where there are two equally operative causes and one is an EOT event the contractor should get an EOT, although the cost/damages consequences may differ in these circumstances.

Concurrent delays are also considered in more detail in Chapter 10.

K. WORK-AROUND MEASURES

A contract sometimes contains an express obligation on the contractor to employ 'work-around measures' in order to avoid a delay. This is sometimes seen as a simple obligation to mitigate. Other contracts provide that the contractor's entitlement to an extension of time will be subject to the contractor demonstrating use of all reasonable measures to overcome the delay. For a contractor who wants an EOT, there is effectively little difference between these types of provisions – work-around measures are mandatory.

Alternatively, work-around measures may be required at the option of the employer. In this case, a contract reserves to the employer the right to order work-around measures as an alternative to an award of an extension of time where the contractor's entitlement has been made out. Additional costs associated with any such measures should be for the employer's account.

L.   ACCELERATION/CONSTRUCTIVE ACCELERATION

Contracts also sometimes reserve to the employer the right to order acceleration of works in lieu of granting an extension of time where the contractor's entitlement to an EOT has been made out. The simplest acceleration measures are out-of-hours work and employment of additional labour. Additional costs associated with any such measures would normally be for the employer's account.

Constructive acceleration is not an acceleration order from the employer but a claim by the contractor that it was entitled to accelerate. For example, claims for constructive acceleration are seen when a contractor is ordered to take measures to achieve the original completion date when in fact an extension of time should have been granted.

Even where it is clear that acceleration has been ordered, quantification of acceleration claims under lump sum contracts can sometimes be fiercely disputed – particularly if one part of the project has been accelerated at the request of the employer at the same time as the contractor is in delay on another part of the project, in which case it may not be clear to whose account (or it what proportions) the additional costs should fall.

M.   EXTENSION OF TIME CLAIMS AND DISPUTES

Claims for extensions of time, along with variation claims, are the most common cause of construction disputes. Extension of time claims invariably involve difficult factual questions of causation, as illustrated by the discussion of critical path analysis and concurrent delays above and in Chapter 10.

As in the case of disputed variation claims, formal notification and documentation requirements are also frequently at issue in extension of time claim disputes.

N.   LIQUIDATED DAMAGES FOR PERFORMANCE FAILURES

Liquidated damages may also be agreed not only in respect of delay but also in respect of failure to achieve the specified performance standards. For example, in the case of a power plant where the electricity output does not meet the pre-agreed threshold, liquidated damages may apply as a factor of the guaranteed gross output. In the case of a transportation system, there may be a failure to achieve a pre-agreed running time where the works have failed to pass the tests on completion. The contract will then generally provide the employer with the option of:

- ordering repetition of the tests, if the perception is that the failure has been a one-off because of teething problems. Liquidated damages for delay can be claimed where repetition of the tests delays completion beyond the completion date;
- rejecting the works in their entirety, where the failures are outside agreed tolerances; or
- taking over the works notwithstanding the failure to achieve the pre-agreed standards and claiming liquidated damages on the pre-agreed basis by way of compensation as a 'one off' hit.

While liquidated damages work to the advantage of the employer in avoiding the need to prove its loss and providing a clear route to set-off against sums due to the contractor, as noted above they may also work in favour of the contractor as a limit on liability where the level of liquidated damages is capped.

X. LIABILITY

There are three obvious areas where a contractor is exposed to liability under a construction contract. These are:

- delay in completion;
- failure to meet specified performance standards; and
- defects in the works.

A. LIMITATIONS OF LIABILITY

In common law jurisdictions, the general position is that breach of contract claims are available unless they are excluded by clear words – for example, stating that the remedies provided in the contract are the only remedies available to the parties (or the contractor) to the exclusion of all other remedies at law.

At common law, the contractor is in principle liable to pay damages on account of all the employer's losses that flow 'directly and naturally' from the breach of its obligations.[17] In practice, the contractor's exposure is not as extensive as this, as a result of the agreement of limitations on its liability. Limitations of liability often reflect the availability of insurance to cover particular risks, such as negligent design.

---

17 This is the requirement of 'remoteness' as laid down in the case of *Hadley v Baxendale*[1843–60] All ER Rep 461.

B.   DELAY AND PERFORMANCE

As mentioned above, the parties frequently agree the payment of liquidated damages if the project is delayed. Liquidated damages for delay will generally be calculated on a daily or weekly basis. Liquidated damages for failure to meet performance standards will vary with the degree of the failure. For example, in a construction contract for the design and build of a power station, liquidated damages may be applied at an agreed rate per unit of heat produced over and above the agreed tolerances, and a separate rate for each unit shortfall in power produced. In transport infrastructure contracts, liquidated damages may be imposed if specified noise levels are exceeded, for failure to achieve specified ride characteristics, or if the travel times are longer than specified.

The contractor will look for such liabilities to be capped (in respect of individual performance requirements and/or overall) at a particular level. This has the effect not only of limiting the liquidated damages payable but also of capping the contractor's liability for breaches of the type covered by the payment of liquidated damages. This is because, generally,[18] a liquidated damages clause will be construed as reflective of the parties' intention that liquidated damages are to be the exclusive remedy in respect of a particular head of loss.[19]

C.   DEFECTS

With regard to defects, a limit in monetary terms (usually part of an overall cap on liability) can often be found. More usually, the contractor will have negotiated to exclude particular *types* of loss, most obviously consequential loss. However, in some jurisdictions, the meaning of consequential loss is unclear and may cause surprise as the exclusion of 'consequential loss' may not in fact exclude very much – for example, under English law, consequential loss has been construed to cover only damages which would not, but for the express knowledge of the contractor, be regarded as losses flowing directly and naturally from any breach.[20] It is sensible to identify the particular categories of loss sought to be excluded, rather than relying on general terminology. Accordingly, a contract may provide for liability to be limited to the costs of repair or reinstatement of the works, excluding liability for

---

18   It is, however, a question of construction of the particular contract: *Baese Pty Ltd v R A Bracken Building Pty Ltd* [1989] 52 BLR 130; *Surrey Heath Borough Council v Lovell Construction Ltd and Haden Young Ltd (third party)* [1990] 48 BLR 108.

19   *Cellulose Acetate Silk Co Ltd v Widnes Foundry (1925) Ltd* [1933] AC; *Temloc Ltd v Errill Properties Ltd* [1987] 39 BLR 30. Similarly in some civil law jurisdictions liquated damages is considered a substitute for any loss suffered e.g. Art. 1152 (1) French Civil Code, Art. 1382 Italian Civil Code and Art. 6.1.8.17 (2) the Dutch New Civil Code.

20   In particular, (under English law) an exclusion of 'consequential damages' does not automatically exclude loss of profits or other financial loss, see *British Sugar plc v NEI Power Projects Limited* [1997] 87 BLR 42.

a list of specified losses, such as loss of use, loss of production and/or loss of contract.

D. ENTITLEMENT TO LOSS AND EXPENSE DUE TO DELAY AND DISRUPTION

Construction contracts may entitle the contractor to claim for loss and expense due to delay and disruption. Delay and disruption claims (also called 'prolongation claims') typically accompany claims for extensions of time. Whereas the latter concern the contractor's claim to extend the date by which it must complete the works, loss and expense claims are for the costs associated with having to take longer to achieve completion. Most costs claimed will be no more than the contractor's ordinary costs of being on site and carrying out the project for a longer period than would have been the case had the delay and disruption not occurred.

Prolongation cost items that are commonly claimed include: extra labour time, additional plant hire costs, overheads, profit, financing charges (such as maintenance of bonds or overdrafts), loss of opportunity and altered working conditions. Contracts may expressly exclude some categories of loss, such as loss of profit and loss of opportunity.

In common law jurisdictions, the absence of an express contractual right to claim for delay and disruption costs leaves a contractor with only its common law right to sue for breach of contract where the employer is responsible for a delay to the contractor's works. Where the delay is not the employer's fault then, even where there is a contractual right to an extension of time (for example, exceptional inclement weather), there is generally no corollary right to delay and disruption costs associated with that delay.

The contractor's entitlement to recover loss and expense on account of delay and disruption (if any) is usually subject to a variety of conditions precedent and other procedural requirements. For example, the contract may require that the contractor must first have established an entitlement to an extension of time, mitigated its loss, provided timely notice to the employer of both the delay and the claim, and adequately substantiated the amounts claimed. Typical conditions precedent to claims are addressed in more detail in Chapter 4.

E. A CONCLUSIVE FINAL CERTIFICATE?

Another means by which the contract may limit the contractor's liability for defects is by providing for the final certificate to be conclusive as to the satisfactory execution of works in accordance with the contractual requirements. Contracts for mechanical and electrical work, for example, generally favour a conclusive certificate, save in respect of a narrow class of latent defects. See for example, clauses 39.12 and 39.13 of MF/1 (rev 4)/2000,[21] a standard form recommended by the United Kingdom based Institution of Mechanical Engineers, Institution of Electrical Engineers and Association of Consulting Engineers for use in connection with domestic or international contracts for the construction of electrical, electronics and mechanical plants. In contrast, the standard form civil and engineering contracts (such as ICE and FIDIC) have moved away from conclusive final certificates. The difference in approach lies in the nature of the work. Defects in mechanical and electrical plant are likely to emerge on commissioning or soon thereafter. Defects in buildings or civil works are notoriously difficult to detect – hence the reluctance to exclude claims following issue of the final certificate. In addition, the continued performance of mechanical and electrical work after commissioning is heavily dependent on the operation and maintenance regimes, thereby quickly reducing the responsibility of the original supplier for any failures.

XI. PROJECT SECURITY

A. TYPES OF SECURITY

The construction contract usually provides for security to be given to support the contractor's (and very occasionally the employer's) obligations. This may be tangible security (the creation of a cash reserve by way of retention) or third party security, either by way of parent company guarantee or third party security in the form of bank guarantees or bonds. The most common forms of tangible and third party security are considered briefly below.

B. ADVANCE PAYMENT BONDS

Where the contractor receives a sizeable advance payment in order to mobilize or to order long lead-time equipment, it is common to secure repayment of the advance by way of deduction from sums earned over the life of the contract. It is also common for this repayment obligation to be backed by a third party guarantee in the form of an 'on-demand' bond. An on-demand bond is one which can be called by a simple written demand without proof of breach by the contractor or of loss or damage suffered by the employer.

In theory, an advance payment bond should be a reducing bond, initially securing the whole of the advance but reducing in size as the advance gets repaid by way of deduction from the value of work carried out. However, this level of

---

21　Institution of Electrical Engineers, *Model Form of General Conditions of Contract including Forms of Tender, Agreement, Sub-contract, Performance Bond and Defects Liability Demand Guarantee for use in connection with home or overseas contracts for the supply of electrical, electronic or mechanical plant – with erection*, (ImechE / IEE, 2000).

sophistication is not often seen, with the result that the contractor carries the cost of a bond for the whole of the advance for the life of the project. While this obviously gives the employer additional security (and may be in lieu of a performance bond or lead to a smaller performance bond) this may be an unnecessary project cost if it affords more security than is required.

C.   RETENTION FUNDS

The employer generally has the right to retain a percentage – usually 3 to 5 per cent – of the value of works certified due in each payment certificate up to practical completion to form a retention fund. The retention fund is available to the employer to meet valid claims against the contractor, including claims for defects. If there are no outstanding claims, typically half of the retention fund will be released on certification of completion and the balance on certification of making good defects or expiry of the defects liability period, whichever is the later.

In part to improve contractor cash flow the practice has evolved – and is now common in major infrastructure projects – of the contractor providing a bond to cover all or part of the retention in lieu of the deduction of such sums by the employer. Since a retention fund is a substitute for a cash sum, against which the employer has unlimited recourse, there is no reason why the retention bond should not be on-demand. In theory this should, until completion, be an increasing bond, mirroring the build-up of the retention fund it replaces but this is rarely the case in practice.

D.   PERFORMANCE BONDS AND GUARANTEES

The purpose of performance bonds and guarantees is to secure the due performance by the contractor of its obligations under the contract. In practice, the real value of these bonds is where the contractor becomes insolvent.

Performance bonds and guarantees may be 'on demand' (described above) or 'on default'. An on default bond operates as a guarantee and requires proof of default on the part of the contractor. Contractors often strongly resist providing on demand bonds due to the risk of wrongful calls. A degree of comfort may be given by requiring the form of demand to be signed by a senior officer of the holder of the bond, who certifies a bona fide belief that the circumstances justifying the call on the bond have arisen.

Performance bonds are frequently sought to a value of up to 20 per cent of the contract price. In many circumstances, the guarantee will be in addition to an advance payment bond (which may or may not reduce so as to have real value over the contract period) and a retention fund or bond. The costs of obtaining bonds and guarantees will be included as part of the contractor's tender and so effectively the employer will be paying for the benefit of such added protection. In addition, in some jurisdictions (for example, Germany), bonds and guarantees must be shown as liabilities on the contractor's balance sheet, which discourages contractors from offering them.

E. PARENT COMPANY GUARANTEES

A parent company guarantee provides the employer with security from the parent company of the party to the original construction contract. Parent company guarantees are usually required from the ultimate parent company or another parent company in the chain – depending on the financial stability of the entities in question. If the ultimate parent company is foreign to the jurisdiction where the project is being constructed, this may influence the choice of guarantor and/or affect the documentation relating to the guarantee.

Whenever a parent company guarantee is requested, the group structure needs to be reviewed to assess the value of the additional security being provided. If the parent company is itself a substantial operating company with its own revenue stream and assets, the guarantee may be of considerable additional benefit. However, if the parent company merely owns shares in operating and asset owning subsidiaries, the value may not be significant, as any claim on the parent company is effectively subordinated to the claims of all the creditors of all the subsidiaries. Despite this, parent company guarantees are frequently requested as the guarantee is not usually limited to a specified percentage of the contract sum (as is a bond) and is usually of longer duration than a bond. The entity providing the guarantee may, however, seek to negotiate a limit to the period of time during which the guarantee may be called.

Sometimes construction contracts require the contractor to provide a parent company guarantee and a bond (which may or may not actually be necessary). Normally the two do serve different purposes. The bond is for a limited sum and time period (often up to completion only) and protects against the risk that the contractor may fail to complete due to insolvency or otherwise. The parent guarantee usually lasts longer and mainly provides comfort in relation to defects that may arise.

F. PAYMENT GUARANTEES

A payment bond or guarantee may be sought by the contractor from an insubstantial employer. Commonly, such bonds or guarantees will cover a fraction of the contract sum, for example, 10–20 per cent. This is generally sufficient to protect the contractor in relation to work it has carried out but which has not been paid for. The contractor can then take the decision whether to walk away from the project and not to incur the costs of any additional works where there is a significant risk of non-payment by the employer, the bond or guarantee having been exhausted.

Provision of a bond of this nature will give rise to an additional project cost. Therefore, alternative methods of providing security for the contractor can and have been devised. For instance, in the Channel Tunnel project (see above), the payment mechanism provided for the contractor to make monthly estimates of future costs two or three months ahead. Subject to scrutiny by the employer, this sum was placed into an escrow account and was available to meet actual costs incurred when the contractor had carried out the work. While this mechanism obviously has a funding cost to the employer, it may be more desirable to provide comfort to the contractor in this way than to purchase third party security in the form of a bond or guarantee.

G.   OTHER FORMS OF SECURITY

Project security may also be provided in other forms such as:

- letters of credit – used, for example, in relation to the purchase of high cost materials or other supplies); and
- trust funds – for example, in the case of the financing of the Iraq Trans Saudi pipeline where the proceeds of oil sales were paid into a trust account earmarked for payment to the contractor on presentation of documents and invoices certified by the Central Bank of Iraq.[22]

XII.   PROJECT FINANCED CONSTRUCTION PROJECTS

A.   INTRODUCTION

The essence of project finance is that there is no, or only limited, recourse to any participant other than the project company. The equity and debt 'cushion' provided by the shareholders of the project company will often be insignificant. To a fundamental extent, therefore, the lenders will call the tune on key terms of the construction contract as well as the conduct and resolution of disputes. In these projects, the interests of the lenders/other financiers are paramount as, whatever the wishes of the project company, its sponsors or the contractor, the construction contract has to be 'bankable' otherwise the project simply will not fly.

The contractor is often (though not invariably) one of the project company's shareholders, or perhaps two or more of the project company's shareholders (or their operating subsidiaries) acting together in a joint venture. The contractor will be looking for a swift exit with a profit and wherever possible to leave risks in the project company. In this context the project company and the lenders' interests are normally aligned as neither will accept risks to be left in the project company that cannot be managed and they will insist on a complete back-to-back pass through

---

22   See *Saipem SpA & Ors v Rafidain Bank & Ors* [1994] CLC 252.

of construction risks. The extent to which this has or has not been achieved in the drafting can give rise to disputes.

Most of the discussion in this chapter applies equally to project financed projects. The difference is really the presence of an additional interest (the lenders) and therefore an additional layer of contract administration and dispute management. Some of these differences are addressed further below.

B.   THE LENDERS' ENGINEER

Lenders will usually appoint their own independent engineer with responsibility for reviewing the technical specification and reporting to the lenders on project progress with rights to inspect the works and off-site facilities. The lenders' engineer is also generally given responsibility for certifying the completion of milestones, the trigger for payments to the contractor under the construction contract and the trigger for the drawdown of financing under the loan agreements.

An employer will not want to have a mismatch between certification under the construction contract and drawdown under the loan documents. Therefore, a solution often seen is that the lenders' engineer's certification is a condition precedent to certification under the construction contract. An alternative is to provide for joint inspection by the lenders' engineer and employer's representative, with a requirement to have due and proper regard to any representations made by the lenders' engineer as to whether or not to issue a certificate under the construction contract. The employer then takes the risk of mismatch in certification under the loan and construction contracts.

C.   LENDERS' STEP-IN RIGHTS

Lenders' step-in rights are another important investment protection mechanism. Lenders to a project company will usually have step-in rights under the construction contract to provide them with the opportunity to rectify any project company breaches or otherwise perform the construction contract where the project company is unable or unwilling to do so. Typically, before a contractor is entitled to terminate the construction contract for project company breach, the project company is allowed a specified period in which it is required (or entitled) to rectify the breach, after which the lenders have their own separate period in which they have the opportunity to remedy the breach.

Lenders will normally have similar step-in rights under the project company's concession agreement or head contract with the government or other authority for which it is carrying out the project. Again, these operate so that the government/authority's ability to terminate the concession agreement/head contract is subject to the lenders having the opportunity to rectify the project company's breach or otherwise perform the project company's obligations.

Lenders' step-in rights are often, but not always, limited to the extent necessary to rectify the breach.

D. VARIATION ORDERS

Lenders are also typically concerned to control the issue of variation orders and contracts may provide for their participation in the authorization procedure (either directly through the lenders' engineer under the construction contract or indirectly by agreement with the project company).

E. LIQUIDATED DAMAGES

A lender's liquidated damages 'wish-list' may be for liquidated damages to cover up to six months' interest payments on the project debt to be paid/deducted at a rate sufficient to cover interest payments.[23] Contractors, however, will resist accepting full risk of debt service costs and delay damages are generally capped at about 15–20 per cent of the total contract value. The contractor's price will reflect the level of exposure – for example, liquidated damages with a cap above 20 per cent of the contract value will generally translate into an inflated price.

F. PROJECT SECURITY

In project-financed projects, lenders look for a guaranteed method of payment of debt service. This tends to mean that the performance bond is seen not simply as protection against the contractor's ultimate insolvency but, as importantly, as a means of providing instant cash to cover any period where the project is not revenue producing. This has consequences for the type of bond or guarantee. As the real purpose of bonds in a project-financed project is not so much to meet the demands of a strongly capitalized employer, but to meet the peculiar demands of a project company dependent on the continuing support of lenders, the norm is for performance bonds to be on-demand. These fall away on completion when the operational phase starts, with revenue stream coming on line to meet debt service payments. In some markets (such as the United Kingdom PFI market) the need to balance the lenders' desire for near immediate cash against the surety's concerns about wrongful calls has been met by 'adjudication bonds'. These are bonds which respond on the delivery of an adjudicator's determination that sums are due to the

---

23 Whilst lenders think in this way, liquidated damages should be determined by reference to the loss suffered as a result of the delay, not the debt service level. In most projects profits will exceed the debt service level, but not in all and not necessarily during and shortly after first start-up.

beneficiary. As with adjudication generally, the adjudicator's decision is not final and while the surety is obliged to pay the sum found to be due by the adjudicator immediately, the amount finally due (whether more or less) will be determined or agreed later. Bonds of this nature are not often used in international projects.

Chapter 3
Dispute Avoidance and Resolution

I. INTRODUCTION

Why are dispute resolution provisions important? This is not a rhetorical question. Parties frequently spend significantly less time and effort drafting and negotiating the dispute resolution provisions in their contracts than they do on the commercial and financial terms. Indeed, such provisions are often found towards the end of the contract amongst provisions like notice and delivery/service requirements, and parties sometimes simply use boiler plate dispute provisions or precedent language from another contract, without considering whether that language is appropriate for the agreement/project in question. In other cases, they may fail altogether to include dispute resolution provisions in their contract.

A lack of attention to dispute resolution provisions may be caused by a variety of factors, including a reluctance by the parties to acknowledge that problems might arise in their relationship in the future and/or pressure to finalize the negotiations and execute the contract, which may prevent parties from undertaking a considered review and analysis of what type of dispute resolution provisions would be most appropriate for the contract/project in question.

Whatever the reason, a party may find that it is seriously disadvantaged as a result of having spent little (or no) time when the contract was prepared addressing the appropriateness of the dispute resolution provisions. This is particularly true if the contract is international in nature, as a party may find itself exposed to disputes being resolved in the local courts of foreign jurisdictions. The laws and procedures of that jurisdiction may be very different from those the party is familiar with, or would wish to apply to its dispute.

II.   DISPUTE AVOIDANCE

Owners, employers and contractors involved in international construction projects generally all want to see work delivered on time and within budget. If lenders are financing the project, they too will want work completed on time and within budget, to ensure that the project will generate the anticipated revenues during the finance repayment period. Avoiding disputes by identifying, investigating and discussing problems at an early stage is one means of achieving these objectives, and parties have increasingly recognized that developing and using multi-tiered dispute resolution provisions tailored to the project in question can assist by:

- allowing parties the flexibility to resolve low value and/or less important problems more swiftly, and with less drain on cost and management time, than would be possible in litigation or arbitration;
- enabling parties to keep the project going, whilst avoiding or minimizing disruption to the project and its completion within the required time; and
- ensuring that there are appropriate means available, if necessary, for resolving disputes that: (i) require a more rigorous and comprehensive review of the circumstances giving rise to, and consequences flowing from, the events in question and their impact on the parties' contractual and legal rights; and/or (ii) may be capable of being resolved only by means of the litigation or arbitration process.

Parties may decide to create their own bespoke tiered dispute resolution provisions (used, for example, for the construction of the Channel Tunnel during the 1980s). Alternatively, many of the standard form contracts used on international construction projects now include multi-tiered dispute resolution provisions. For example, each of the most recent editions of FIDIC's three main standard form contracts (the Red Book, Yellow Book and Silver Book) provides for disputes (including disputes related to the engineer's determinations of claims, or, in the case of the FIDIC Silver Book, determinations made by the employer or its representative) to be determined in the first instance by binding adjudication. This is followed by ICC arbitration in the event that one party is dissatisfied with the adjudication determination.

To take another example, the ENAA model forms for power plant construction and process plant construction provide for disputes to be resolved by mutual consultation between the parties, followed by (in certain specified cases) reference to an expert, whose decision is final and binding unless either party refers the dispute to ICC arbitration. Again, the AIA *A201 – 1997 General Conditions of the Contract for Construction* also provide for tiered dispute resolution provisions whereby claims (with the exception of those relating to hazardous materials) are in the first instance referred to the architect for his decision. A party dissatisfied with that decision may, within 30 days of receiving the decision, provide notice that it intends to refer the matter to arbitration. Prior to arbitration, the parties are required to endeavour to resolve the claim by mediation (in accordance with the Construction

Industry Mediation Rules of the American Arbitration Association (AAA)). If mediation does not resolve the dispute, the matter may be arbitrated in accordance with the AAA's Construction Industry Arbitration Rules.

Settling any doubts about the enforceability of tiered dispute resolution clauses, the English courts have enforced provisions whereby parties have contractually agreed to submit to some form of alternative dispute resolution (ADR) as a stage of their dispute resolution processes. In particular, the Courts have required parties to undertake a particular stage of their dispute resolution processes (e.g. mediation, or review by a dispute board) before allowing them to proceed with court or arbitration proceedings that are intended to finally resolve the dispute.[1] Courts in the United States have also enforced contractual provisions requiring parties to engage in an agreed ADR process prior to pursuing other remedies.[2]

Although multi-tier dispute resolution provisions are now the norm in major international construction projects, such provisions need to be tailored to each individual project as they do not all require the same number, or types, of tiers. Including more tiers than the nature of the project requires may be counterproductive, in that it: (i) unnecessarily increases the cost and time required to resolve disputes; and/or (ii) hinders final resolution of the matter by unduly complicating and/or prolonging the dispute resolution process.

Determining which tiers of dispute resolution procedures may be appropriate to a particular project will depend on a number of factors, including:

- the size of the project from a monetary perspective;
- the scope and expected duration of the project;
- the jurisdictions in which the parties are based;
- the location of the project;
- whether the parties have an existing (or developing) long-term, ongoing relationship; and
- whether there are multiple parties (e.g. sub-contractors, lenders) and/or multiple agreements involved in the project.

For example, if a project involves a long-term concession, the parties may want to include senior management discussions and some form of ADR as procedures in their dispute resolution provisions, as these can provide a contractual framework that supports the ability to maintain and improve the parties' ongoing working relationship. On the other hand, if an agreement is simply for the construction of a facility, with no ongoing operating or maintenance relationship between the parties, parties might consider it less useful to include mandatory management discussions and ADR provisions. They may instead want to ensure that disputes can be quickly

---

1 See *Cable & Wireless plc v IBM United Kingdom Ltd* [2003] BLR 89 and *Channel Tunnel Group Ltd v Balfour Beatty Construction Ltd* [1993] AC 334.
2 See *HIM Portland, LLC v Devito Builders, Inc.*, 211 F.Supp.2d 230 (1st Cir. 2002) and *Kemiron Atlantic, Inc. v Aguakem International, Inc.*, 290 F. 3d 1287, 1290 (11th Cir. 2002).

referred to adjudication so that the dispute can be resolved relatively swiftly with the parties obliged to implement the decision (and get on with completing the project) pending final resolution of the dispute through arbitration or litigation.

Whatever tier(s) of dispute resolution procedures parties include in their contracts, they should also include express provision that the parties are obliged to continue to perform their obligations pending resolution of any dispute.[3]

The primary advantages and disadvantages of the various forms of dispute resolution procedures are discussed briefly below. The procedures are discussed in greater detail in later chapters.

III.   OPTIONS FOR TIERED DISPUTE RESOLUTION PROCEDURES

A.   MANDATORY DISCUSSIONS

Concern about the adversarial nature of construction contracts has led to the industry's operations being examined in a number of jurisdictions. For example, reports in the United Kingdom such as the Reading report, *Trusting the Team*,[4] and the Latham report, *Constructing the Team*[5] (a joint industry and government review of the construction industry), advocated dispute *avoidance* in the first instance and called for problems to be identified early.[6]

In particular, these reports encouraged participants in the construction industry to avoid resorting to any formal dispute resolution procedures, by taking steps to resolve problems at the lowest possible level within their own organizations as quickly as possible. Only if no solution can be found is the problem then referred to higher levels of management. Interestingly, this concept conflicts with the introduction of the compulsory adjudication provisions of the UK Construction Act 1996 (discussed in detail later in this chapter) and other jurisdictions which provide a right to refer disputes to adjudication *at any time*. The result of this is that either

---

3   However, as discussed later in this chapter, in some jurisdictions a contractor may be entitled to suspend performance of its obligations in situations where it has not been paid monies that it claims are owing to it under the contract.
4   John Bennett and Sarah Jeyes, *Trusting the Team: the best practice guide to partnering in construction* (Thomas Telford, 1995).
5   Michael Latham, *Constructing the Team: Final Report of the Government/Industry Review of Procurement and Contractual Arrangements in the UK Construction Industry* (HMSO, 1994).
6   See also: Construction Industry Review Committee, *Construct for Excellence* (2001), a report on construction practices in Hong Kong, available at: http://www.wb.gov.hk/archives/index.aspx?langno=1&nodeid=35 (accessed 27 October 2005); and Construction Industry Council and Department of Environment, *Building our Future Together: Strategic Review of the Construction Industry* (Stationery Office Dublin, 1997), a report on an investigation into the construction industry in the Republic of Ireland.

party can always leapfrog the negotiations process and immediately refer disputes to adjudication.

Of course, parties are always free to negotiate at any time, so why should they include specific provisions to do so in their contract? One main reason is that if such a provision were not included, some might perceive a party's request to negotiate as a sign of weakness.

The primary advantages of contractually specifying that potential disputes must in the first instance be referred for mandatory discussions at senior management level (e.g. managing directors or chairmen) are that:

- it requires parties to turn their minds to a problem early, perhaps before the problem becomes substantial in nature;
- involving senior management may overcome any deadlock over the problem or entrenchment of positions that may have developed at the project management or operational level;
- senior management may be aware of broader issues, or interests, related to the dispute, the project and/or the parties' relationship, which may be relevant to that particular party's approach to the problem and its possible resolution; and
- if senior management are able to resolve the problem, it is a much faster and cheaper way of resolving the dispute than commencing court or arbitration proceedings.

The primary disadvantages of mandatory senior management level discussions are:

- there may be no means of forcing another party's representative to engage in a meaningful and effective discussion. Under English law, for instance, such agreements to negotiate are unenforceable as: (i) the agreement lacks the necessary certainty, so that the courts have insufficient objective criteria to determine whether a party had complied with its obligation to negotiate; and (ii) the courts refuse to police the circumstances in which a party may withdraw from such negotiations.[7] Successful operation of such provisions in these cases is, therefore, largely dependent on the goodwill of the parties;
- inevitably, such discussions will not result in determination of legal questions of principle, but will be aimed at securing a commercial compromise. Accordingly, the outcome may not prove to be as useful in resolving future disputes of a similar kind as a binding determination imposed in formal dispute resolution proceedings; and
- one party may try to thwart the process by refusing to make itself available to participate in the discussions. However, the risk that dispute resolution procedures will be stymied in this manner can be avoided by ensuring that the

---

[7] *Courtney & Fairbairn Ltd. v Tolaini Brothers (Hotels) Ltd* [1975] 1 All ER 716; *Smith (Paul) Ltd v H & S International Holding Inc* [1991] 2 Lloyd's Rep 127; *Walford v Miles* [1992] 2 AC 128.

contract: (i) includes a limit on the time for holding such discussions; and (ii) allows disputes to be referred to the next stage once that period of time has expired, even if the management level discussions have not occurred.

Finally, it may be appropriate on certain projects to create a standing board of senior management representatives that remains in place for the duration of the project/contract, rather than arranging ad hoc meetings whenever a dispute arises. This technique is often adopted where partnering or alliancing arrangements are in place (see Chapter 4) or the contract is for a substantial period of time. The board members will usually:

- meet regularly to try to identify and resolve problems before they develop into disputes; and
- when necessary, engage in mandatory discussions to attempt to settle disputes.

B.   ADR – Alternative Dispute Resolution

Other means of resolving disputes (in addition to adjudication, arbitration or litigation) are available if principal-to-principal negotiations fail. Typically this involves the parties engaging a neutral third person to help them reach an acceptable settlement. Such procedures, generally known as ADR, aim to encourage the parties to acknowledge the weaknesses of their own case and the strengths of their opponent's case, and to recognize the wider commercial implications of the dispute. ADR is not necessarily intended to decide on the merits of the issue or to lay blame at one party's door. It is essential to recognize that this procedure does not attempt to achieve a comprehensive review of facts or legal or technical issues – the aim is simply to reach a commercial solution. For this reason (among others), ADR is often considered to work best where the parties lack an understanding of each other's case and where the parties have a genuine desire to remain in a long-term relationship. In consequence, ADR will often prove more successful at the start of a project than at the end, when there may be no ongoing commercial connection or prospect of an ongoing relationship.

Terminology used to describe the varying forms of ADR is not universal, and can sometimes be interchangeable. However, the main types of ADR used on international projects are as follows.

1.   Mediation and Conciliation

The parties may engage an impartial third party to act as mediator or conciliator. There is no consistently recognized distinction between the processes of mediation and conciliation. To the extent that there is a difference, a conciliator may be more likely to issue a non-binding recommendation to the parties. Because of this lack of differentiation, the term 'mediator' is used in this book to include conciliators.

Dispute Avoidance and Resolution 61

A mediator does not make decisions for the parties, but instead meets with the parties to assist them to negotiate their own solution to the dispute. The parties can agree whether the mediator should consult with them separately and/or jointly. The mediator cannot compel the parties to reach a settlement, but he may take a very persuasive approach.

As discussed in more detail in Chapter 6, the procedure for a mediation or conciliation will generally involve:

- appointment of a suitably qualified mediator, in whom both parties have confidence;
- exchange of a short summary of each party's case, appending only key documents;
- a relatively brief meeting (e.g. one to two days), with the opportunity to address the mediator in the absence of the other party;
- an opportunity for the parties to negotiate face to face, with or without the mediator present; and
- the mediator's assistance in drawing up any agreement reached by the parties.

The advantages of mediation over litigation and arbitration are that it is:

- swift, with the whole procedure designed to produce agreement within weeks/months rather than months/years, thus providing an opportunity to avoid an antagonistic drawn-out dispute that ties up management resources and damages on-the-job relations;
- relatively cheap, principally as a result of the length of time taken and the lesser emphasis on the proof of rights;
- flexible, as mediation enables the parties and the mediator to take account of matters other than the strict contractual and legal positions such as the future commercial relationship of the parties. For this reason mediation is sometimes described as being an 'interest based', as opposed to 'rights based', method for resolving disputes; and
- confidential, with any offers to resolve the dispute made on a without prejudice basis, so that the offer may not be disclosed in any subsequent formal dispute resolution procedure.

However, there are also disadvantages, including:

- the risk that proposing mediation where there is no contractual requirement to ask for ADR will be perceived a sign of weakness: the thinking (historically) being that a claimant with a strong case will want to proceed straight to a binding determination;
- they have no teeth – the mediator cannot require either party to take any steps such as disclosing relevant documents damaging to its case;

- the outcome is non-binding – either party can refuse to reach an agreement or reject any recommendation that the mediator may make.
- there is no effective precedent value of a result in mediation, whether for future similar disputes between the same parties or in relation to disputes with third parties, such as sub-contractors;
- a frank early disclosure of each party's case can be damaging as it gives the other party the opportunity to receive and evaluate information that may not otherwise have been available until a much later stage in formal proceedings. Such early disclosure may permit a party to prepare its case better for arbitration or litigation. Accordingly, sometimes a party may agree to a mediation without ever having any real expectation of reaching a settlement;
- if it fails, it may be a waste of time and money;
- it will not produce a legally 'correct' result – the solution always represents a compromise; and
- the non-investigatory nature of the procedures may make them ill-suited to complex, technical disputes.

In addition to mediation, other types of ADR (also discussed in more detail in Chapter 6) include:

2. Mini- Trial/Executive Tribunal

Here a panel is created, usually consisting of a senior executive of each party, with a neutral chairman such as a retired judge or other senior lawyer. The parties make limited representations to the panel, which is responsible for reaching a negotiated settlement. The chairman may merely ensure fair play or may become involved in the process – much as a mediator would – to assist the parties to resolve the issue. This form of ADR may involve a more detailed examination of the parties' legal positions than mediation, and is considered by some to be best employed when there has been exchange of pleadings and disclosure of documents.

3. Early Neutral Evaluation

As an alternative to requesting assistance in the reaching of an agreement, the parties can obtain an independent evaluation of both the technical and legal issues of their case by an independent expert or in the context of existing proceedings perhaps a judge or arbitrator. The intention is to provide the parties with a realistic view of the prospects of success, thereby encouraging settlement. However, if this process takes place close to the scheduled trial date, substantial costs will already have been incurred by the parties. That said, it may still be preferable to attempt to resolve disputes through this means than proceed to a full trial. The person who provides the evaluation will receive written submissions in advance, which will often (though not invariably) be supported by short oral submissions by the parties at a hearing.

This provides an opportunity to question the parties. Once the evaluation has been given, he is disqualified from being involved in the case further if it should continue.

C. FACTORS TO CONSIDER IN DRAFTING ADR PROVISIONS

If parties decide to adopt mediation or conciliation as part of agreed dispute resolution provisions, they have the option of conducting the procedure through an institution, or on an ad hoc basis.

Mediation and conciliation services with standard procedures are offered by institutions throughout the world, such as:

– the Centre for Effective Dispute Resolution (CEDR);
– the International Chamber of Commerce (ICC);
– the American Arbitration Association (AAA);
– the Chartered Institute of Arbitrators (CIA);
– the Hong Kong International Arbitration Centre (HKIAC); and
– the London Court of International Arbitration (LCIA).

Such institutions usually have a standard form of procedure and a method of appointing the neutral third party (mediator or conciliator) in the absence of agreement between the parties. For example, CEDR, based in London, has for a number of years specialized in providing mediation services in both domestic United Kingdom disputes and international disputes. There are also a number of specialized mediation service providers in the United States and other parts of the world. For example, in the last few years the ICC (based in Paris) has introduced its own set of rules (the ICC ADR Rules[8]) dealing with ADR procedures. Additionally, the United Nations International Trade Law branch (UNCITRAL) has published Conciliation Rules that parties can agree to adopt. Aspects of common mediation rules are considered further in Chapter 6.

Alternatively, parties can develop their own rules concerning the selection of the neutral third party and the conduct of the ADR procedure. Indeed, many companies have their own bespoke ADR procedures, which are introduced into their contracts as the first stage of the dispute resolution process, sometimes with an option to leapfrog straight to litigation or arbitration if both parties agree that ADR is not appropriate.

---

8   Note that the ICC refers to its ADR Rules as 'amicable dispute resolution' rules, as compared to 'alternative dispute resolution' which is what the term ADR is frequently understood to mean.

Finally, it should be remembered parties can always agree to participate in an ADR process for any disputes that have arisen, even if there is no ADR provision in their contract.[9]

D.   ADJUDICATION AND THE USE OF DISPUTE REVIEW BOARDS

In construction projects it is often important that the parties have a means of achieving binding decisions to resolve their disputes on an interim basis, so that work on the project can continue whilst the parties await the outcome of protracted litigation or arbitration proceedings (which can take months, often years, to complete). Many parties in international construction projects achieve this objective by including adjudication in their dispute resolution provisions.

Adjudication can take different forms. For example, it can involve a single person or a panel deciding the dispute on an interim basis. The decision maker(s) will often be technical specialists or experts in their fields. The use of panels of experts to act as interim decision makers has become increasingly popular in international construction projects.

In part, this has evolved form the role of the 'engineer' as decision maker in the first instance under various standard forms of construction contracts. For example, earlier editions of the FIDIC contracts provided that: (i) disputes would in the first instance be determined by the person appointed as engineer under the contract; and (ii) his decision would be binding upon the parties until such time as it was reversed by arbitration held in accordance with the ICC Rules of Arbitration (ICC Rules).

Sometimes contractors developed cynicism as to the engineer's impartiality. This is perhaps not surprising, as frequently a dispute will concern a decision made by the engineer in the administration of the contract, for example as to the amount of additional payment due to the contractor in respect of variations. In addition, the engineer is, of course, on the pay-roll of the employer. A contractor's cynicism is likely to be even more acute where the contract provides for disputes to be determined in the first instance by the employer's representative, who is a member of the employer's own organization. Lenders to projects also increasingly had concerns over the potential lack of independence of the decisionmaker and pressed for alternative means of decision making in the first instance.

As a result panels of experts with particular skills, expertise and familiarity with the type of project or industry in question began to be used during the 1980s and 1990s. Often these panels were chosen by the parties and appointed at the outset for the duration of the project. Panel members were paid (with the costs shared equally by the parties) regardless of whether or not any disputes were actually

---

9   As discussed above, the advantage of including mandatory ADR provisions in a contract is that it removes what can (at least historically) sometimes be a perceived stigma associated with being the party to suggest using ADR.

referred to them. Effectively, the parties created a standing panel of decision makers who would be available should the parties need to call upon them.

International construction projects where such panels have been used include:

- the Channel Tunnel Project, which had a standing panel (the Disputes Review Board) of five members, of whom three would form a tribunal to make an interim binding decision concerning a particular dispute;
- the Hong Kong Airport Project, which had a standing Dispute Review Group, formed of six members, plus a convenor, who came from various areas of expertise expected to be relevant to disputes that might arise on the project. Depending on the nature and complexity of a particular dispute, one or three members of the Group would be selected to decide a dispute;
- the Ertan Project in Sichuan Province, China (a hydroelectric power plant involving a dam, tunnels and underground power house complex), which had a standing, three-member Dispute Review Board to make interim binding recommendations concerning disputes arising out of the various civil engineering contracts forming part of the project work; and
- the Channel Tunnel Rail Link Project currently underway in the United Kingdom, which has two standing panels that determine disputes based on their subject matter. The technical panel (comprised of engineers) deals with construction-related disputes, and the finance panel deals with disputes concerning the financial provisions of the concession agreement.

In the United States, the concept of using a standing panel of impartial, qualified people to provide non-binding recommendations for resolving disputes developed during the 1970s and 1980s, when such boards were used on various construction projects, including the second bore of the Eisenhower Tunnel (Colorado), the Mount Baker Ridge Highway Tunnel (Washington), the Chambers Creek Tunnel (Washington) and the El Cajon Hydroelectric Project (Honduras). By the mid-1990s, it was reported that such boards had been used on:

- 68 projects that were completed as of January 1994;
- 98 projects that were ongoing as of January 1994;
- 162 projects that were planned as of January 1994; and
- 60 per cent of tunnel and underground projects bid in the United States in 1993.[10]

Historically, such boards were used in the United States largely for tunnel projects, but by the mid-1990s they were being used on other types of heavy construction work (for instance, highways), and building and process contracts, involving a

---

10 Matyas, Robert, A. A. Matthews, R. J Smith, and P. E. Sperry, *Construction Dispute Review Board Manual* (McGraw-Hill, 1996), pp. 10–11 and Appendix A.

variety of employers, including American Telephone & Telegraph, the Hawaiian Department of Transportation and the International Monetary Fund.[11]

Certain institutions and organizations now include adjudication provisions in their standard form contracts or terms and conditions applicable to international construction projects. For example, in the mid-1990s FIDIC introduced adjudication as a possible tier of dispute resolution in its various standard form contracts, and each of the current editions of FIDIC's Red Book, Yellow Book and Silver Book now include adjudication as the primary means of dispute resolution.[12] The World Bank also includes adjudication provisions in its Standard Bidding Documents – Procurement of Works (May 2005), which is the standard form of contract used on large scale civil works projects funded by the World Bank where the estimated costs are more than USD 10 million.[13] Additionally, the World Bank and other multilateral development banks and international financing institutions have prepared a Standard Bidding Document for Procurements of Works and User's Guide (July 2005) which such organizations intend to use for their own standard bidding documents relating to projects that they decide to finance. That document also includes adjudication provisions that have been accepted by various development banks (for instance, the Asian Development Bank and the European Bank for Reconstruction and Development) for use in their standard bidding documents.[14] Those adjudication provisions are based in part upon the adjudication provisions developed and adopted by FIDIC.

The AAA also publishes its own Construction Industry Disputes Review Board Procedures that parties can adopt. Additionally, the ICC introduced its own Dispute Board Rules in September 2004 (ICC DB Rules).[15] The ICC rules provide for 'Dispute Adjudication Boards' (panels that make binding interim decisions), 'Dispute Review Boards' (panels that make non-binding recommendations that become binding if neither party expresses dissatisfaction within 30 days of receiving the recommendation) and 'Combined Dispute Boards' which usually issue recommendations as a Dispute Review Board but may in certain limited circumstances make interim binding determinations as a Dispute Adjudication Board (e.g. if one party requests that the panel do so, and the other party does not object).

The adjudication procedures published by the World Bank, FIDIC, the AAA and the ICC are considered in more detail in Chapter 5. A table comparing the key features of each procedure is also set out in Chapter 5.

---

11 Ibid.
12 See clause 20 of each of the current editions of the Red Book, Yellow Book and Silver Book (First Editions, 1999).
13 See http://www.worldbank.org (accessed 27 October 2005).
14 For example, see the Asian Development Bank's Standard Bidding Document (November 2004), available at: http:// www. adb. org/ Procurement/ prequalification- bid- documents. asp (accessed 27 October 2005).
15 Available at: http://www.iccwbo. org/ drs/ english/ dispute_ boards/ all_ topics. asp (accessed 27 October 2005). For more information see chapter 5.

The interim binding nature of the adjudication panel's decision results from the parties contractually agreeing that they will implement and abide by the adjudicator's decision until such time as the dispute is finally resolved. The decisions are not enforceable by a summary procedure in the same way as arbitral awards or judgments. Instead, a claim for breach of contract has to be brought against a defaulting party. The enforcement and grounds for attack of such decisions is considered further in Chapter 5.

Common features of adjudication panels in international construction projects include:

- a panel consisting of three (or sometimes five) members, with each party choosing a panel member (whose appointment is approved by the other party) and the parties, in consultation with the other panel members, selecting a third person to act as chair of the panel. Sometimes a sole adjudicator will be appointed instead;
- a requirement that panel members be (and remain) independent of the parties and impartial. The panel members chosen by the parties do not act as advocates for, or representatives of, the party that nominated the member;
- provision that the panel's discussions be confidential in respect of any non-parties or subsequent litigation or arbitration proceedings;
- a requirement that the panel members keep themselves regularly informed of the progress of the project, for example by reading progress reports and other materials provided jointly by the parties, and making regular visits to site (e.g. every three months);
- frequently, a requirement that panel members have expertise and skills in the type of project/industry in question;
- provision for the panel to take the initiative in investigating and ascertaining the facts and law related to the dispute; and
- a requirement that the panel render decisions (usually written) within a relatively short period of time (often within three months of a dispute being referred to the panel).

A number of different terms may be used to describe adjudication panels. For example, the FIDIC standard form contracts refer to 'Dispute Adjudication Boards', whilst the World Bank refers to 'Dispute Review Boards' and Dispute Boards. These panels all make interim binding decisions.

The terms 'Dispute Board' and 'Dispute Review Board' may have different meanings to parties from different jurisdictions. Thus it is important to understand the effect that the board's decision will have on the parties. This determines whether the proposed panel or board is an adjudication panel imposing a binding interim decision on the parties (as provided for in the standard form contracts of FIDIC, the World Bank and Dispute Adjudication Boards created pursuant to the ICC DB Rules), or simply makes a non-binding recommendation (as is common in the United States and with Dispute Review Boards created pursuant to the ICC DB Rules). It is expected that the parties' acceptance and implementation of the non-binding recommendation will follow from their confidence in the board in question, the theory being that recommendations by a board comprised of impartial members with the appropriate expertise and understanding of the project in question should be persuasive to the parties. It has also been suggested that in some cases a party will be encouraged to accept and implement a non-binding recommendation where the contractual provisions establishing the board and its powers and procedures provide that its recommendations are admissible, to the extent permitted by law, as evidence in subsequent litigation or arbitration proceedings concerning the dispute in question. For example, Article 25 of the ICC DB Rules provides that a recommendation made by a Dispute Review Board is admissible in any subsequent judicial or arbitral proceedings between the parties.

Reference to adjudication panels/boards in the remainder of this chapter are (unless indicated otherwise) to panels/boards making binding interim decisions.

E. ADVANTAGES AND DISADVANTAGES OF ADJUDICATION

The advantages of adjudication can be summarized as follows:

- it provides the parties with a dispute resolution process that is relatively swift and less costly than court or arbitration proceedings (of course such proceedings may be avoided only temporarily if a party is dissatisfied with the decision of the adjudication panel);
- it allows the parties to continue with their project/contract while the dispute is being resolved. This can be particularly important on major construction projects where the parties will normally all want to ensure that work continues, particularly in light of the significant costs that may be incurred if work is stopped for only a relatively short period of time; and
- it provides the parties with the opportunity to select decision makers with appropriate skills, technical expertise and experience for the project/contract in question. This provides the benefit of having decisions made by technically qualified people who will understand and be familiar with the often complex technical issues that arise in international construction projects. At the same time, the parties have the comfort of knowing that if the panel of experts makes a flawed decision (on a factual or legal basis), it will still be possible to have a court or arbitral tribunal re-examine and possibly set aside the panel's decision.

The drawbacks of adjudication (and possible means of overcoming certain drawbacks) are:

- although the parties gain the benefit of having decisions made by people with technical expertise, sometimes such panels make decisions that are more

'technical' than 'judicial' in nature. That may not be appropriate for the final determination of the dispute. Similarly, panel members with technical expertise may not be best suited to deciding mixed questions of fact and law. (These disadvantages are to an extent alleviated by the fact that the panel's decision may later be reviewed and possibly set aside in arbitration or litigation);
- panels are commonly required to provide decisions within a relatively short period of time (typically three months or so). This means the panel may not be able to conduct an in-depth and rigorous analysis of all the issues (factual and legal) that the parties consider to be relevant to their dispute;
- the panel does not have the ability (absent agreement of the parties) to require disputes arising under different contracts to be joined or consolidated with disputes arising under the contract in question, even if the same parties are involved in both disputes or the disputes arise out of the same events or circumstances (e.g. disputes in relation to a main contract and a sub-contract dispute). Parties may attempt to overcome the risk of inconsistent decisions by including provisions in their contracts for joinder and/or consolidation of related disputes before one panel. Alternatively, they might enter a stand alone 'umbrella' dispute resolution agreement addressing all of the contracts and allowing related disputes to be heard before one adjudication panel;
- the decision of the panel is enforceable only as a matter of contract, not as a judgment of the court or award of an arbitral tribunal. It can therefore be more time consuming and costly to enforce the panel's decision in the event that one party does not comply with it;
- although the panel's decisions may have persuasive value, they are not binding precedents for similar disputes that might arise between the parties in future; and
- the panel's proceedings may prove difficult to control due to the wide freedom generally given to the panel to act as experts and investigate the facts. For example, under English law, experts are not bound to follow rules of natural justice and may make their own investigations and reach decisions without referring back to the parties for their comments, and there is no statutory framework such as exists for arbitration, (namely, the English Arbitration Act 1996), which applies to check perceived procedural unfairness. These issues are addressed in more detail in Chapter 5. Such risks can be mitigated to an extent by drafting procedures designed to provide parties with greater control over, or at least involvement in, the process of fact finding and determination. For example, as a minimum it is prudent to provide that all communications that each party has with the panel be simultaneously copied to the other party. Also, occasionally provision is made for a panel to provide a draft of its decision for comment by the parties.

F. FACTORS TO CONSIDER IN DRAFTING ADJUDICATION PROVISIONS

Assuming parties want to include adjudication (i.e. an interim binding decision) as a tier of dispute resolution, they will need to consider and address a number of issues including:

- whether the parties want a single person, several people (typically three) or one or more people selected from a panel (typically five to ten people) to act as adjudicator(s);
- whether the parties want to create their own bespoke adjudication provisions, and if so:
    - what type of powers they want to give to the panel (e.g. do the parties want the panel or board to also be able to make non-binding recommendations and if so, in what circumstances)? There is no firm precedent here to assist in making this decision. For instance, the FIDIC Red Book provides that binding decisions will generally be made but permits parties to agree jointly to refer a matter to the Dispute Adjudication Board for its non-binding opinion (see clause 20.2). This option is not, however, found in the basic versions of the FIDIC Yellow or Silver Books. Another approach (as discussed above) is found in the ICC DB Rules which allow parties to create a 'Combined Dispute Board' which primarily makes non-binding recommendations, but can in specified circumstances make binding determinations;
    - will the panel be able to render a decision on the basis of a majority opinion, or will it need to be a unanimous decision?
    - what type of procedures relating to the adjudication do the parties want to include in the contract (e.g. will the parties have a right to request a hearing)?
    - the allocation of the costs and expenses of the panel;
    - the extent to which the panel will have the power to take the initiative in ascertaining and investigating the facts and law related to the dispute;
    - whether the panel may appoint its own advisers to assist on matters of legal interpretation or areas outside its expertise (and if so, whether the prior consent of the parties is required in advance of the panel making any such appointment);
    - the timescale for the panel to provide its decision;
    - whether the panel will have powers to award interest and/or costs;
    - confidentiality;
    - whether the panel members will be barred from acting as witnesses in any subsequent litigation or arbitration;

- the panel's powers in the event that one party refuses to participate in a reference. For instance, can the panel proceed to a determination if a party is absent or refuses to participate?; and
- will the sole adjudicator or panel be appointed from the time that the parties' agreement is entered into (or shortly thereafter) until the end of the construction of the project (i.e. a standing adjudicator/panel), or will an adjudicator or panel only be appointed only if and when a dispute arises (i.e. ad hoc)? The primary advantage of standing panels is that whenever a dispute does arise the members of the panel will have a high degree of familiarity with, and understanding of, the project and its progress and the panel will generally be readily available to render a decision within a relatively short period of time.[16] Practically, a standing panel may also be able, if so desired by the parties, to act as an informal sounding board when issues first arise and before they are formally referred to dispute resolution.

Alternatively, the parties might decide to adopt existing adjudication provisions, such as those that have been developed by FIDIC, the World Bank, the AAA or the ICC (see Chapter 5 for a comparison of the key features of the institutional rules). Other factors to consider in deciding whether to include adjudication in your dispute resolution provisions include:

- the fact that although the parties can initially select panel members with expertise from a range of disciplines, it is of course possible that the panel may lack the expertise to deal with a particular dispute that arises. One way to address this is to provide that the panel has the power to obtain external expert advice if required; however, the associated costs will have to be borne by the parties;
- the risk that the panel members may find it difficult to approach fresh disputes with an entirely uncoloured view of the merits, in light of information received in relation to previous disputes on the project;
- the fact that the use of a standing panel can be costly, particularly if the project is of a significant duration; and
- the fact that the increasing popularity and use of standing panels may result in it becoming difficult to find suitably qualified people who are available to accept relatively long term appointments as panel members.

Finally, as with the other methods of dispute resolution we have previously discussed, it is important that parties who include adjudication provisions in their contract also include either litigation or arbitration as the final tier of dispute

---

16   For example, each of FIDIC's Red Book, Yellow Book and Silver Book provides for decisions to be made within 84 days of a dispute being referred to the Dispute Adjudication Board (see clause 20.4).

resolution, unless they want the determination by the adjudicator(s) to be final and binding. Additionally, it may be prudent to provide that any dispute regarding non-compliance with a panel's decision will be referred directly to arbitration (or litigation) to avoid having first to have recourse to the panel.

G.   STATUTORY ADJUDICATION

1.   The United Kingdom Experience

The adjudication procedures described above involve a contractually agreed form of dispute resolution. In the United Kingdom, radical legislation in the late 1990s (the UK Construction Act 1996), introduced mandatory fast-track adjudication for construction contracts carried out in England, Wales, and Scotland.[17] As a result, even if the parties have decided, for whatever reason, that adjudication is not an appropriate form of dispute resolution for their contract or project, if that project falls within the ambit of Part II of the UK Construction Act 1996, they may nevertheless have adjudication imposed on them in respect of certain types of dispute. This is also the case in some other jurisdictions, including New Zealand, Singapore and a number of the Australian states, which have also introduced legislation imposing compulsory adjudication on aspects of the construction industry. In addition, compulsory adjudication in the construction industry is being considered in Hong Kong, South Africa and some states in the United States.

Adjudication was not, of course, a new concept to the United Kingdom construction industry – families of standard form construction contracts in the domestic market, such as those published by the Joint Contracts Tribunal (JCT), included adjudication provisions in relation to payment disputes in their sub-contracts prior to the enactment of the UK Construction Act 1996. Adjudication was also seen on larger projects (e.g. the Channel Tunnel) with the use of a panel of experts as discussed above.

What was radical about Part II of the UK Construction Act 1996 was mandatory regulation by Parliament, focused on a specific industry and imposing procedures on parties to construction contracts that they might not wish to adopt. This occurred against the backdrop of, amongst other matters, the Latham report (referred to above). Compulsory adjudication was one of the pivotal recommendations in that report. The primary policy consideration was the belief that adjudication would provide a swift, cheap mechanism for sub-contractors to pursue claims, thereby preventing perceived abuse by contractors (and their clients).

Essentially, Part II of the UK Construction Act 1996 provides that fast-track adjudication (decisions are to be made within 28 days of a dispute being referred to adjudication) is available for all 'construction operations' carried out in England,

---

17   The UK Construction Act 1996 was extended to Scotland pursuant to the *Scheme for Construction Contracts (Scotland) Regulations 1998* (SI 1998/687).

Scotland or Wales (with certain limited exceptions). The term 'construction contract' is defined to include agreements for carrying out 'construction operations', and both 'construction contract' and 'construction operations' are broadly defined in the UK Construction Act 1996. In particular, these terms capture many activities that parties might not initially consider to be core construction activities. For example, 'construction contracts' is defined to include contracts for project management, construction management, architectural, surveying and advisory services. 'Construction operations' is defined to include, amongst other matters, providing labour for the carrying out of construction operations and external or internal cleaning of buildings and structures carried out in the course of constructing, altering, repairing, extending or restoring a building structure.

If an activity within an agreement falls within the statutory definition of an agreement for 'construction operations', Part II of the UK Construction Act 1996 provides that any party is entitled to refer a dispute arising in relation to those construction operations to adjudication *at any time*. In the case of multi-tiered dispute resolution provisions, this enables a party to 'leapfrog' over other dispute processes (e.g. mandatory senior management discussions or ADR) that might be provided for in the relevant agreement.

The extremely broad scope of Part II of the UK Construction Act 1996 is often surprising to parties, even if they have a general familiarity with English law. Indeed, the adjudication provisions in Part II of the UK Construction Act 1996 would have applied to the construction contract for the Channel Tunnel, at least in so far as those parts of the works carried out in England, if the UK Construction Act 1996 had been in force when the tunnel was constructed.

It is not possible for parties to contract out of the application of Part II of the UK Construction Act 1996. However, Part II of the UK Construction Act 1996 is territorial in that its application is limited to construction operations carried out in England, Scotland or Wales (and possibly offshore of those countries).[18] For example, it would not apply to construction projects in the Middle East, even if the parties selected English law as the governing law for their contract.

Part II of the UK Construction Act 1996 sets out a series of eight minimum requirements that must apply to any adjudication relating to 'construction operations' falling within the ambit of the UK Construction Act 1996.[19] These are:

1. Each party shall have the right to give notice *at any time* of its intention to refer a dispute to adjudication.
2. A timetable must be provided with the object of securing the appointment of the adjudicator and referral of the dispute to him within seven days of such notice.

---

18 The English courts have not yet expressly considered whether Part II of the UK Construction Act 1996 applies to oil and gas facilities located offshore of the United Kingdom. It appears, however, that it may not, based on the language in Part II and other relevant legislation.
19 S. 108(5).

3. The adjudicator is required to reach a decision within 28 days of referral or such longer period as is agreed by the parties *after* the dispute has been referred.
4. The adjudicator is allowed to extend the period of 28 days by up to 14 days with the consent of the party by whom the dispute is referred.
5. The adjudicator has a duty to act impartially.
6. The adjudicator must be able to take the initiative in ascertaining the facts and the law.
7. The decision of the adjudicator is to be binding until the dispute is finally determined by legal proceedings, arbitration or agreement. Alternatively, the parties may agree to accept the decision of the adjudicator as finally determining the dispute.
8. The contract must provide that the adjudicator is not liable for anything done or omitted in the discharge or purported discharge of his functions as adjudicator unless the act or omission is in bad faith (and any employee or agent of the adjudicator is similarly protected from liability).

In the event that parties fail to provide for adjudication in their agreement (or they do provide for adjudication but fail in their contractual provisions to provide for any one of these eight compulsory requirements), a statutory scheme of adjudication set out in the UK Construction Act 1996 (the Scheme) will automatically apply to the parties and their dispute.[20]

The adjudication process required by the UK Construction Act 1996 is extremely fast and the adjudicator must reach his decision within 28 days unless both parties agree to extend the time after the dispute has been referred or the claimant agrees to an extension of 14 days. It is also weighted in favour of the claimant. For example, the claimant can prepare its submissions at leisure and then refuse an extension of time, putting great pressure on the respondent to prepare its defence both quickly and effectively.

The Scheme provides a framework for the adjudication but does not set out detailed procedural steps, for example, specifying when the respondent must deliver its submissions. It also does not provide that a party is entitled to a hearing. Other concerns include the fact that the UK Construction Act 1996 and the Scheme do not expressly contemplate adjudication by a panel. Further, while the Scheme does provide for joinder of related disputes if all parties agree, that may prove difficult to achieve once a dispute has arisen and has to be determined within the applicable short timetable.

There are an increasing number of organizations in the United Kingdom with their own adjudication procedures, including the Construction Industry Council Model Adjudication Procedures (CICMAR), Technology and Construction Court Solicitors Association (TeCSA) (formerly known as the Official Referees' Solicitors Association), the JCT, the ICE and the Technology and Construction Court Barristers Association (TeCBar). Some may prefer to incorporate one of these organization's

---

20 S. 108(5).

procedures (possibly with amendments)[21] or to prepare self-standing adjudication procedures tailored to meet the requirements of the particular project (and the eight minimum requirements set out in Part II of the UK Construction Act 1996) rather than relying on the Scheme.

The provisions of Part II of the UK Construction Act 1996 apply only so far as the contract relates to 'construction operations'.[22] This can result in some aspects of a contract being subjected to the UK Construction Act 1996, but not others. For example, in a turnkey contract for civil and process plant works, the latter part of the works will generally be excluded from the definition of construction operations by virtue of section 105(2). Similarly, a services contract which includes, as one element, maintenance of a building, will only partly come within Part II of the UK Construction Act 1996.

Different dispute resolution provisions may therefore apply, depending on the category of the works concerned. This situation is clearly unattractive as it may result in jurisdiction disputes over the correct method of dispute resolution. Clients and contractors alike may therefore prefer to adopt adjudication procedures that are compliant with the UK Construction Act 1996 for all disputes arising under such a composite contract. This avoids the prospect of different procedures operating depending on the nature of the underlying dispute. Indeed, in cases where it is impossible to distinguish between the elements of the dispute falling within and outside the UK Construction Act 1996, this may be a necessity.

2. Excluded Activities and Contracts

A number of specific activities are excluded from Part II of the UK Construction Act 1996, including:

- drilling for, or extraction of, oil or natural gas and the extraction of minerals;
- assembly, installation or demolition of plant or machinery on a site where the primary activity is nuclear processing, power generation or water or effluent treatment;
- 'supply only' contracts for manufacture or delivery to site of:
  - building or engineering components or equipment;
  - materials, plant or machinery; or
  - components for systems of heating, lighting, air conditioning, ventilation, power supply, drainage, sanitation, water supply or fire protection, or for security or communication systems.

---

21   It is important where an organization's adjudication procedural rules are adopted, to ensure that those rules address the eight mandatory requirements in S. 108 of Part II of the UK Construction Act 1996.
22   S. 104(5).

However, 'supply and fit' contracts which also provide for installation of such equipment are *not* excluded.

Additionally, certain classes of construction contract are excluded by the Construction Contracts (England & Wales) Exclusion Order 1998 (SI 1998/648) (Exclusion Order), which excludes concession-style agreements entered into with the government or a public body under the private finance initiative provided certain requirements set out in the Exclusion Order are met.

The policy decisions behind certain of the exemptions were debated in Parliament. The exemptions for nuclear processing and power generation, for example, were justified on the basis that projects of this kind are generally for an identified client and well-funded. Relationships in these industries in the United Kingdom, where there are a relatively small number of long-term players, are generally perceived to be good and generally there are not the same problems in relation to non-payment down the contractual chain as there are on civil and building projects. The distinctions drawn in parliamentary debates, however, have not always been well reflected in the legislation. Design obligations in relation to plant or machinery for power generation, for example, are not excluded.

It would be prudent for a party who intends to engage in construction projects in England, Scotland or Wales, and is not familiar with the provisions in and application of Part II of the UK Construction Act 1996 to obtain specialist advice concerning the application of the UK Construction Act 1996. Ideally, such advice should be sought at the time that the dispute resolution provisions in the contract in question are being drafted and negotiated.

H.   THE FINAL TIER: LITIGATION OR ARBITRATION?

In major international construction projects, parties typically select arbitration, rather than litigation, as the final tier of dispute resolution. There are a variety reasons for this, including those discussed below.

Arbitration avoids recourse to the courts of any one jurisdiction, which is often important when the project involves parties from different countries (with the project/facilities in question possibly being located in yet another country). Arbitration is often seen, therefore, as providing a 'neutral' forum for deciding disputes before an impartial tribunal.

In addition, the enforcement of arbitration awards is greatly facilitated by the New York Convention,[23] which provides for what is effectively a fast track procedure for enforcing arbitral awards as compared to judgments from foreign courts (see Chapter 13 for more detail). A copy of the New York Convention appears at Annex 7.[24]

---

23   The Convention on the Recognition and Enforcement of Foreign Arbitral Awards (1958).
24   See also http://www.uncitral.org/uncitral/en/uncitral_texts/arbitration/NYConvention.html (accessed 27 October 2005).

Arbitration also provides the parties with the freedom to choose an 'arbitration friendly' venue for the 'seat' (or legal place) of the arbitration. Ideally the arbitration law of the seat will provide that the local courts are to exercise restraint in interfering with the conduct of an arbitration and provide a supporting role, intervening only in limited circumstances (for example, to check a lack of procedural fairness or bias on the part of an arbitrator). One indication of whether a jurisdiction that is being considered as the seat of the arbitration is 'arbitrationfriendly' is its adoption of the UNCITRAL Model Law on International Commercial Arbitration, adopted by the United Nations in 1985 (the UNCITRAL Model Law).[25] A copy of the UNCITRAL Model Law appears at Annex 8.

The objective of the UNCITRAL Model Law is to have numerous jurisdictions adopt it (subject to any special or unique requirements of the particular legal system in question) and by doing so to create an arbitration law that is recognized and applied internationally. The UNCITRAL Model Law is premised on the principle that local courts in the place where an international arbitration is being held should support, but not unduly interfere with, the arbitral process. Accordingly, the UNCITRAL Model Law provides, among other things, that:

- the courts play a limited role in the arbitral process (e.g. ordering interim relief; subpoenaing witnesses or documents; deciding challenges to the impartiality or independence of arbitrators);
- courts support the arbitral process by ordering a stay of court proceedings brought in breach of a valid arbitration agreement;
- any inherent jurisdiction of a court to interfere in arbitral proceedings be excluded;
- nationality should not preclude a person from acting as an arbitrator; and
- the parties all receive equal treatment and be given a proper opportunity to present their respective cases.

To date, over 40 jurisdictions have adopted the UNCITRAL Model Law, including Australia, Bahrain, Canada, Egypt, Germany, Hong Kong, Mexico, Singapore, Spain and some of the United States (Texas, California, Oregon, Illinois and Connecticut). However, in some cases a jurisdiction has adopted the UNCITRAL Model Law but made significant amendments to it, for example, permitting anti-arbitration injunctions to be made (as in India) or adding substantive grounds for review of arbitral awards (as in Egypt). Accordingly, if you are not familiar with the local laws of the place that you are considering selecting as the seat of your arbitration, it is prudent to obtain advice (prior to selecting that jurisdiction as the seat of the arbitration) from advisors who are both qualified in, and

---

25 United Nations Commission on International Trade Law Model Law on International Commercial Arbitration, available at: http://www.uncitral.org/uncitral/en/uncitral_texts/arbitration/1985Model_arbitration.html (accessed 28 July 2005).

experienced with, the laws of the jurisdiction under consideration. This is advisable even if the jurisdiction has adopted the UNCITRAL Model Law.

The seat of the arbitration is also relevant to enforceability of the award – ideally the country selected will be a signatory to the New York Convention. Again, this should be checked prior to agreeing the location of the seat.

Arbitration provides a final and binding award, generally not subject to any appeal. There are exceptions to this principle, depending on the law of the seat of the arbitration. In England, for example, under the English Arbitration Act 1996 there remains a limited right of appeal to the courts on a question of law. (Grounds for challenge and appeals from an arbitral award are discussed in more detail in Chapter 13.)

Arbitration also provides the parties with the advantage of privacy – arbitral hearings are conducted in private. It is often assumed that parties to an arbitration (and the tribunal members) are bound to keep information concerning the arbitration, and any resulting award, confidential. However, the laws of different jurisdictions may treat the existence of an implied obligation of confidentiality differently. Accordingly, if an arbitration agreement (or the procedural rules of a selected institution) does not expressly impose an obligation of confidentiality, national laws governing the arbitration become relevant to determine whether such an obligation exists. For example, in *Esso/BHP v Plowman*,[26] the Australian High Court rejected the prevailing English law view that parties are subject to a general implied duty of confidentiality.

Parties should also be aware that national laws may consider there to be exceptions to an implied general duty of confidentiality. For example, in *Hassneh Insurance Co of Israel v Mew*,[27] the English courts held that a party (who was not subject to an express confidentiality obligation) could disclose an award and the reasons behind the award to a third party against whom it had a claim, in order to establish the basis of the party's right to claim against that third party.

Additionally, even where institutional arbitration rules impose a duty of confidentiality on the parties, some institutions may nevertheless publish certain limited information (redacted or edited to preserve the parties' anonymity) concerning arbitral awards. For example, the ICC publishes summaries of arbitral awards, although it will not provide copies of the award (Article 28 (2) of the ICC Rules). Article 27(8) of the AAA's International Arbitration Rules (AAA Rules) provides that the AAA/International Centre for Dispute Resolution (ICDR) may publish or otherwise make publicly available selected awards, decisions or rulings (edited to conceal the parties' names or other identifying details), unless agreed otherwise by the parties. Article 32(5) of the UNCITRAL Arbitration Rules provides that the consent of the parties is required in order for an award to be made public.

A final advantage of arbitration is that the parties are free to choose their tribunal by agreement. This, in the authors' view, is a very significant advantage over

---

26 [1995] 128 ALR 391.
27 [1993] 2 Lloyd's Rep 243.

litigation. Parties can make a selection based on a number of criteria, including their own past experience of an arbitrator, knowing how he is likely to approach legal and procedural issues. For example, their case may be more likely to find favour before a commercially minded arbitrator as opposed to a technical, 'black-letter' academic lawyer.

However, arbitration (like all other forms of dispute resolution) can have its drawbacks. In particular:

- it is not necessarily a faster or cheaper means of resolving disputes than litigation. In fact arbitration may in some cases be more costly as parties must pay the fees and expenses of the arbitral tribunal, as well as administrative institutional costs if the arbitration is administered by an institution. Additionally, the members of the arbitral tribunal may have limited availability (due to other commitments) to hold and conduct hearings. This can be particularly difficult if there is a three person arbitral tribunal and it is necessary to find times when all three members of the panel are available. As a result, some arbitral institutions are actively seeking to address the issue of the time taken to complete an arbitral proceeding. For example, the ICDR (the international division of the AAA) claims that by careful selection of arbitrators and management of the process, certain arbitrations can be started and completed within approximately one year;
- arbitral awards (which are based on contractual agreement) are not binding on non-parties to the contract, and generally arbitrators do not have power to join or consolidate related proceedings unless all of the relevant parties involved agree to do so. As indicated above, this can be cured by providing for joinder and consolidation in all relevant contracts (or having all parties enter into an umbrella dispute resolution agreement). Alternatively, a degree of protection can be obtained by suitable selection of institutional arbitration rules. In particular, Article 22.1(h) of the LCIA Rules provides that an arbitral tribunal has the power (unless agreed otherwise by the parties) to order that a third party be joined in the arbitration as a party, provided that both the third party and person(s) applying to have the third party added to the proceedings agree to the joinder. The ICC Rules do not go as far, although Article 4(6) provides that the ICC International Court of Arbitration may, at the request of a party, include in existing proceedings claims made in a subsequent Request for Arbitration filed in connection with a legal relationship between the same parties as are involved in the existing proceedings, provided that the Court has not yet approved or signed the Terms of Reference in the existing proceedings;
- enforcement of the award will be problematic in countries that are not signatories to a treaty recognizing international arbitral awards (including the New York Convention). Moreover, the practical difficulties of enforcing an award in a hostile environment should not be overlooked. It is prudent to seek local law advice in respect of prospects for enforcement of an award in jurisdictions where the other party's principal assets are located.

Some major international construction projects do include litigation rather than arbitration as the final tier of dispute resolution. The primary advantages of litigation, as compared to arbitration, are that:

- state courts often have the power to order that related proceedings be joined or consolidated and heard together (even if the parties do not agree to do so), thus minimizing the risk of inconsistent decisions being made;
- state court judgments may bind third parties; and
- state courts often also have other powers that an arbitral tribunal does not have, including the power to order interim relief and to subpoena witnesses.

The choice of arbitration versus litigation to resolve disputes on international construction projects will depend to an extent on the particular nature of the project. Generally, however, the main disadvantages of choosing litigation rather than arbitration to resolve international construction disputes are that:

- it is often more time consuming and costly to enforce a court judgment in a foreign jurisdiction than an arbitral award, particularly if the foreign jurisdiction has ratified the New York Convention;
- use of national courts may give a party a 'home court' advantage;
- court proceedings are usually public;
- Parties usually are not able to ensure that the judge(s) who will decide their dispute is either experienced in, or knowledgeable about, construction-related matters;[28]
- generally, parties must follow prescribed rules of procedure that cannot be adapted to fit any specific needs of the dispute or particular objectives of the parties; and
- the dispute may be subject to appeal(s) in relation to any pre-trial rulings made by the court and/or the trial court's judgment on the merits of the claim.

Attached as Annex 1 is a table that briefly summarizes the primary advantages and disadvantages of arbitration and litigation.

---

28 However, some jurisdictions may have specialist courts. For instance, construction related disputes in England and Wales are generally heard by the Technology and Construction Court.

I. FACTORS TO CONSIDER IN DRAFTING ARBITRATION PROVISIONS

Parties who opt for arbitration rather than litigation must consider a number of issues when drafting the arbitration provisions. In particular, such parties will have to:

- decide whether to have an institutional or ad hoc arbitration;
- choose the seat and language of the arbitration;
- determine the number of arbitrators (one person or a panel of three);
- determine whether any rights to appeal an arbitral award should be preserved;
- consider the ability to seek interim and injunctive relief;
- address multiple party and/or multiple contract situations; and
- consider the impact of limitation periods if the contract provides for multiple tiers of dispute resolution procedures.

These are each addressed briefly below.

1. Institutional Versus Ad Hoc Arbitration

An arbitral tribunal can be appointed, and the arbitration conducted, either in accordance with institutional rules adopted by the parties in their contract (e.g. the arbitration rules of the LCIA,[29] the Arbitration Institute of the Stockholm Chamber of Commerce,[30] the AAA[31] or the ICC[32]), or on an ad hoc basis.

An ad hoc arbitration is one where the parties either: (i) draft bespoke rules for the appointment of the arbitral tribunal and conduct of the arbitration (which allows them to tailor the rules to their particular requirements); or (ii) adopt model rules such as the UNCITRAL Arbitration Rules,[33] which are the best known and most widely used rules developed for ad hoc arbitrations.

The principal disadvantage of a purely ad hoc arbitration (i.e. one that does not provide for an appointing authority or institutional support or model rules of procedure) is that, practically, its effectiveness depends on the willingness of the parties to co-operate in reaching agreement on the appointment of the arbitral tribunal and conduct of the arbitration. In particular, there will be no institution to police the proceedings before the arbitral tribunal is appointed or to rule on applications to remove arbitrators. Thus, if one party does not wish to co-operate, it can result in the early administration of the arbitration being difficult and time consuming. Indeed, costly court applications may be required in order to ensure

---

29 Available at: http://www.lcia-arbitration.com (accessed 27 October 2005).
30 Available at: http://www.sccinstitute.com (accessed 27 October 2005).
31 Available at: http://www.adr.org (accessed 27 October 2005).
32 Available at: http://www.iccwbo.org/index_court.asp (accessed 27 October 2005).
33 Available at: http://www.uncitral.org/uncitral/en/uncitral_texts/arbitration.html (accessed 27 October 2005).

that a tribunal is appointed, though where the ad hoc arbitration is conducted in accordance with the UNCITRAL Arbitration Rules, this can be avoided since the Secretary General of the Permanent Court of Arbitration at the Hague can be asked to designate an appointing authority.

In contrast, choosing to conduct the arbitration in accordance with institutional rules means that there will be a professional organization available to administer and supervise the proceedings. Of course, institutions require parties to pay for such administrative and supervisory services. The institutions will also usually have a schedule of the fees and costs that are payable to the members of the arbitral tribunal. The ICC, for example, fixes the amount of the ICC's administrative expenses and the fees of the arbitrator(s) in all cases ad valorem on a sliding scale by reference to the amount in dispute. Other institutions (e.g. LCIA) simply have a fixed schedule of fees and costs.

The other main advantage of adopting institutional rules for an arbitration is that the procedural rules developed by the institution in question will usually have been 'tested', both in previous arbitrations and in international enforcement proceedings. This should provide some comfort to a party concerning the conduct of the proceedings and its ability ultimately to enforce an arbitral award.

Of course, before adopting any set of institutional rules, the parties need to review the rules prior to agreeing to use them. Such a review is necessary to ensure that the procedures prescribed by the rules are acceptable to the parties and/or appropriate for potential disputes arising in relation to the project/contract in question.

Parties will sometimes want to modify an institution's procedural rules. The parties' ability to modify institutional rules and have the institution in question agree to administer the modified procedural rules will depend in large part on the particular procedural rules, and the attitude of the institution, in question. For example, both the LCIA Rules and the ICC Rules expressly contemplate that certain rules can be modified by the parties[34] and the Construction Industry Arbitration Rules published by the AAA provide that parties may modify the procedures provided for in the rules[35] (but only with the consent of the arbitrator(s) if the modification is proposed to be made after the appointment of the arbitrator(s)). There are, however, clearly limits. It is difficult to see, for instance, how any institution could agree to parties attempting to disapply its fee structures or the ICC (for instance) agreeing that its process of review of awards before issue should not take place. Accordingly, when major changes are being contemplated to an institution's rules it is sensible to approach that institution before contracting in order to make sure that the amended procedures will be acceptable to it.

---

34 See for example, Art. 23 (1) (Conservatory and Interim Measures) and Art. 32 (Modified Time Limits) of the ICC Rules, and Art. 15.1 (Submission of Written Statements and Documents), Art. 22.1 (Additional Powers of the Arbitral Tribunal) and Art. 28.4 (Award of Arbitration and Legal Costs) of the LCIA Rules.
35 Rule 1(a).

2. The Seat of the Arbitration

As mentioned above, the seat of the arbitration is generally the place where the arbitration hearings are held (unless agreed otherwise by the parties and arbitral tribunal), and the selection of the seat is important as it determines what additional procedural rules may apply to the arbitration by virtue of the local arbitration laws. The seat of the arbitration is also important to the enforceability of the award, so parties should consider whether there are reciprocal enforcement arrangements between the jurisdiction in which the arbitration is seated and the likely place(s) of enforcement (in particular, whether both jurisdictions have ratified the New York Convention).

Other factors that should be considered in deciding the seat of the arbitration include the convenience of the venue for the parties, witnesses and arbitrators, and the geographical neutrality of the venue if the parties are from different jurisdictions. Ideally, the parties should specify the seat of the arbitration in the arbitration clause (versus waiting to select and agree the location of the seat after a dispute has arisen).

3. The Language of the Arbitration

The language of the arbitration should also be addressed in the arbitration clause. Factors to consider when choosing the language of the arbitration include the language(s) of the contractual documentation, the law governing the contract, and the languages spoken by the parties and any likely witnesses.

4. Number of Arbitrators

A primary advantage of choosing a panel of three arbitrators is that it provides more opportunity to obtain the appropriate mix of cultural or linguistic attributes and/or technical expertise that may be relevant to an international project. Additionally, having three arbitrators rather than one may provide parties with some level of comfort that it is less likely that all three arbitrators might fail to understand, or take account of, all of the relevant factual and legal issues in reaching their decision. Of course the additional costs involved in having a three person tribunal may not be warranted if the contract in question is not of a particularly significant value.

5. Right to Appeal

Parties may choose arbitration over litigation because it generally results in a decision that is final, as the laws of many jurisdictions permit arbitral awards to be appealed only in limited circumstances, for example where there is fraud or the arbitral tribunal did not have jurisdiction to determine the dispute. Under the English Arbitration Act, parties can contract out of their right to appeal an award on a point of law, although they cannot contract out of the right to challenge an award on the basis of substantive lack of jurisdiction of the tribunal or serious irregularity affecting the tribunal, the arbitral proceeding or the award.[36]

Parties will need to consider whether to include provisions in their contract that expressly exclude the parties' right to appeal any arbitral award. If the parties have chosen to apply the arbitration rules of an institution, the institution's rules often deal expressly with the ability of the parties to seek recourse concerning an award. (For example, both the LCIA Rules and the ICC Rules provide that parties, by adopting the rules, waive their right to recourse insofar as such a waiver can be validly made[37]). The AAA Rules provide that awards are 'final and binding' on the parties (Article 27(1)).

The effectiveness of such an exclusion (whether in the parties' contract or the institutional rules agreed to by the parties) will need to be considered in the context of the law of both the seat of the arbitration and where the award is likely to be enforced.

6. Continuing Performance

Standard form contracts commonly used in international construction projects frequently provide that the parties shall continue to perform their obligations under the contract despite a dispute being referred to the dispute resolution procedures.[38] As mentioned above, if parties are using a bespoke contract (or their standard form does not include such a provision), the parties will usually want to include such a provision, in order to ensure that work continues whilst disputes are being resolved. However, the laws of the jurisdiction chosen to govern the contract and/or the law of the place of performance may nonetheless entitle the contractor to suspend work in certain situations, for instance for non-payment. See for example, section 112 of the UK Construction Act. Such rights also exist under French[39] and Dutch law.[40] Some standard form contracts also include such suspension rights.[41]

Employers, owners and lenders will also generally want to ensure that the contractor cannot treat a reference of a dispute to the dispute resolution procedures as an event entitling the contractor to an extension of time and/or additional costs (e.g. on the basis that failure by the employer to accept the contractor's claim is an act or omission causing delay to the works). Parties may want to expressly provide for this in their contracts.

---

36 Ss 67–69.
37 Art. 26.9 LCIA Rules; Art. 28(6) ICC Rules.
38 See, for example, clauses 20.4 of FIDIC Red, Yellow and Silver Books respectively.
39 Art. 1184 of the French Civil Code and the general principal *'exception d'inexécution'*.
40 § 6:262 of the Dutch Civil Code (*Burgerlijk Wetboek*).
41 See, for example, clause 16 of the FIDIC Red Book.

7. Interim and Injunctive Relief

Parties need to consider whether it is necessary, or desirable, to include express provisions allowing the parties to seek interim or interrogatory relief from national courts in aid of arbitration or in connection with arbitration proceedings, and if so, how the courts might treat such a provision. This will depend on the law governing the contract, the law of the seat of the arbitration and the law of the jurisdiction(s) where the parties might seek to obtain such a remedy. For example, in England the courts will not treat such a provision as supplanting arbitration as the agreed method of dispute resolution, but as entitling them to act only to support the arbitration in circumstances where the arbitral tribunal does not have the necessary power to act in urgent cases.[42] This may not be the situation in other jurisdictions, and therefore specialized advice will usually be required. In addition, the arbitration rules of certain institutions include provisions that contemplate court applications prior to the constitution of the tribunal or, in exceptional cases, after the tribunal is in place.[43]

8. Multiple Parties or Contracts

As discussed above, on major international construction projects there are often more than two parties involved in the project and/or more than one contract related to the project. In such situations, another issue that must be considered when drafting arbitration provisions is joinder or consolidation of disputes. For example, a main contractor with a series of sub-contracts or back-to-back supply agreements may want to provide for disputes under the sub-contracts to be consolidated with disputes under the main contract (and vice versa) in situations where the decision concerning a dispute under one contract has direct effects on the other contract(s). The advantage of being able to consolidate disputes in such circumstances and having them heard by one arbitral tribunal is that it avoids the cost and time of arguing the same issues in different arbitrations. It also avoids the risk that different arbitral tribunals could render inconsistent decisions on the same facts.

However, employers and owners often resist such provisions, on the basis that they do not want to become embroiled in disputes further down the contractual chain.

The arbitration rules of the LCIA and ICC permit a tribunal to order joinder or consolidation, but in very limited circumstances.[44] The Construction Industry Arbitration Rules published by the AAA allow for joinder or consolidation, if that is provided for in the parties' agreement or the applicable law (Rule 7).

If joinder or consolidation or similar provisions are agreed, they need to be addressed in each of the various relevant contracts. Alternatively, as mentioned above, parties may agree to enter into one, standalone 'umbrella' dispute resolution

---

42 *Channel Tunnel Group Ltd v Balfour Beatty Construction Ltd, supra.*
43 Art. 25.3 LCIA Rules; Art. 23(2), ICC Rules.
44 Art. 22.1(h) LCIA Rules; Art. 4(6) ICC Rules.

agreement that deals with disputes arising under the various contracts and addresses, amongst other matters, when and how consolidation of disputes can occur under the relevant agreements. Both the preparation and drafting of consolidation provisions in various contracts, and drafting of a separate umbrella dispute resolution agreement, will usually be very complex in nature and usually requires specialist advice.

It will also be necessary to ensure that any confidentiality provisions in the contracts in question are not compromised as a result of the consolidation/joinder provisions or stand alone dispute resolution agreement.

One possible alternative to joinder or consolidation in circumstances involving sub-contracts is to provide for a procedure whereby:

- the sub-contractor is entitled to make representations and present submissions in the name of the contractor in the proceedings against the employer or owner;
- the sub-contractor must comply with all relevant time limits in presenting such submissions and evidence, and shall be entitled to copies of the evidence and submissions supplied by the employer or owner only where they are relevant to issues raised in the dispute between the contractor and sub-contractor (the ability to do this will need to be addressed in any confidentiality provisions in the main contract in order to allow the contractor to pass such information along to the sub-contractor); and
- it is agreed that the outcome of the dispute proceedings under the main contract will be binding both as between the employer/owner and the contractor and as between the contractor and the sub-contractor.

Such procedures are sometimes referred to as 'name-borrowing'. The parties will need to seek advice on the enforceability of procedures like this under the law(s) applicable to their contracts.

Parties will also need to consider how to deal with the appointment of arbitrators when more than two parties will, as a result of consolidation or joinder, be involved in particular arbitration proceedings. Of course even if there is only one contract, but more than two parties, it will be necessary, when drafting, to consider and address the appointment of the arbitral tribunal in such situations. One particular issue that arises is the balancing of parties' rights to participate in appointing the arbitral tribunal and the need to balance the number of parties. Some of the institutional arbitration rules expressly address this issue, in part as a result of the decision in 1992 of the French Supreme Court in *Siemens v BKMI and Dutco* (on which see Chapter 8).

9. Limitation Periods

Finally, in preparing multi-tiered dispute resolution provisions, the parties will usually provide that the various processes/proceedings provided for in each tier must be completed before a party is entitled to refer a dispute to the next tier (e.g.

by requiring completion of ADR proceedings before arbitral or judicial proceedings can be commenced). For instance, as noted above, the courts in England and the United States will generally require parties to follow agreed procedures and not permit a party to leapfrog to the next stage if the chosen method of dispute resolution does not suit their purposes.[45] This may cause difficulties if a limitation period may expire prior to the time that a party will be able to commence arbitral or judicial proceedings as the final tier of the dispute resolution procedures. A solution to this problem is to include a provision in the dispute resolution clause which allows that in such circumstances arbitration (or litigation) may be commenced on the condition that the arbitral/judicial proceedings then be stayed until such time as the parties have completed all of the earlier sequential processes/proceedings provided for in their dispute resolution clause.

10.   Drafter Beware

Generally, parties will want to be able to rely on the arbitration provisions that were included in their agreement. The enforceability of an arbitration clause will depend on a number of factors (including the law applicable to the contract) that need to be considered when drafting the arbitration clause. There are certain things that you should or should not do when drafting your clause, in order to avoid some of the main pitfalls that can arise concerning the enforceability of arbitration clauses, including the following:

- do not equivocate – make it clear whether or not arbitration is the exclusive remedy. Do not, for example, include language that can lead to a situation where it appears that the parties have not in fact made a choice between litigation or arbitration (e.g. 'In the case of arbitration, the X Rules of Arbitration apply; in case of litigation, any dispute shall be brought before the courts of Y'). Similarly, avoid using shorthand (e.g. 'Arbitration, if any, by ICC Rules in London') as it may lead to ambiguity that can later be exploited by your opponent;
- before you agree to adopt the rules of an institution, check the rules to confirm that they are acceptable to you, and ensure that you correctly identify the institution (e.g. 'Disputes to be resolved through arbitration by the AAA', not the 'AA'). Also, it is generally best to adopt that institution's model arbitration clause. A selection of model clauses is set out in Annex 2. Amending such model clauses is often unnecessary, and can be unwise unless you have taken specialist advice. Moreover, as discussed earlier, there is a risk that an institution may later refuse to administer the arbitration if you have amended its model clause;

---

45   See *Channel Tunnel Group Ltd v Balfour Beatty Construction Ltd, supra* and *Kemiron Atlantic, Inc. v. Aguakem International, Inc, supra*. In England this is subject to the application of Part II of the UK Construction Act 1996 (discussed above).

- if you are designating an appointing authority, ensure that authority's willingness to accept the responsibility, otherwise you may find that the arbitration agreement becomes inoperable because the designated authority is unwilling to undertake the role you envisaged. That could lead to having to litigate your dispute, possibly in a jurisdiction that you would not have chosen; and
- it is usually best to avoid trying to combine the procedural rules of various jurisdictions. For example: 'Disputes relating to the validity or interpretation of this Agreement shall be governed by the Arbitration Rules of X institution in Country A, in accordance with the laws of the European Union and federal laws applicable therein'. There is a risk that the laws of one country may contain mandatory provisions that are inconsistent with the laws of another country that is referred to, which could lead to intractable problems in practice.

IV. BITS AND MITS

If an international construction project involves a state party, access to international remedies directly against that state might be available under a treaty. As a result, the party might not be restricted to actions being brought against the state or state agency under the agreement in question. Such treaties can be bilateral investment treaties between two states (BITs) or regional multilateral investment treaties (MITs) (e.g. the Energy Charter Treaty and the North America Free Trade Agreement).

In a BIT, both of the states that are parties to the treaty in question assume certain obligations in respect of investments made in their respective territories by each other's nationals. BITs can give an investor a treaty right to commence arbitration directly against the state for a breach of any of those treaty obligations. MITs provide similar protections.

Whether or not a party can have recourse to a BIT or MIT will depend very much on the specific BIT or MIT in question and the particular facts of the case, and a detailed consideration of BITs and MITs is beyond the scope of this work. However, the types of issues that need to be considered include:

- determining if there is a relevant BIT or MIT in place between the state of which the investor is a national and the state where the investment is made. If the investor is a company, the company may need to be incorporated in and have its registered office in the state which has entered into the treaty with the host state;
- determining what constitutes an 'investment' under the relevant treaty. For example, investment may be broadly defined to include:
  - moveable and immoveable property;
  - direct, and sometimes indirect, interests such as shares and other interests in companies;
  - monetary claims and contractual rights;

- copyrights and intellectual property; and
- business concessions and other public law rights;
- identifying the type of protection provided by the treaty. For example, does the treaty provide for:
    - fair and equitable treatment;
    - protection and security of investments;
    - 'most favoured nation' treatment;
    - equal treatment with nationals of host state;
    - no expropriation (or measures tantamount to expropriation) without prompt, adequate and effective compensation?

The treaty may give an investor the right to commence arbitral proceedings directly against the host state. For example, BITs often provide for an investor to have recourse to international arbitration (often after a specified 'amicable settlement' period) under the auspices of the International Centre for the Settlement of Investment Disputes (ICSID) (provided the two contracting states in question are parties to the 1965 Washington Convention, which established this arbitration institution), or the UNCITRAL Arbitration Rules. The Energy Charter Treaty offers investors of member states a choice of arbitration under the auspices of ICSID, the Arbitration Institute of the Stockholm Chamber of Commerce or the UNCITRAL Arbitration Rules.

V.   RELATED PROVISIONS

Parties drafting and negotiating dispute resolution provisions in agreements concerning international construction projects also need to address separate, but related, provisions, such as:

- the law governing the agreement;
- state immunity; and
- if litigation is selected to resolve disputes, provisions concerning jurisdiction and service abroad.

Each of these topics raises its own unique issues, and depending on the nature and location of the project, and the nature and nationalities of the contracting parties, can require extensive consideration of various public international law and conflicts of law questions which are beyond the scope of this work. However, at a minimum, the following issues may arise.

A.	GOVERNING LAW

When selecting the substantive law to govern to the contract (i.e. is the law that will apply to determine the substantive rights, obligations and remedies under the contract), the parties should ask themselves the following questions.

- Will the local courts/government in the location of the project be able to interfere with or review the contract terms, even if the parties have chosen the law of another jurisdiction as the governing law?
- If contracting with a state or state agency/emanation (such as a state bank or state owned oil marketing organization), will it be possible for it to agree that the law of another jurisdiction shall govern the agreement? If so, are the national courts of the state (and jurisdictions where proceedings might be brought) likely to uphold and enforce such contractual provisions?
- Will certain areas of local law be applied to the contract on the basis that they are mandatory, even if the law of another jurisdiction governs the contract? For example, the laws of many jurisdictions provide that parties cannot contract out of local law provisions relating to bankruptcy, licensing and administrative consent requirements (such as planning and environmental consents) or property rights.
- Is the law of the chosen jurisdiction sufficiently developed to be used adequately in connection with the project? For example, currently the laws in some areas of Eastern Europe and the former Soviet Union may not be adequately developed to address all of the legal issues that might arise in relation to what are commonly known as public private partnership (PPP) or privately financed concession (PFI) style infrastructure projects.
- Do the parties want to consider including 'international' concepts into their governing law provisions? For example, parties will sometimes provide that the law of a chosen jurisdiction and 'generally accepted principles of international law' shall govern the agreement (although this may lead to problems of interpretation). Similarly, parties in some industries (for instance, oil and gas) frequently agree that in addition to complying with the applicable law, parties shall perform their obligations in accordance with 'good international industry practice'.

It is advisable to obtain specialist local law and public international law advice in order to properly consider and address these and other issues relating to the choice of the governing law for the contract.

B.	STATE IMMUNITY

Foreign states and their agencies/emanations are generally immune both from suit and from enforcement against state assets. Accordingly, if your counterparty in an international construction project is a state or its agency/emanation, then state immunity issues will need to be considered and addressed. In particular, it is necessary to ensure that the contract contains effective waiver of immunity provisions both in respect of proceedings being brought against the state or its agency/emanation and in respect of immunity against enforcement of an arbitral award or judgment made against that party.

C.   JURISDICTION AND SERVICE ABROAD

If litigation is chosen as the final tier of dispute resolution for an international project, a jurisdiction clause should be included in the contract.

A jurisdiction clause is intended to determine which national courts will hear a dispute between the parties and apply the law chosen as the governing law. It is, of course, possible to give jurisdiction of a dispute to the courts of one country (or alternatively a number of countries) that is different from the country whose law is chosen by the parties to govern their contract.

Drafting and negotiating jurisdiction clauses involves addressing a number of issues that can often be complex. These include whether the courts of a chosen country should have exclusive or non-exclusive jurisdiction over disputes and how courts in other jurisdictions (e.g. where assets may be located or the parties' project may be based) will approach the enforceability of the jurisdiction clause. Specialist legal advice may also be needed in respect of conflicts of law and/or public international law issues that arise in connection with jurisdiction clauses.

Additionally, it may be prudent to provide for an agent to accept service of court proceedings within the jurisdiction of the relevant courts if litigation is to be the final tier of dispute resolution and the parties are from various jurisdictions. Such provisions can help avoid potentially costly and time consuming efforts that might otherwise need to be undertaken to effect service of court proceedings on the defendant. They will be critical where there is no means of otherwise effecting service on the prospective defendant. For example, if service is required by means of service through diplomatic channels, and those channels have broken down, proceedings to enforce legitimate contractual rights will effectively be stymied. This occurred when sanctions were imposed on Iraq following the invasion of Kuwait in 1990 and diplomatic relations were suspended between England and Iraq. It was therefore impossible, in the absence of an agent for service clause, to serve court proceedings on Iraqi state entities, as the English rules of service required service through diplomatic channels.

The same service issues do not arise in relation to arbitration proceedings. For example, under the ICC Rules all that is required to commence the arbitration is delivery of the Request for Arbitration to the Respondent: no specific manner of service is stipulated. However, it may still be advisable to appoint an agent for service of process in order to avoid any difficulties that might otherwise arise if one party seeks assistance from a local court in support of the arbitration (e.g. to obtain interim relief) and cannot serve relevant court processes on the other party.

VI.   DRAFTING CHECKLIST

As discussed above, there are many factors and issues that need to be considered and addressed when drafting dispute resolution provisions for use in international construction projects. Set out in Annex 3 is a non-exhaustive checklist of some of the more important of these factors/issues.

Chapter 4
Claims Administration

I. INTRODUCTION

Ideally, a construction contract should operate as a valuable project management tool. It should include provisions which facilitate dispute avoidance and claims handling, in order to prevent disputes escalating through formal dispute resolution procedures. Effective procedures will ideally lead to early solutions, lower costs and fewer delays to the project overall. If there is the opportunity to have input in the drafting stage of the construction contract it is valuable to involve the project managers or contract administrators. It is frequently the case that a contract drafted by lawyers without any input from those who will be running the project does not accurately reflect the procedures which are used on the ground. Frequently, comments are expressed when disputes arise that practices simply do not follow the contract procedures, which are unduly cumbersome or simply inappropriate for the project in question.

This chapter looks at a number of tools for effective claims management including:

- 'early warning' provisions;
- the requirement for provision of particulars;
- conditions precedent to claims;
- rights of access to information;
- audit rights;
- 'keep working' provisions;
- identification of communication/reporting lines; and
- partnering and alliancing techniques.

II. EARLY WARNING PROVISIONS

A contract may include provisions requiring the contractor to give notice of actual or potential difficulties, so as to enable the parties to address solutions at the earliest opportunity. The idea is to encourage joint problem solving. Such arrangements may be formalized by an agreement to refer problems to a project board, for example, which is a popular technique in partnering arrangements (see further below). An example of such a clause in the FIDIC Silver Book is as follows:

> 'The contractor shall promptly give notice to the employer of specific probable events or circumstances which may adversely affect or delay the execution of the Works'.[1]

The ENAA's standard form for power plant construction, requires the contractor to monitor progress of all activities specified in the project's programme and to supply a monthly progress report to the employer, indicating 'where any activity is behind the program giving comments and likely consequences and stating the corrective action being taken'.[2]

An early warning requirement is one of core clauses of the Engineering and Construction Contract (ECC). Both parties, the contractor and the employer's project manager, are obliged to notify the other once either of them becomes aware of any matter which could increase the total contract price, delay completion or impair the performance of the works.[3]

The contract may provide for consequences for failing to provide such notice, for example, barring the opportunity to pursue claims in respect of such events. If so, the contractor must take great care to ensure notice is given to the employer in compliance with such a clause. For example, in the ECC, the sanction for failure by the contractor to give early warning of a compensation event is to reduce the payment due to him for any related claim to the level that he would have been entitled to if he had given early warning and allowed cost saving actions to be taken.[4]

III. REQUIREMENTS FOR PROVISION OF PARTICULARS

From the employer's perspective the construction contract will ideally require a contractor to provide accurate and contemporaneous information in respect of potential claims. Third parties, such as project insurers, financiers, governmental authorities (for example health and safety or environmental officers, depending on

---

1 Clause 8.3; the Yellow Book and Red Book contain similar provisions.
2 Clause 18.3 ENAA Power Plant Model Form 1996.
3 Clause 16.
4 Clause 63.5.

the nature of the event in question) may be interested in such information. Such clauses usually require information to be provided in relation to each of the following:

- the event which gives rise to the claim;
- its contractual relevance in terms of the risk allocation agreed between the parties and the agreed treatment of such risk;
- the consequences of the event in terms of the time and cost impact to the project;
- whether it is possible to mitigate the consequences of the event, for example, by adopting work-around measures or by a variation order in circumstances where, for example, a particular product is unavailable to permit the use of a suitable alternative.

An example, again from the FIDIC Silver Book, is the requirement to give notice of force majeure:

'notice shall be given within 14 days after the Party became aware, or should have become aware, of the relevant event or circumstance constituting Force Majeure'.

IV. CONDITIONS PRECEDENT TO CLAIMS

Requirements to give particulars may act as conditions precedent to the entitlement to pursue the claim. In such cases, if the notice or particulars required are not provided within a certain time period the claim is barred. The question whether a requirement to give notice of a claim or particulars in relation to the claim is a condition precedent is one of the true construction of the agreement. Some are unambiguous. For example, the FIDIC Silver Book requires a contractor to give notice within 28 days of an event or circumstance which may give rise to a claim. If it fails to do so it loses the right to claim time and/or cost, and the employer has no liability in that respect. Clause 20.1 of the FIDIC Red and Yellow Books contain identical provisions. A less black and white approach is to be found in clauses 53(1) & (4) of the ICE D&C. There, if a contractor fails to notify the employer in writing of any claim within 28 days, it shall only be entitled to payment 'to the extent that the Employer's Representative has not been prevented from or substantially prejudiced by such failure in investigating the said claim'.

Do such clauses have to be reasonable? Under English law there is no general overarching principle of reasonableness. The only exception to this is where one party is contracting on the other's standard terms of business. In these circumstances the Unfair Contract Terms Act 1977 (UCTA) subjects to the test of reasonableness clauses which restrict liability in respect of breach (including clauses which require the party to make a claim within a certain time-limit).[5] The test of reasonableness

---

5 *Unfair Contract Terms Act 1977, s. 3(2)(a).*

for the purposes of UCTA is that 'the term shall have been a fair and reasonable one to be included having regard to the circumstances which were or ought reasonably to have been, known to or in the contemplation of the parties when the contract was made'.[6] The time for determining the reasonableness of the term is the time when the contract was made. The determination is therefore unaffected by the size of the loss or damage suffered except to the extent that such events were or ought reasonably to have been in the contemplation of the parties at the time the contract was made.

In *T Mackley & Co Ltd v Gosport Marina Ltd*,[7] the English High Court held that clause 66 of the ICE Conditions of Contract (6th edition) required a decision of the engineer on a dispute as a condition precedent to the entitlement of a party to the contract to refer the dispute to arbitration.

It is often the case on major projects that the terms and conditions put forward by the employer as part of its tender package are negotiated between the parties. Accordingly, it cannot be said that the contractor is contracting on the employer's standard terms and conditions of business. Similarly, the standard forms such as FIDIC, ICE and AAA do not constitute one or the other party's standard terms of business. Accordingly, on major projects UCTA has limited application to conditions precedent to pursuing claims.

Civil law codes may impose a general requirement of reasonableness on contractual terms. Generally, however, such principles are applied with restraint.

V. RIGHTS TO ACCESS TO INFORMATION

A critical issue for an employer is access to accurate contemporaneous information about the status of the project. If lenders are involved there will generally be a requirement for regular progress reports. There may in addition be obligations to notify insurers in respect of certain events.

There are various different sources of information addressed in construction contracts, for example:

- status reports to be provided on a monthly or other specified periodic basis;
- site meetings again to take place on a specified periodic basis, generally weekly (a contractual clause may specify who is to attend such meetings);
- provisions for shared databases, for example in relation to design information;
- requirements for updated programmes to be provided on a regular basis. for example, the FIDIC silver book requires revised programmes to be submitted

---

6 *Unfair Contract Terms Act 1977, s. 11(1)*.
7 [2002] BLR 367.

where the previous programme is inconsistent with the actual progress of the contractor's obligations.[8]

There is a good general clause in the FIDIC Silver Book,[9] which specifies the nature of information required to be provided:

- a detailed description of progress;
- photographs;
- the stage in manufacture of main items of plant;
- the contractor's personnel and equipment on site;
- quality assurance documents and test certificates;
- a list of variations and notices of claim;
- safety statistics and hazardous environmental-related incidents; and
- a comparison of actual against planned progress and of events/circumstances which may jeopardize completion and the measures which have been taken to overcome delay.[10]

Clause 7.3 of the Silver Book also permits the employer to have full access to site to be able to inspect/test and to check progress of manufacture of plant and materials. The FIDIC Red and Yellow Books contain identical provisions.[11]

VI. AUDIT RIGHTS

Rights to audit costs which have been incurred or which will be incurred as a result of orders or contracts placed with suppliers and/or subcontractors are critical in any cost plus or target cost contract. Invariably, the contractor's entitlement to recover costs under these contracts will be subject to a test of such costs being reasonably and properly incurred (see Chapter 2). There may, in addition, be the negative condition that the cost should not be incurred as a result of a wrongful act or omission by the contractor.

The employer will want to have access to detailed records and accounts, including invoices, orders, day work records, records of payment and charges made.

---

8   Clause 8.3; for similar provisions in other standard forms see FIDIC Red Book (clause 8.3), FIDIC Yellow Book (clause 8.3) and ENAA Process Plant Model Form 1992 (clause 18.2). However, under some standard forms the contractor is obliged to produce a revised programme only if the employer requests one e.g. ENAA Power Plant Model Form 1996 (clause 18.4) and ICE D&C (clause 14.4).
9   Clause 4.21.
10 In addition, clause 1.12 FIDIC Silver Book provides the employer with a general right to request information in order to verify contractual compliance.
11 For other examples of provisions giving the employer or his representatives rights of access to the works, see the ENAA Process Plant Model Form 1992 (clause 23.8), ENAA Power Plant Model Form 1996 (clause 23.8), ICE D&C (clause 37), AIA A201-1997 General Conditions of the Contract for Construction (clause 3.16.1) and EEC (clause 28.1).

It is important that the contract should include requirements for records to be retained and a contractor should in turn impose identical requirements on its sub-contractors to ensure that it complies with its obligations under its contract. In some circumstances, contracts provide for an independent audit. This is increasingly common in partnering arrangements, where the philosophy of co-operation and the ethos of goodwill and mutual trust are seen to be undermined where the employer retains the right to conduct the audit itself. In such circumstances, the costs of the independent audit are generally shared between the parties.

VII. 'KEEP WORKING' PROVISIONS

Construction contracts may include an express requirement for the parties to continue to perform their obligations under the contract despite the existence of a dispute. The contract may expressly say that there is no right to suspend work or to terminate the contract. These rights may, however, be inconsistent with local law. For example, under English law there is a statutory right to suspend work for non-payment which cannot be excluded by contract.[12]

In the context of variations, the employer may have the right to require a contractor to proceed with the variation despite the time and cost consequences not having been agreed in advance. On a fixed price lump sum contract, this will put the contractor at a severe disadvantage in circumstances where the employer does not accept that the work in question is a variation at all but is instead within the original scope of work and therefore that there is no entitlement to additional payment. A relaxation to this position that is sometimes negotiated is that the employer shall pay a proportion of the contractor's costs actually incurred in carrying out the variation pending the resolution of the dispute.

On a cost plus contract, such issues are less controversial as the contractor is entitled to recovery of its reasonably and properly incurred costs in any event. The issue becomes more complex, however, where the contract includes targets and adjustments to those targets are required where variations are ordered.

Where a dispute does arise, ideally there will be an appropriate dispute resolution mechanism which will provide for early resolution of such disputes (such techniques are addressed in Chapter 3 above).

VIII. PROVISION FOR COMMUNICATION/REPORTING LINES

Construction contracts invariably provide for one formal point of contact on the side of the employer and the contractor, naming a specific representative for each party. Ideally clear lines of authority will be set from the outset identifying who

---

12   S. 111 UK Construction Act 1996.

can sign off on site instructions, variation orders and amendments to the contract. The topic of contractual correspondence is addressed further below.

IX.   PARTNERING AND ALLIANCING

Becoming increasingly popular in United States, United Kingdom and Australia is the use of partnering and alliancing techniques.[13] These are aimed at establishing joint working and co-operation with varying degrees of formality. At one end of the spectrum is a non-binding partnering charter, whereby parties agree to abide by a set of behavioural principles with the intent that they shall conduct their relationship in a spirit of mutual trust and co-operation and shall strive to avoid disputes by early communication of issues and shared problem-solving. In the United Kingdom it is common for such partnering charters to be drawn up following a facilitated workshop, in order that the parties' key objectives can be jointly agreed. A partnering board or committee may be set up to address in the first instance problems that cannot be resolved at project level. If disputes cannot be resolved, the parties would then proceed through the formal dispute resolution procedures.

Increasingly in the United Kingdom parties are looking to embody partnering principles in their contractual arrangements on individual projects. Standard form contractual documentation is now available.[14] Parties have the option of entering into a single contract incorporating all terms relevant to partnering as well as the delivery of the project or, alternatively, to use two contracts, an agreement embodying the partnering elements that sits alongside the traditional works contract. In these circumstances, it is usual to amend the works contract to reflect the terms of the partnering agreement. There would then be the option, if the partnering agreement were to be terminated, for the parties to revert to the traditional works contract.

It is common for such arrangements to include:

- procedures for information sharing and joint working, including, for example, shared databases, secondments and arrangements for office sharing;
- procedures to break down formal hierarchies to create a flatter structure for direct communication, for example, between sub-contractors and the employer and its consultants;

---

13 For a discussion on partnering and alliancing in the construction industry including a review of the historical development of partnering, structures for partnering and alliancing and typical contractual arrangements, see Sally Roe and Jane Jenkins, *Partnering and Alliancing in Construction Projects* (Sweet & Maxwell, 2003).

14 For example, see: Institution of Civil Engineers, *The NEC Partnering Option* (Thomas Telford, 2001); and Association of Consultant Architects and Trowers & Hamlins, *PPC 2000 The ACA Standard Form of Contract for Project Partnering (Amended 2003)* (Association of Consultant Architects, 2003).

- the creation of a core group or team comprising representatives from all, or key, members of the partnering arrangements to whom key issues are referred, including, for example, the approval of a variation to the project and consequences in terms of time/cost and any impact on performance criteria;
- the calculation of project costs and any pain share/gain share payments;
- the exclusion and admission of partners;
- dispute resolution; and
- the creation of an integrated project team, identifying the best person for each role from the partners. 'Man-to-man marking' is avoided to reduce costs.

Alliancing is said to represent a more 'hard-headed' approach to partnering because, as well as the behavioural and organizational issues addressed above, it involves profit-sharing and often risk-sharing schemes. A distinctive feature of an alliance is an incentivized contract which is central to the partnering relationship, with sanctions applying for failure to achieve key performance indicators and targets. The scheme will usually link the rewards of all alliance members to the overall outcome of the project and not their individual performance. The fundamental aim is to encourage co-operation and reduce the potential for adversarial conflict. With incentivization arrangements linked to targets, the same tensions in relation to the award of variations as arise in fixed price lump sum or target cost contracts will apply. It is therefore generally the case that the alliance board has the right to decide whether a variation should be admitted to the project, and the consequences upon the risk-sharing arrangements.

Surveys carried out in both the United States and the United Kingdom report significant cost savings as a result of adopting partnering arrangements. For instance, an analysis of project performance for partnering projects in the United States Army Corps of Engineers carried out in December 1992 by David Charles Western concluded that there is a 9 per cent improvement in cost and 8 per cent improvement in time on partnered projects. Again, estimates published in the United Kingdom by the Reading Construction Forum indicate that cost savings of 2 to 10 per cent are typically achieved with project specific partnering and savings of 30 per cent are realistic over time with long-term or strategic partnering.[15] The benefits to be gained will turn largely on the type of partnering arrangements employed and the degree of party integration and risk-sharing. As the greatest opportunity for influencing out-turn costs and quality of the project arises at the early planning and design phases, generally speaking it is considered to be most effective to adopt partnering techniques from the outset.

It is clear from the above discussion in relation to partnering techniques that they are designed to avoid the traditional claims-orientated adversarial approach adopted on traditional construction projects. It is common, for example, to suspend the requirement to give notice of claims and to submit full and detailed particulars to support a claim within a given timeframe during the currency of the partnering

---

15   Bennett and Jeyes, *Trusting the Team*.

arrangements. Avoiding the time and costs associated with the preparation and review of claims materials is seen as one of the principal benefits of adopting partnering arrangements.

Parties need to consider, however, what will happen if the partnering arrangement falters and the parties resort to claims pursuant to the terms of the works contract. Ideally, there will be express provision for such circumstances and the consequences of termination of the partnering arrangements. In the absence of such express provision an employer may argue that the contractor is out of time to pursue claims which arise in respect of circumstances during an early part of the project when the partnering arrangements were in place. In response, a contractor may be able to demonstrate that the employer has waived its right to receive such notices and particulars on the grounds that the parties were voluntarily following the partnering arrangements. It is possible that the arrangements may stop short of a formal variation of the contract, depending on the manner in which they are documented. For example, if there is merely a partnering charter which is expressed to be non-binding, an argument that there has been a formal variation of the contract may not succeed. Instead, however, the contractor may be able to resort to an argument based on waiver by estoppel. There is limited case law in the United Kingdom regarding the effect of a non-binding charter, but one case which came before the courts in 1999[16] concerned a partnering charter which included an obligation for the parties to work together in a spirit of mutual co-operation and trust. This charter was acknowledged to provide standards by which the parties were to conduct themselves and against which their conduct and attitudes were to be measured. The Judge, his Honour Judge Humphrey Lloyd, observed that where a charter is in place the parties are unlikely to adopt a rigid attitude to formation of a contract or to be concerned with compliance with contractual notice procedures for the pursuit of claims. Indeed, the Judge went further and stated:

> 'An arbitrator (or Court) would undoubtedly take such adherence to the charter into account in exercising the wide discretion to open, review and revise, etc which is given under the [construction contract]'.[17]

To avoid arguments over waiver or estoppel parties may wish to provide that, if the partnering arrangements are terminated, the contractor may then have a specified period of time which is, in all the circumstances, a reasonable period, to provide particulars of any claims in relation to events which have taken place during the currency of the partnering arrangements. Where the partnering agreement includes risk sharing provisions, increased costs and delays sustained during the currency of the partnering agreement may be required to be taken into account when calculating an 'exit fee' or payment for release from the partnering arrangements.

---

16   *Birse Construction Limited v St David Ltd* [1999] BLR 194. The decision was appealed but there was no discussion on the effect of the charter in the Court of Appeal.
17   Ibid. p. 203.

X. EFFECTIVE CONTRACT MANAGEMENT

Effective contract management may assist in minimizing disputes that do proceed to formal dispute resolution. Both parties will be concerned to track progress and delay on the project and monitor increased costs. In any formal dispute, contemporaneous correspondence and documentation will be critical to establishing entitlement to any claims brought by the contractor and the legitimacy of defences raised by the employer. Documents will need to be stored either in hard copy or, increasingly, electronically, to ensure a party is in the best possible position to proceed to formal dispute resolution. Obvious key documents are:

- the most recent version of the contract and any amendments to the contract agreed between the parties;
- progress documents such as programmes, method statements, information release schedules;
- on site records for example site diaries, equipment lists, timesheets, day work sheets, site instructions;
- correspondence and information exchanged between the parties, including minutes of meetings, which will ideally be signed to reflect agreement to the terms of the meetings;
- internal correspondence; and
- variation orders and instructions.

In other words, any document of potential relevance to the contract, its interpretation, performance and claims.

Increasingly, electronic filing techniques are used and it is sensible to provide for unique document identification numbers. Scanning may be used, or storage of information on shared databases or intranets. Care should be taken when selecting information to be included in a shared database, as it may be inappropriate for some internal communications or commercially sensitive information to be disclosed to the other party to the contract. From the perspective of managing disputes it is very helpful to include as part of the data entry process, a means of tagging key issues for later search and retrieval of the documents. This requires considerable thought in advance of setting up any such database as to the comprehensive list of issues that should be included in order to achieve consistency in data entry. Issues can, however, later be added as and when they arise, although there will be the risk that relevant documents already on the database relating to such issues will not be captured.

Consideration should be given to security and the control of access to confidential documents. It may be possible to grant wider access in 'read-only' format to ensure that those accessing the documents are not able to amend or change the documents in any way.

Contractors in particular need to be aware of time periods set out in the contract. As noted above there may be notice requirements, which operate as conditions precedent to claims. Generally it is sensible for the parties to:

- track notice periods;
- track the performance of obligations within an agreed time limit;
- track payment within agreed time limits; and
- track claims time limits/requirements for the provision of particulars.

Thought also needs to be given to a document retention policy. On a major project which will last for a number of years, standard corporate policy may result in the destruction of materials relevant to the early stages of the project after a specified period of time. It is sensible where a project is ongoing, however, and there is the possibility of disputes arising to preserve all documents in relation to the project, as it may be that events taking place at the beginning of the contract will have relevance directly or indirectly to final claims. Of course, the objective of introducing conditions precedent to claims and requiring that particulars of events likely to give rise to claims be served promptly is to avoid 'stock-piling' of claims and to enable the events and their consequences to be addressed contemporaneously. The aim on the part of the employer is to avoid global or rolled-up claims where the argument is made that it is too difficult because of the complexity of the interaction of the different events in question to track the consequences flowing from each. The contractor then seeks to recover a portion, or even the entirety, of its costs as a global claim not specifically allocating heads of cost to individual claims.

It is sensible for the contract correspondence to be written with an eye to the possibility of future formal dispute resolution. Oral communications or instructions should be confirmed in writing, for example. 'Warning' letters should be written where an action or omission is proposed by the other party to the contract which may be prejudicial or result in breach of contract. Issues to address in relation to contract correspondence include:

- Does the writer have authority to speak on behalf of the company?
- Is the letter addressed to the correct party/representative of the party?
- Has the letter gone through any internal review procedures?
- Is the language appropriate (business-like, factual, firm)?
- Does the letter avoid unnecessary concessions?
- Does the letter explicitly preserve rights where appropriate?
- Is the correspondence supported with facts as opposed to opinions?
- Are there clear references to contractual provisions and preceding relevant correspondence or documentation?

Care should be taken to avoid the waiver of rights. Of course, it is not uncommon in long-term projects to make accommodations with the other party for functional reasons. Care should be taken, however, to ensure that one-off concessions will not operate to waive contractual rights generally. The first step is to check the terms of the contract for provisions in relation to waiver. Some contracts specifically address this issue, providing that failure to pursue a right on one occasion should not be deemed a waiver of rights for the future. Explicit reservations of rights in correspondence is sensible, however. It should not be assumed that merely stating 'without prejudice' on a letter will be sufficient.

There is the further point that under most sophisticated legal systems the conduct of the parties in implementing the agreement can affect the rights and obligations under the agreement even if there is not an express written variation. Acquiescence to conduct in breach of contract over an extended period where there is reliance on such acquiescence, for example, may lead to a party being estopped from pursuing the relevant rights in the future. It may be sensible for the in-house legal team to do regular check-ups on contract correspondence to address such issues.

Great care needs to be taken in relation to emails, which encourage a more informal style of communication and may be difficult to control. The risk is that such communications may be required to be disclosed in later litigation or arbitration. Very broad rules of discovery[18] apply in Anglo-American systems, requiring the production of all categories of documents which are relevant to the issues in dispute which are within the control of either of the parties. Protection is afforded in these systems to legally privileged documents but there is no automatic protection for documents merely because they are confidential or internal, including, for example, informal email exchanges at a project level or far more formal documents such as board minutes.

Civil law systems have a far more restricted approach to disclosure, for example in France and Holland parties are not generally required to disclose categories of documents other than those they wish to present to support their case. The trend in international arbitration is towards the fusion of these two approaches (see the discussion in Chapter 11 below).

---

18 See the discussion in Chapter 11.

Chapter 5
Adjudication and Dispute Review Boards

I.   INTRODUCTION

Chapter 3 discussed the use of dispute review boards as a means of interim dispute resolution, and the advantages and disadvantages of such procedures were considered. Such procedures are becoming increasingly common and well-established on international construction projects. As a form of expert determination they provide a relatively fast-track procedure as a useful alternative to arbitration or litigation during the life of the project.

This chapter looks at typical procedures for a reference before a dispute review board, assuming that the parties have agreed that the board should make a decision (as opposed to merely a non-binding recommendation). After a general discussion of such procedures, there is a brief commentary on the institutional rules with a comparison of the key features provided in a table. The chapter then discusses how to approach a reference before a dispute review board tactically, including the manner of preparation of submissions and representation at the hearing. Finally, the effect of the decision, how it may be enforced and how it may be attacked are considered.

II.   COMMENCING THE REFERENCE

It is assumed for the purposes of this discussion that the parties have decided that there should be a standing panel of three experts and accordingly the dispute review board (referred to below as the DRB) has been established at the outset of the project.

Assuming that a dispute has now arisen and negotiations to resolve the dispute have failed, typically a reference is commenced by either party giving a notice to refer the dispute to the president of the DRB with a copy to other party. The notice to refer is typically a brief document setting out:

- the nature and a brief description of the dispute;
- the names and addresses of the parties involved; and
- the nature of the relief sought.

Where selection is to be made among a standing panel of, say, five experts, with two experts sitting with the president to determine a particular dispute, such notice to refer should provide sufficient information to enable the president to select the most suitable experts from the standing panel to hear the reference. (Of course, parties may have agreed that experts should be selected in strict rotation, in which case there would be no discretion regarding the selection of the panel members when a dispute arises). Similarly, where there is no standing panel and an ad hoc panel is to be created it would be at this stage, following the service of the notice to refer, that the procedure for appointing the panel members would be followed.

Where the members of the DRB are fully constituted from the outset, the rules may require the referring party to deliver with the notice to refer its written statement of case, detailing the grounds upon which its claim is based and providing supporting documentation upon which it wishes to rely. Where a DRB is to be constituted, this document may be required to be delivered within a specified number of days following the constitution of the DRB. Alternatively, the rules may require the parties to exchange written submissions simultaneously rather than providing for a sequential process. The simultaneous exchange of submissions provides the opportunity for a quicker time frame, although of course, parties will need a proper opportunity to respond to the submissions from the other side. Depending on the overall time frame for the reference it is common to see a period of up to 28 days for the submission of responses to the other party's submissions.

III.   THE POWERS OF THE DISPUTE REVIEW BOARD

A typical feature of a DRB is that its members are given full power to take the initiative in ascertaining the facts and the law. In particular, the DRB may have the express power to:

- request clarification or additional information from either or both of the parties;
- make such site visits and inspections as it considers appropriate;
- convene meetings upon reasonable notice to the parties at which both parties shall be entitled to be present;
- appoint its own advisors to advise on matters of legal interpretation or expertise outside the area of expertise of each of the members on which the parties are not agreed;

- open up, review and revise any decision, approval, recommendation or determination made, notice or certificate given by the employer and/or the employer's engineer or representative; and
- make use of the specialist knowledge of each of the members.

The freedom of the DRB members with regard to the procedures they may adopt and the use of their own expertise is a crucial difference from arbitration. Experts do not act in a judicial capacity. They are not required to follow the rules of natural justice and there is no objective standard of fairness that must be complied with. There is no statutory code for the conduct of expert references for the parties, the courts or an arbitral tribunal to refer to as exists for arbitration. Indeed, the absence of remedies for procedural irregularity has been described as the 'crucial difference' between expert determination and arbitration:

> 'An arbitration award may be set aside because the procedure fails to conform to the statutory standard of fairness which is closely derived from the principles of natural justice: no such remedy is available to invalidate an expert's decision. An expert can adopt an inquisitorial, investigative approach and need not refer the results to the parties before making the decision. An arbitrator needs the parties' permission to take the initiative and must refer the results to the parties before making the award'.[1]

Of course, an expert will be required to follow any methodology and instructions prescribed by the parties, whether in their agreement or the terms of reference of a specific dispute. Accordingly, if they have agreed that certain procedures are to apply (for example, the exchange of documents, the holding of a hearing) an expert will be required to comply with such procedures.

IV. DO THE RULES OF NATURAL JUSTICE APPLY TO DISPUTE REVIEW BOARDS?

If the agreement is silent on procedural issues, for example, whether a hearing should be held, will rules of natural justice (meaning due process of law) be imported into the procedure? The English courts have said:

> 'The principles of natural justice are of wide application and great importance, but they must be confined within proper limits and not allowed to run wild.
> ... [The expert] must throughout retain his independence in exercising [skilled professional] judgment; but provided he does this, I do not think that, unless the contract so provides, he need go further and observe the rules of natural

---

1 John Kendall, *Expert Determination*, 3rd edition, (Sweet & Maxwell, 2001), para. 1.1.2.

justice, giving due notice of all complaints and affording both parties a hearing. His position as an expert and the wide range of matters that he has to decide point against any such requirement. ... It is the position of independence and skill that affords the parties the proper safeguards, and not the imposition of rules requiring something in the nature of a hearing. For the rules of natural justice to apply, there must be ... something in the nature of a judicial situation; and this is not the case'.[2]

The English courts appear to accept that in view of the pressures and constraints of the adjudication timetable under the UK Construction Act 1996, which requires that a decision should be reached within 28 days of referral to an adjudicator, they should recognize that Parliament must have intended the rules of natural justice could not apply in full to adjudication.[3] Similarly, it has been accepted that the distinction must be preserved between the role of an adjudicator on the one hand and that of a court or an arbitrator on the other:

'...one has to recognize that the adjudicator is working under pressure of time and circumstances which make it extremely difficult to comply with the rules of natural justice in the manner of a court or an arbitrator. Repugnant as it may be to one's approach to judicial decision-making, I think that the system created by the Act can only work in practice if some breaches of the rules of natural justice which have no demonstrable consequence are disregarded'.[4]

A related line of argument which has been addressed by the courts is whether the European Convention of Human Rights should apply to statutory adjudication. Article 6 of the Convention lays down a number of essential prerequisites for the valid determination of civil rights and obligations. The hearing must be fair, the hearing and judgment must be public, the hearing must take place within a reasonable time and the tribunal must be independent and impartial and established by law.

---

2   *London Borough of Hounslow v Twickenham Garden Developments Ltd* [1970] 3 All ER 326, 346–7.
3   See *Macob Civil Engineering Ltd v Morrison Construction Ltd* [1999] BLR 93. See also *Glencot Development and Design Co Ltd v Ben Barrett & Son (Contractors) Ltd* [2001] BLR 207, 218 where an inevitable compromise is recognized 'it is accepted that the adjudicator has to conduct the proceedings in accordance with the rules of natural justice or as fairly as the limitations imposed by Parliament permit'.
4   *Balfour Beatty Construction Limited v Lambeth London Borough Council* [2002] BLR 288, para. 28 and see also *Discain Project Services Ltd v Opecprime Development Ltd* [2001] BLR 285, para. 35 where it was recognized that some deviations from the standards required of impartiality and fairness may be disregarded.

The Convention has been held not to apply to statutory adjudication on various differing grounds, including:

- that the award is provisional and therefore does not involve final determination of the rights of the party;[5]
- that an adjudicator is not a public authority under the Convention and is not bound by the Human Rights Act 1998;[6] and
- that even if the adjudicator were so bound, the process of adjudication includes court proceedings to enforce the decision and accordingly the requirement for a public hearing is met.[7]

There have been instances, however, where adjudicators' decisions have been set aside on the basis of a want of fairness in the treatment of the losing party (for example, where one party is consulted on submissions made by the other without a reciprocal process).[8]

Parties may wish to include more specific drafting in bespoke procedures for references before DRBs to produce specific minimum safeguards which a court may be slow to import. For example, they may wish to provide that the DRB shall convene a meeting at the request of either party to give the parties the opportunity to make oral submissions before the DRB reaches its decision, or that all communications with one party must be copied to the other at the time of sending.

V. INSTITUTIONAL RULES

By way of illustration, it is useful to look at rules that have developed for the use of DRBs either as an adjunct to the standard forms or promoted by organizations for projects in which they are involved. The FIDIC contract provisions and the rules of the World Bank, AAA and ICC are addressed below. A comparison is presented in tabular form further below.

A. WORLD BANK DISPUTE REVIEW BOARDS

The use of DRBs, initially on tunnelling projects and eventually extending to other heavy civil engineering works and building construction projects, had been widespread for decades in the United States domestic construction market and was considered highly successful. As a result, in 1994, the World Bank began consultation with various interested parties, including FIDIC, on the use of DRBs

---

5 *Elanay Contracts Ltd v The* Vestry [2001] BLR 33.
6 *Austin Hall Building Ltd v Buckland Securities Ltd* [2001] BLR 272.
7 Ibid.
8 *Discain Project Services Ltd v Opecprime Development Ltd supra.*

on international construction contracts. In 1995, the World Bank published a revised edition of its *Standard Bidding Documents – Procurement of Works* (SBDW). The SBDW is the standard form of contract that must be used on any large scale civil works project financed wholly or partly by the World Bank. The 1995 revised SBDW for the first time extended the use of DRBs to international projects.

The SBDW incorporates FIDIC's Conditions of Contract for Works of Civil Engineering Construction (the 'old Red Book'), subject to a set of amendments contained in the Conditions of Particular Application prepared by the World Bank. In particular, Condition 67 of the old Red Book, which provides for settlement of disputes at the first instance by the employer's engineer, was replaced.

The amended Condition 67 introduced the obligatory use of a dispute review board comprising three members for the resolution of disputes at the first level where the contract is valued at USD 50 million and above. For contracts valued at less than USD 50 million, the parties had the option of using either a three-member DRB, a single person acting as a Disputes Review Expert (DRE) or an engineer independent of the employer.

The decision of a DRB/DRE was termed a 'recommendation'. A recommendation was final and binding only if neither party had given notice to the other of its intention to commence arbitration within 14 days of receipt of the recommendation. If notice of arbitration was given on time, then the parties were not bound by the recommendation. The DRB/DRE proceedings were to be governed by the appropriate set of rules appended to the SBDW; these were the *Dispute Review Board Rules and Procedure* and the *Rules and Procedure for the Function of the Disputes Review Expert*.

A revised edition of the SBDW was published in May 2000. It made significant changes to dispute resolution mechanism on World Bank financed projects. The mandatory use of a three-member DRB on contracts above the USD 50 million threshold was maintained. However, the option of using an independent engineer to resolve disputes on contracts below this threshold was removed. Disputes at first instance were to be referred to either a DRB or a DRE.

In addition, the recommendation of a DRB or DRE was binding and was required to be given effect to unless or until revised by an arbitral award. If neither party gave notice of its intention to commence arbitration within 14 days of receipt of a recommendation, it became final and binding.

The current version of the SBDW was published in May 2005. It is based on the Master Procurement Document for Procurement of Works & User's Guide approved by the multilateral development banks (MBDs) and other international financial institutions in October 2004 as part of their measures to harmonize the process of obtaining development funding worldwide.

The relevant changes introduced in the 2005 version of the SBDW are:

– the replacement of the old Red Book with the new FIDIC Conditions of Contract for Construction MDB Harmonized Edition 2005, a harmonized version of the Red Book agreed between FIDIC and the MDBs;

- the replacement of the terms 'DRB' and 'DRE' with Dispute Board (DB); and
- the removal of the requirement for a three-member DRB where the contract sum is valued at USD 50 million and above.

It is left to the parties to agree in the contract whether a DB will comprise one or three members. In absence of the contract stating the number of members, the DB shall comprise three members.

The dispute resolution clause, *General Conditions of Dispute Board Agreement* and the *Procedure Rules* for DBs are set out in section VII of the current version of the SBDW, and are reproduced at Annex 4.

The table further below provides a comparison of the main features of the rules and procedures applicable to the World Bank's DB proceedings with those of other institutions.

B. FIDIC DISPUTE ADJUDICATION BOARDS

FIDIC first introduced the concept of a dispute board into its contracts with the publication in 1995 of its conditions of contract for design-build and turnkey projects (the 'Orange Book'). FIDIC also issued supplements to the old Red Book in 1996 and the old Yellow Book (conditions of contract for electrical and mechanical works, 1987 edition) in 1997, extending the use of dispute boards to these contracts.

Prior to the publication of the Orange Book, the first tier for resolving disputes between the employer and the contractor under FIDIC contracts was the engineer. In the event either party was dissatisfied by the engineer's decision, the dispute could then proceed to arbitration. However, the role of an engineer (who was remunerated by the employer) acting as impartial adjudicator of disputes between the employer and contractor became increasingly untenable. On the one hand, contractors were distrustful of engineers remunerated by the employer impartially deciding disputes. On the other hand, employers from less developed countries were similarly suspicious of engineers who were usually from the same country as the contractor.[9] FIDIC therefore decided to follow the lead of the World Bank and adopted the use of dispute boards, which FIDIC called dispute adjudication boards (DABs). FIDIC provided that the findings of a DAB (termed a decision) are binding on the parties and are to be implemented forthwith, unless or until the decision has been revised by an agreement between the parties or subsequent arbitration.

The DAB provisions under FIDIC are part of the contract's dispute resolution clause and are applicable whenever a FIDIC contract is used,[10] unless the parties have amended or deleted the provisions. A set of procedural rules to govern the

---

9   See John Bowcock, *What FIDIC has to offer and plans for the future,* available at: http://www1.fidic.org/resources/ contracts/bowcock97.asp (accessed on 26 July 2005).
10 See clauses 20 FIDIC Red, Silver and Yellow Books respectively.

DAB and the general conditions of a tripartite agreement between the employer, contractor and DAB members is annexed to the contracts.

A comparison of the main features of the rules and procedure applicable to FIDIC DAB proceedings with those of other institutions is shown in the table further below.

C.   AMERICAN ARBITRATION ASSOCIATION (AAA) DISPUTE REVIEW BOARDS

The AAA published its *Dispute Review Board Guide Specifications* (DRB Guide Specifications) on 1 December 2000. The DRB Guide Specifications provide the rules and guidelines for the DRB. The AAA also issued a Three-Party Agreement (a model contract providing for the rights and duties of the employer, contractor and DRB members) and an AAA roster of experienced persons from which DRB members can be selected. The DRB Guide Specifications is a stand-alone document, which can be incorporated into any contract. However, it is a guide and parties are advised to make sure it fits in with the rest of the contract, especially the other dispute resolution provisions.

According to the AAA, its DRB Guide Specifications draws on the latest DRB models, including those used on the Boston Central Artery Tunnel Project, the Puerto Rican Tran Urbano Project and the Golden Gate Bridge Retrofit Project in California. Whilst the use of DRB has been prevalent in the United States for decades, the main users have generally been parties involved in public sector construction. The AAA has developed the DRB Guide Specifications to make DRB more accessible to the private construction sector.

A peculiar feature of the DRB Guide Specifications is the heavy involvement of the AAA. The AAA believes that although parties can choose to administer the process themselves, the involvement of an institution such as itself enhances the sense of neutrality. The AAA's role includes providing lists of potential DRB members, acting as a liaison between DRB and parties, including scheduling meetings and site visits, organizing payment of the DRB's fees and disbursements and communicating minutes of meetings and the DRB's recommendation to the parties.

Another feature of the DRB Guide Specifications, which intending users should approach with caution, is the process of nominating members for the DRB. It has the potential to become a drawn-out process if one party objects to the nominee of the other. A party is allowed 14 days to nominate a replacement and the other party has a further 14 days to reject the replacement and is not required to disclose the reason for non-acceptance. This process will continue, 'until two mutually acceptable members are named'.[11]

---

11   Specification 1.02.C.5.c DRB Guide Specifications

See table further below for comparison of other features of the AAA's DRB Specifications with other institutional rules. A copy of the AAA's DRB Guide Specifications appears at Annex 5.

D.   INTERNATIONAL CHAMBER OF COMMERCE (ICC) DISPUTE BOARDS

The ICC published its dispute board documents in September 2004. A copy appears at Annex 6. The documents comprise a set of Dispute Board Clauses (ICC DB Clauses), Dispute Board Rules (ICC DB Rules) and a Model Dispute Board Member Agreement (DBMA). The objective of the ICC's documents is to provide parties intending to use dispute boards in their contract with a comprehensive and flexible framework. According to the Chairman of the ICC Task Force charged with drafting the rules,[12] the main factor differentiating the ICC's approach from others is that the ICC's system is intended to be adaptable and applicable to medium/long term contracts in any industry anywhere in the world.

As noted in Chapter 3, parties intending to use the ICC documents have three options. The first is to opt for a Dispute Review Board (DRB). A DRB attempts a consensual approach to resolving disputes and issues a 'recommendation'. A recommendation becomes final and binding only if neither party expresses dissatisfaction with it or the recommendation is upheld by arbitration or by the courts. The second option is the use of a Dispute Adjudication Board (DAB), which approaches dispute resolution in a less consensual manner. A DAB issues a 'decision' which must be implemented forthwith. The decision becomes final if neither party files a notice of dissatisfaction. If dissatisfaction is notified, the matter will be finally decided by arbitration or through the courts. The third option is a hybrid of a DRB and DAB called a Combined Disputes Board (CDB). The CDB will normally issue a recommendation but if either party requests a decision and the other does not object, the CDB will issue a decision.

Another distinct feature of the ICC rules is the power given to the ICC under Article 21 to review a decision by a DAB or CDB before it is communicated to the parties. However, this power is exercisable only if the parties have expressly authorized it in the relevant clause.

In addition, parties are encouraged under Article 17(3) to seek to resolve the matter amicably even though DB proceedings are ongoing.

The table below provides a comparison of the main features of the rules and procedure under the ICC dispute boards system with those of other institutions.

---

12   Pierre M. Genton, in a presentation at an ICC/FIDIC conference in Paris 29 – 30 April 2004.

VI. COMPARISON OF RULES AND PROCEDURES FOR DISPUTE ADJUDICATION BOARDS/DISPUTE REVIEW BOARDS

	WORLD BANK	FIDIC	ICC	AAA
Ad hoc / Standing	Standing; to be appointed by the date stated in the contract.	Standing;[1] to be appointed by date stated in Appendix to Tender and/or Ad hoc;[2] to be appointed within 28 days of notice of intention to refer dispute.	Standing unless otherwise agreed. To be established at time of entering the contract.	Standing. It could take up to 84 days[3] from date contract entered into before the board is in place.
Size of Panel	One or three members	One or three members	One or three members	One or three members

	WORLD BANK	FIDIC	ICC	AAA
Time limits	Decision to be given within 84 days of receipt of the reference of the dispute, parties may agree longer period. 28 days to give notice of dissatisfaction with the decision and intention to commence arbitration.	Decision to be given within 84 days of receipt of the reference of the dispute, parties may agree longer period. 28 days to give notice of dissatisfaction with the decision.	Recommendation/decision to be given within 90 days of commencement of proceedings, parties may agree longer period. 30 days to give notice of dissatisfaction with the recommendation or decision.	Recommendation to be given within 14 days of hearing,[4] parties may agree to extend this period. 14 days to give notice of acceptance or rejection of the recommendation.

	WORLD BANK	FIDIC	ICC	AAA
Procedure – Exchange of statements Right to hearing,	DB may conduct hearings and request exchange of statements. Parties must be given ample opportunity to present their case. There are no express restrictions on the use of lawyers at hearings but the DB has final say on conduct of the proceedings and can refuse audience to any persons other than the representatives of the parties.	DAB may conduct hearings and request exchange of statements. DAB can adopt an inquisitorial procedure. The is no express mention on the use of lawyers, but the DAB has the right to refuse admission or grant audience to anyone other than the parties' representatives.	Parties entitled to request hearing or the Board may order. Unless otherwise agreed or DRB/DAB/CDB orders otherwise exchange of statement of case and response. Parties must appear in person or through authorised representatives, but they may be assisted by advisors/lawyers.	A hearing is to be scheduled within 7 days of receipt of response to the referral request. Parties are to exchange of statements of case. DRB may limit the exchange of any other documents, evidence or exhibits. Expressly limits the participation of lawyers in the proceedings.

Adjudication and Dispute Review Boards

	WORLD BANK	FIDIC	ICC	AAA
Jurisdiction	Resolution of disputes formally referred to it by either party.	Resolution of disputes formally referred to it by either party. Parties may agree to seek an opinion on a matter on an informal basis from the DAB or a member thereof.	Resolution of disputes formally referred to it by either party. Resolution of disputes on an informal basis provided all parties agree.	Resolution of formally disputes referred to it by either party. DRB may facilitate the resolution of problems and claims before they become formally referred disputes.

	WORLD BANK	FIDIC	ICC	AAA
Joinder of disputes	No provision.	No provision.	No provision.[5]	Yes – DRB can consolidate disputes by notifying the parties in writing. The onus is on the contractor to provide the DRB with all documents in disputes involving subcontractors and to ensure subcontractor attends and assists in the proceedings.

Adjudication and Dispute Review Boards

	WORLD BANK	FIDIC	ICC	AAA
Monitoring Project Progress (mandatory site visits etc)	At intervals of not more than 140 days but not less than 70 days. At site visits, DB is expected to become acquainted with progress of works and any problems. Produce a report after each visit.	The Red Book only: at intervals of not more than 140 days but not less than 70 days. DAB to use site visit to become acquainted with progress of works and any problems. Produce a report after each visit.	Schedule meetings to keep it informed about the contract and site visit if relevant. Minimum of 3 site visits per annum if site visits are relevant. At meetings and site visits DB is to review performance of contract, provide informal assistance with respect to any disagreement and produce a written summary.	Review monthly progress reports submitted to it by the parties. Site visits at least every 3 months unless the DRB and parties agree otherwise. Each visit to comprise a field inspection of the works and an informal round-table discussion on status of the works, disputes and claims. Produce minutes of meeting to be given to parties by the AAA.

	WORLD BANK	FIDIC	ICC	AAA
Effect of Decision / Recommendation	Binding and must be implemented forthwith unless and until revised by amicable settlement or an arbitral award. Becomes final if no notice of dissatisfaction is given within 28 days of receipt of decision.	Binding and must be given effect to promptly unless and until revised by amicable settlement or an arbitral award. Becomes final if no notice of dissatisfaction is given within 28 days of receipt of decision.	A recommendation is binding and final unless a notice of dissatisfaction is given within 30 days. If notice is given there is no obligation to comply with recommendation until dispute finally determined. A decision is binding and must be given effect to unless and until revised by arbitration. If neither party has given a notice of dissatisfaction within 30 days of receipt of the decision, then parties agree (insofar as such agreement can be validly made) to continue to comply with the decision and not to contest it.	Recommendation is non-binding.

Adjudication and Dispute Review Boards

	WORLD BANK	FIDIC	ICC	AAA
Next tier	56 days from date of notice of dissatisfaction to attempt a amicable settlement of the dispute. ICC arbitration, unless otherwise agreed, if amicable settlement negotiations fail.	56 days from date of notice of dissatisfaction to attempt a amicable settlement of the dispute. ICC arbitration if amicable settlement negotiations fail.	Arbitration or litigation.	No provision, but parties are free at any time during DRB proceeding to refer the dispute to the AAA for mediation or any other ADR method.

1. FIDIC Red Book.
2. Under FIDIC Silver and Yellow Books parties have the option of maintaining a standing DAB.
3. Or longer, if either party objects to a nomination. See Specification 1.02.C which gives parties in each case 14-day timeframes to accept/reject alternative nominations to the Board until 'mutually acceptable members are named'.
4. Whilst the rules provide time limits for responding to referrals, it is left to the DRB to fix the date of hearing, which, the rules recommend, should be around the time of the next site visit; see Specification 1.04.D
5. But note Art. 15(4), though not an express joinder provision, permits the DB to adapt the rules to apply to multiparty situations where there are more than two parties to the contract.

VII. FINDING THE BEST APPROACH

A. PREPARATION OF SUBMISSIONS

The key procedural objective for the parties in using the DRB is to clarify the dispute or issues as economically and quickly as possible. Rather than identification of the issues through pleadings the parties are (typically) encouraged to present their respective cases in a narrative form. For example, the ICC DB Rules provide that the Statement of Case should include a clear and concise description of the nature and circumstances of the Dispute and a presentation of the referring party's position on the issue(s) in dispute, along with any support for its position such as documents, drawings, schedules, and correspondence.

B. DISCLOSURE OF DOCUMENTS

Whilst the DRB may request further materials, whether of its own motion or upon the suggestion of the other party, there are (typically) no fixed requirements for disclosure of documentary evidence beyond the need, as a matter of proof, for each party, to submit documents upon which it relies and to respond to requests for further information made by the DRB. No doubt the DRB would draw inferences if a party refused to provide documents it was requested to produce.

C. HEARINGS

Under the commonly used rules referred to above there will be hearings. Sometimes, as discussed above, the rules express a bias towards hearings being scheduled for the next regular site visit. An example is ICC DB Rules Article 19.2 which states:

> 'Hearings shall be held during scheduled meetings or site visits, unless the Parties and DB otherwise agree'.

The DRB is given freedom in relation to the procedure to be followed at the hearing stage. Article 19.8 of the ICC DB Rules suggests a default running order namely:

- presentation of the case by the parties;
- identification by the DRB of any issues which need to be explored further;
- clarification by the parties concerning the issues identified by the DRB; and
- responses to the clarification only to the extent that new issues are raised by the other party.

Generally external lawyers take a low profile in hearings before the DRB. It may be that there are certain issues of legal principle which are central to the dispute, in which case external lawyers may play a role in presenting the party's case on those issues – alternatively an in-house lawyer may be used. Certainly it is unusual to have formal witnesses as one would see in an arbitration or court proceedings. Instead, each party's case is generally presented by representatives of the party who has first hand knowledge of the relevant issues – such as the site engineer, employer's representative, contract administrator or programmer. The essential aim of the procedure is to achieve a more streamlined and informal process than arbitration. Generally the tribunal would take an inquisitorial approach, asking questions of those presenting the case. It would be very unusual, and in the authors' experience, unknown, for cross-examination of those presenting to be permitted.

Ideally the members of the DRB will have the relevant technical expertise and accordingly there will not be the same necessity to 'educate' as may be required in respect of arbitrators who are lawyers with limited experience of the technical issues in dispute.

VIII. THE DISPUTE REVIEW BOARD'S DECISION

The DRB will have a specified period of time within which to reach its decision. It is prudent to provide for a period of extension with the consent of the parties, otherwise it may be argued by the losing party that the decision is of no effect if it is delivered outside the prescribed time. Again, depending on the rules, the DRB's decision may need to be unanimous, or a majority decision only. For example, the ICC DB Rules require three-member DBs to make every effort to reach a unanimous decision. Where this cannot be achieved, a majority decision is permissible. It is useful to provide a power to correct any clerical error or mistake in the decision outside of the time limit for delivering the decision.

The decision may also address the question of costs in relation to the reference. Generally, each party bears its own costs in relation to any reference before a DRB, with the reasonable costs and expenses of the DRB being borne by the parties in equal shares. It is unusual, but of course possible, to provide for the DRB to allocate its costs in unequal shares and to order the losing party to pay some or all of the winning party's costs.

A. THE STATUS OF A DRB DECISION

The DRB decision is not an arbitral award capable of enforcement under the New York Convention, nor does it have the status of a court judgement. Instead, the decision is binding only as a matter of contract between the parties on the basis of their agreement to be bound by the terms of the DRB rules. Accordingly, its effect will be governed by those rules. It may, for example, should the rules so provide, constitute only a recommendation which is not binding on the parties at all (see table above). More typically the decision will be binding unless and until reversed in arbitration on service of a notice of dissatisfaction or referral to arbitration by either party. It is generally the case, however, that the rules will provide that the decision does have to be complied with in the meantime. Indeed this is the very essence of the value of DRB proceedings – they provide a means of resolving disputes in the first instance to enable the project to continue even if, following arbitration, the result may be reversed or modified in some way.

B.   ENFORCING A DRB DECISION

The appropriate method of enforcing a DRB decision is therefore an ordinary action for breach of contract. What is the proper forum for such a claim? This, of course, will depend again on the drafting of the dispute resolution clauses. Ideally, there will be a provision expressly providing that disputes with regard to the enforcement of the DRB's decisions do not need to be referred in the first instance to the DRB but instead may leapfrog straight to the agreed procedure for final determination of disputes, most usually (on international projects) arbitration but possibly litigation in the designated courts.

On what grounds may a defendant to such action seek to resist enforcement?

What is clear is that the arbitral tribunal faced with a claim for breach of contract will not be revisiting the merits of the claim. This may, of course, be the subject of another set of arbitral proceedings, assuming a notice of dissatisfaction or referral to arbitration has been served. In the enforcement proceedings there may, however, be arguments available to the defendant to resist enforcement.

Potential grounds include:

1.   No Jurisdiction Over an Issue

If the DRB purports to decide matters which, on a true construction of the referral documentation, were not referred to it then the decision is outside its jurisdiction and will not be enforced. The argument is essentially that the DRB has asked itself the wrong question.[13] Alternatively, it may be said that the very nature of the decision is one which the DRB is not contractually permitted to make. For example, the DRB may have issued a provisional decision where it is permitted only to issue a final determination (albeit subject to reversal at the next tier of dispute resolution). Further, it may have applied its own assessment of what is fair and reasonable in the circumstances and not strictly applied the terms of the contract. In short, a DRB charged with applying the contract is not permitted the freedom to act as *amiable compositeur* to arrive at its own assessment of what it perceives to be a fair result. In the authors' experience this may be a real temptation for a DRB, and if the parties

---

13   See, by analogy, case law on the enforcement of an expert's decision, for example, *Shell United Kingdom Ltd v Enterprise Oil plc* [1999] 2 Lloyd's Rep 456 and *Jones v Sherwood Computer Services plc* [1992] 2 All ER 170.

consider this to be an attractive option, they should consider building in the flexibility to use the DRB in this way.

2. Lack of Fairness

As discussed above, enforcement of a DRB's decision may be refused if there has been a lack of fairness in the procedures adopted. The authors are not aware of any decided cases in the United Kingdom on this issue. However, by analogy with case law in England on the status of an adjudicator's decision, examples may include where:

(i) assistance is sought from a programming specialist, but the parties are not given an opportunity to comment on the final report prepared by him;[14]
(ii) one party is consulted upon submissions made by the other without a reciprocal process;[15]
(iii) the failure to make available to one party information obtained from the other party and various third parties;[16] and
(iv) use of an analysis different to that advanced by the parties without informing the parties of the proposed methodology and without seeking their observations on its suitability.[17]

---

14 *RSL (South West) Ltd v. Stansell Ltd* [2003] EWHC 1390.
15 *Discain Project Services Ltd v. Opecprime Development Ltd supra.*
16 *Woods Hardwick Ltd v. Chiltern Air Conditioning* [2001] BLR 23.
17 *Balfour Beatty Construction Limited v Lambeth London Borough Council supra.*

Chapter 6
Forms of ADR

I. INTRODUCTION

The obvious question is why, in a work on construction arbitration, is there a chapter on mediation, conciliation and other forms of ADR. The answer is that this book is intended to provide a guide through the real world of construction arbitration and is not an abstract treatise on the practice of arbitration in a vacuum. Arbitration is not is not an end in itself but a means to an end. In achieving the objective (settlement of a dispute) it will often be sensible to consider whether that objective is best served by pursuing arbitration to the exclusion of all other modes of dispute resolution. If not, whether a two-pronged approach is a more sensible, practical and cost efficient way to proceed. For that reason, and because in the real world the possibility of resolving disputes in a more consensual and less confrontational way will be encountered, this chapter sets out to identify the forms of ADR which are available, the circumstances in which they might be considered and how they interface with the arbitration process itself.

So how, when disputes are to be resolved by arbitration, do the parties come to resolve their differences by ADR? To start with, it is now commonplace for judges to take an active role in seeking to settle court cases by the process of court encouraged mediation. In England and Wales, for instance the Woolf Report recommends increased use of ADR, and parties to litigation are required to state whether ADR has been considered as part of the pre-trial review.[1] An unreasonable refusal to do so may later result in cost penalties if the litigation proceeds.

---

1   Lord Woolf MR, *Access to Justice: Final Report to the Lord Chancellor on the civil justice system in England and Wales* (HMSO, 1996).

Similarly, parties involved in litigation in other jurisdictions may be required (by the relevant courts and/or applicable civil procedure rules) to consider attempting to resolve their disputes through ADR. For example:

- in Hong Kong, parties in a construction dispute are required (pursuant to a Practice Direction issued by the court) to indicate at their pre-trial directions hearing whether any attempts have been made to resolve the dispute by mediation;[2]
- in Canada, parties to a civil dispute being heard in the Federal Court may be subject to a court order requiring a proceeding or issue in a proceeding to be referred to a 'DR conference' whereby a judge or prothonotary may conduct a mediation, early neutral evaluation or mini-trial (described below);
- in the United States, the rules of civil procedure applying to disputes before the Federal District Courts and the United States Courts of Federal Claims provide for consideration of ADR at pre-trial conferences as a means of assisting in the resolution of the dispute;
- in proceedings brought in certain courts in Australia, parties are required to consider (as part of their case management) using mediation to attempt to resolve their dispute;
- in Spain, judges are required to invite the parties to a preliminary hearing which, among other things, has the aim of giving the parties the chance to reach an amicable settlement;
- in Sweden, where the dispute is amenable to out of court settlement, the courts must take all appropriate measures to allow the dispute to be resolved amicably;
- in civil court proceedings in Germany, oral hearings must be preceded by conciliation hearings, unless an attempt at settlement has already taken place or it is evident that the conciliation hearing has no prospect of succeeding; and
- in Greece, disputes that fall within the jurisdiction of the court of first instance cannot be heard unless there has first been an attempt at conciliation.

While there is, as yet, no great groundswell of opinion that arbitrators should be adopting a similar approach, there are murmurs that this should be part of their function. However, the lack of any such standard practice means that if any form of ADR is to be attempted it generally has to happen at the instance and with the agreement of the parties but normally not with the involvement of the arbitrator.

Parties' agreement to participate in ADR may come about in one of two ways. First, as discussed in Chapter 3, the dispute resolution provisions of the construction contract may require the parties to attempt ADR as part of the dispute resolution process. At one time it was considered that such clauses, as a matter of English law and possibly some other common law based legal systems, were unenforceable and could be simply and safely ignored by a party which felt disinclined to seek to

---

2 However, refusal to mediate does not necessarily lead to adverse costs sanctions in Hong Kong.

attempt to resolve their disputes in this fashion. Any doubts about this were laid to rest as a matter of English law when Coleman J, a judge of the Commercial Court in London, required a reluctant party to participate in a mediation even though it said it did not wish to and argued that it was not obliged to.[3] As a result, any dispute resolution provision governed by English law which includes a requirement to mediate (or attempt some other form of ADR) must now be regarded as binding. Therefore, a reluctant party can also now be compelled, by an application to either an arbitrator or a court, to participate in a mediation. This does not, of course, imply that the chosen ADR technique must result in a concluded settlement or even that it must be pursued to the bitter end if it is obvious that it will fail. But at the very least the steps agreed by the parties for participation in the ADR procedure must be attempted.

The second way an attempt at ADR might come about, even in the absence of any requirement in the underlying contract, is at the instance of the parties or, increasingly, their advisers. Where ADR has taken hold – principally in the United States and in England – it is now commonplace for both litigators and arbitration specialists alike to regard mediation and other ADR techniques as one of the tools by which clients' disputes may be resolved.

However it comes about, the two questions which need to be addressed at this stage are (i) what form of ADR should be attempted and (ii) who should be the neutral. Obviously the two questions are interdependent, with the answer to the second largely depending on the answer to the first. If, however, the parties and their advisers are unable to determine the most appropriate form of ADR it occasionally happens that they can reach agreement on a neutral that they respect and then leave the final form of the procedure to him. More often, though, in such circumstances the assistance of one of the recognized ADR administering bodies is employed, taking brief details of the nature of the dispute and then recommending both a procedure and (if necessary) a neutral.

II. TYPES OF ADR

ADR sits between the forms of principal-to-principal negotiation that take place both at site and project management levels every day of every week on construction projects and the more formal systems of dispute resolution processes, including expert determination, adjudication, Dispute Review and Adjudication Boards and, of course arbitration and litigation themselves. Yet even within this narrow part of the dispute resolution spectrum, ADR techniques span a considerable gap. The principal common feature is the involvement of an independent neutral whose objective is not necessarily to determine the parties' respective rights and obligations, but to facilitate an effective resolution of disputes which have arisen.

---

3 *Cable & Wireless plc v IBM United Kingdom Limited supra.*

While this may involve consideration – sometimes quite careful and detailed consideration – of rights and obligations, as often as not this is in the context of encouraging the parties to consider what the alternatives to reaching a binding negotiated settlement might entail. In the context of the common or garden construction case, this contemplation of what might take place if they do not reach a negotiated settlement will almost inevitably lead the parties to realize that it would entail considerable expense and diversion of management time from their principal business. As such, anything other than a negotiated settlement at as early a stage as is possible is destructive to the parties. For this reason, if for no other reason, early settlement should be encouraged in all possible circumstances. The neutral's job is to facilitate this. In order to achieve this desirable outcome, the neutral's skills can be deployed in a number of different ways, the principal of which are discussed below.

Of course, there is no obligation on any of the parties to reach a settlement as a result of any form of ADR procedure and the neutral has no powers to compel the parties to enter into an agreement or to hold them to any particular agreement that may or may not have been made in the course of the ADR procedure. Only when a formal agreement has been reached and recorded by the parties in writing[4] is the dispute settled. Nor is the settlement agreement produced in the course of a mediation immediately enforceable, unlike an arbitration award.

This last point has been regarded by some as a distinct weakness in the process and was considered at some length in the course of the agreement of the new UNCITRAL Model Law on International Commercial Conciliation.[5] Considerable discussion took place over many sessions on whether a settlement agreement reached in the course of a mediation should be accorded the same sort of status as an arbitration award, perhaps after signature by the mediator. In the end, however, the agreed form of the Model Law went only so far as to state that any agreement reached is binding and enforceable (which it would presumably be in any event) whilst leaving it to adopting states to specify whether and if so how such an agreement should be enforced within their own territories.[6] There is not (and there is not expected to be) any equivalent to the New York Convention by which settlement agreements arising out of ADR procedures will be internationally enforceable.

---

4   Writing is not, perhaps, strictly necessary but is required by the principal standard rules for mediation.

5   See the UNCITRAL Model Law on International Commercial Conciliation with Guide to Enactment and Use 2002, available at : http://www.uncitral.org/uncitral/en/uncitral_texts/arbitration/2002Model_conciliation.html (accessed 18 November 2005).

6   Ibid., paras 87 – 92.

A.   MEDIATION AND CONCILIATION

At the least judgmental end of the scale of ADR procedures lie mediation and conciliation. Unfortunately there are no hard definitions of either of these terms and as a result they may either be used more or less interchangeably (according to some) or be distinguished (say others) by the willingness of a 'conciliator' to adopt a more rights based, judgmental approach than a mediator. For present purposes, no real distinction is intended to be drawn between the process of mediation and conciliation, it being recognized that the degree to which the neutrals concerned are willing to take a more interventionist, judgmental approach has much more to do with the personal temperament and outlook of the neutral than with the label attaching to the process. For that reason the terms 'mediator' and 'mediation' are used without any intention thereby to distinguish the process from conciliation.

Ignoring the semantics, the process of a mediation/conciliation is about the bringing together of the parties by judicious use by the neutral of both carrot and stick, usually, though not invariably, in a non-judgmental way. In other words, the neutral's approach is one which causes the parties to work out for themselves, prompted by the neutral as and when necessary, which elements of a proposed deal are to their advantage and which are not. It is for the parties, not the neutral, to assess and factor into their analysis of the settlement point, the chances of them succeeding and of the costs, particularly the irrecoverable costs, that failure to settle would involve.

Where the contract contains an express requirement to participate in a mediation, the process will start as laid down in the contract. This may either contain quite detailed provisions as to the conduct of the mediation or may be in a short form, often referring to the provisions of one of the recognized ADR centres. Either way, the first significant step is the choice of the mediator. At this point, it is important to remember what it is that the mediator is doing or, rather, not doing. By and large, what the mediator is not employed to do is to determine the respective rights and obligations of the parties. What the mediator is attempting to do is to bring the parties to the point where they realize that it is in their best interests to settle early and cheaply, even if that means making some compromises along the way. For this reason, the choice of mediator is as much about picking the person with the right interpersonal skills as knowledge of the subject matter of the dispute. And it is much more about picking the mediator with the right interpersonal skills than finding the person who will determine what the 'right' answer is.

The ideal mediator will, therefore, have the skill to identify what it is that each party wants to get out of the mediation by way of settlement and how to bring them together, by focussing on the relevant and removing the obstacles to settlement that a deep-rooted dispute often contains. Not least, the mediator must recognize the personal involvement of the individual participants, many of whose actions have contributed to the dispute and whose bonuses, reputations or possibly even careers depend not on getting the most sensible commercial settlement in all the circumstances, but on getting the answer to which they have committed themselves, along with the organization for which they work.

Taking these interpersonal skills as a given, the next issue is the importance of the mediator having some familiarity with the subject matter of the dispute or of the businesses within which the parties operate. Of course, in an ideal world, a mediator would have both the relevant skills and the detailed knowledge, not just of the parties' businesses, but also of the subject matter of the dispute. The reality is that this happens only occasionally and that such 'new age' mediators can rarely be found – or if they do exist they will not have the time to devote to the mediation. Given that the ideal is not attainable, a mediator's understanding of the way the business operates, and the opportunities within that business area, is more important than a knowledge of the specifics of the dispute. Accordingly, the choice of mediator will normally be made on the basis of the mediator's reputation, knowledge of the business area within which the dispute has arisen, and availability. That done, any deficiency in the background knowledge of the mediator can often be made good by the selection of an appropriate 'pupil mediator'. Most ADR institutions require mediators recommended by them to have undergone both theoretical and practical training, as well as undergoing a process of continuing education and practice. For this reason there is, at least at the time of writing, a steady supply of would-be pupil mediators willing to get hands-on experience of mediations. While not having any formal role in the conduct of the mediation – and not being paid for their participation either – a suitably qualified pupil mediator working with the chosen mediator can often add invaluable experience to the process. So too can having lead and co-mediators, with the co-mediator supplying industry – or subject matter – specific input, though obviously at additional cost.

Other factors that may be considered are the likely prejudices of the mediator and the fees. Time and time again the question is asked whether the mediator is a 'black letter lawyer' or a 'commercial man', or whether the mediator is a 'contractor's' or an 'employer's' man. These matters are, in practice, clearly of great significance to the users of the mediation process and for that reason should not be ignored. After all, the process requires the parties to reach agreement, and it is in no-one's interest if one party perceives the mediator as being partial to the position of the other. However, while there is often great pressure and much temptation to influence the process of mediation, much as is (quite properly) done when selecting an arbitrator, the best result must be to select a mediator who does not have a reputation for favouring the position of one party or the other. As for costs, while some mediators charge as much as the most expensive arbitrators, the overall costs of a mediation – measured in days of work rather than weeks or months – means that the question is of marginal relevance.

Once the mediator(s) have been chosen, the next step will often be informal approaches by the mediator to both parties, either together or separately, in order to establish the procedure and to allow the mediator an early opportunity to direct the process in a way which he considers most likely to result in a favourable outcome. Unlike arbitration, there are no restrictions on the parties (or their advisers) talking to the mediator without the other being present, although occasionally some advisers do attempt to stamp their own authority on the process by seeking to dictate that virtually all contact with the mediator is to be joint. Fortunately, most mediators explain why this is not likely to be in the best interests of either party and most often the process continues as the mediator sees fit.

It should be noted that there are, in fact, very few inviolable rules relating to the conduct of a mediation. The single most important one, which is found in virtually all guides and rules of mediation, is that the process is to be without prejudice – i.e. nothing said or done in the course of the mediation is to be relied on as evidence in any subsequent court or arbitration proceedings. The other fundamental rule is that what is said by one side to the mediator in confidence must not be passed on to the other party without consent. How that consent is given varies from mediation to mediation, with the default position generally being that everything said by one party in the absence of the other is to be regarded as confidential unless disclosure is expressly authorized. Occasionally the default is the reverse, with the mediator feeling free to pass on everything said to him unless it is specifically identified as being confidential and not for onward disclosure. Either way can work satisfactorily but it is, of course, essential to know which default position is to apply.

One of the first things that the mediator will do is to circulate a draft mediation agreement.[7] This will deal with the key issues of the use to which information disclosed in the mediation may be put. It will also deal with confidentiality, as well as more mundane matters such as fees, venue for the hearing and its duration. Since in practice (like many construction contracts) the mediation agreement is not normally signed right at the start of the mediation, but only after a number of steps in the mediation may already have taken place, it is generally a good idea if the key issues set out above are agreed in a short written exchange to cover the period from the very start of the process until the mediation agreement is signed, normally immediately before the start of the oral hearing.

The draft mediation agreement – and no doubt the mediator in person – will also make it clear that it is expected that those leading the mediation for each party will have the necessary authority to settle. This is clearly an important factor, and one which causes considerable anxiety in practice. Of course, in the real world nobody representing a corporate body comes authorized to settle at any price, though normally most do come with something resembling authority to settle within a realistic range. However, where it becomes apparent – regardless of what is being said – that those leading a party's delegation to the mediation do not in fact have authority (or cannot easily obtain it via mobiles or Blackberries) the process is likely to founder quite quickly. Especially difficult are cases where third parties such as sub-contractors, professional advisers, insurers or lenders to SPVs have an interest in the outcome. While a degree of understanding is normally present concerning

---

7   For a sample form of mediation agreement, see the website of the Centre for Effective Dispute Resolution at http://www.cedr.co.uk (accessed 17 November 2005).

the difficulties that these situations present, it is nevertheless incumbent on the party having to report to, and obtain approval from, other interested parties to take all necessary steps to have those other parties engaged in the issues and the process. This might extend to having them present on an observer basis to reduce the information lag that would otherwise occur.

Following the introductory and administrative stages of the mediation, it is likely that the parties will be invited to present position papers to the mediator and each other. Typically, mediators prefer these to be fairly short but the parties (and their advisers) tend to make them longer than they need necessarily be, often also attaching many files of supporting papers. In part, this is because the function of these position papers tends to be forgotten. Although the mediation process will involve a review of the matters leading up to the dispute and the parties' respective rights and obligations, the purpose of the mediation is to find a settlement which is as much forward looking as it is an historic review of what went wrong. The real function of these position papers is not, therefore, to set out in great detail all that has gone before. Most, if not all, of this will already be known to the parties and be of only marginal relevance to the mediator. The real functions of the position paper are (i) to educate the mediator about the parties' current positions and the possible avenues for a successful settlement and (ii) to educate the parties' decision makers (who, likely as not, will not previously have had direct exposure to the other party's case) about the key factors they should take into account in agreeing to a settlement on the terms that the other party considers acceptable.

Matters dealt with in position papers may therefore include key strengths in a party's own position or key weaknesses in the other party's (relevant if the matter is not settled and has subsequently to be determined by an arbitrator), but it is just as likely they will include suggested opportunities for each side to improve on their own commercial position in the present project. However, unlike any form of court or even arbitration proceedings, there is no reason why the messages being sent should be restricted to the matter in dispute or even the project or business relationship that gives rise to it. There is no reason why the position papers should not identify instances where the parties might jointly take advantage of future opportunities not readily exploitable by either in isolation, or raise possible adverse commercial consequences to one or both of them if settlement is not reached. In short, a lengthy regurgitation of the contractual correspondence at this stage is a wasted opportunity. The focus should be on getting key issues about the settlement point across to the decision maker of the other party.

Following the exchange of position papers (and possibly reply position papers), experienced mediators will often ask a series of follow-up questions, suggesting areas of the written presentation that might be developed or clarified. After that, there will be some form of mediation hearing. Again, there are absolutely no hard and fast rules as to how such a meeting will be conducted. All that can sensibly be said is that almost all start off with general introductions (mainly for the mediator's benefit, as it would be unusual for the parties not to know each other's representatives) and an opportunity for each party to make an oral presentation to the other party/ies and the mediator. Again, this is a stage in the process that provides great opportunities, and equally one which is often wasted. What is definitely not required at this stage is a (further) detailed recital of the history of the dispute. What the first part of the hearing offers each party is the opportunity to present its reasons why a settlement should be agreed and on what terms. While this may have been done before, this is usually the first opportunity to give it directly to the decision maker of the other party without any filtering or spin added by the decision maker's subordinates and advisers. This is important, because it may well be that those subordinates and advisers have a vested interest in a particular outcome, either because they were part of the problem or because they have become identified with a particular view on the position, from which it is now difficult to resile.

From this point in a mediation onwards it becomes increasingly difficult to predict what procedure will be adopted. More often than not, the mediator will suggest that the parties break out into separate rooms so that the mediator can explore specific issues with each party to improve its knowledge of the issues and the likely area of settlement. At the same time, the mediator will often take the opportunity to make sure that each party is taking a realistic view on the strength of its position and the consequences of not reaching a settlement. Normally the mediator will avoid taking a position on the underlying dispute but will cause the parties to undergo a process of reality testing by which they are made to face up to the key points in their and the other party's cases and the financial and other consequences of not settling. As part of this, the mediator may require each party to produce summaries for their internal use. In addition, specific matters (commercial, technical or legal) may be identified for further discussions in small groups between the parties. This has the benefit of allowing the mediator to isolate and sideline potential stakeholders in the process who might represent an obstacle to success.

Whatever the process, there will come a stage when the mediator feels comfortable enough to attempt to sketch out a settlement that the parties have, in essence, bought into. Not unusually there is still a gap (gulf, more often) between the parties at this stage which needs further work, involving more reality testing and more discussion about the impact that resolving the dispute in a traditional manner will have on the parties' businesses.

At some later stage, one of two things will happen. First, it may become apparent that no further progress can be achieved in the time allotted for the hearing, with the result that there is no purpose in continuing. If that has happened as a result of the need to obtain a higher level of authority or third party consent, the mediator will encourage the parties to obtain the authority/consent immediately but, failing that, seek to maintain the progress of the mediation either by agreeing a date for an adjourned hearing or fixing a date and time for a conference or video-conference call to take matters forward.

The other thing that happens more often than might be expected is that a settlement agreement is reached. In this event the challenge for the mediator and the parties is to document that agreement in an acceptable form, ideally before the parties leave the hearing. In reality it is only relatively simple disputes where the settlement can be fully documented with the resources available at the hearing, even if lawyers are present for both parties. Experience shows, however, that a written record – even if only a heads of agreement – of the settlement should be signed by the parties before they leave the mediation hearing, thereby reducing the chances of the agreement being reviewed – perhaps by others not present at the mediation – and the deal evaporating before being consummated.

B.   MINI-TRIAL

The mediation process described above involves, in any complex case, each party being supported by a large number of people required to provide the necessary factual, legal, technical and financial support to the decision maker. A mediation involving 20 or more people is, as a result, not uncommon. Nevertheless, having this number of people involved is unwieldy and makes the mediator's job more difficult as it is by no means always easy to separate the decision makers from the rhetoric and the interests of other stakeholders and to get the decision makers to focus on what is in the best interests of the parties they represent. One particularly effective way this can be done is by the ADR technique known as 'mini-trial'. In essence, this is an amalgam of a conventional hearing and a mediation as described above. The essential difference between a mini-trial and a mediation is that the decision makers of each party are isolated from their teams and placed with their opposite numbers and the neutral.

Again, while there are no hard and fast rules on how mini-trials are to be conducted, the stages up to the oral hearing are likely to be similar to those in a typical mediation. In other words, there will be selection of a neutral, agreement of the ground rules (without prejudice nature of documents, confidentiality of proceedings etc.) and preparation and exchange of position papers. The difference comes about at the hearing itself. At this point, instead of there being a hearing of the sort previously described, the decision makers and the neutral form a panel to which each party's representatives make presentations. At the conclusion of the presentations, including after any reply submissions and questions from the panel, the panel retires without the presence of any of the party's other representatives. What then takes place is a mini-mediation between the parties' decision makers, moderated and guided by the neutral.

There are a number of advantages to this. For a start, the fact that the decision makers will be isolated from their support staff means that they will be very much better briefed than if they knew they had immediate and unfettered access to their advisers and others in their organization. This enhanced degree of preparation enforced upon them increases the chances of the success of the process. The same effect can often be found in mediations that take place away from both parties' home bases. Even with increasingly sophisticated communications, the extra degree of preparation attendant in going 'off site' produces an increased chance of a settlement. The second advantage is that the neutral gets sole access to the decision makers and can direct their focus onto what is in their organization's best interests, generally speeding up the process by which mediated settlements tend to be reached.

Despite these advantages, the fact is that mini-trials are far less common in Europe (at least) than mediations, although the Centre for Effective Dispute Resolution in England and the Netherlands Arbitration Institute both cater for the process.[8]

C. EARLY NEUTRAL EVALUATION

As discussed in Chapter 3, a further technique – which is by definition an evaluative process – is the process of Early Neutral Evaluation (ENE). Unlike the processes of mediation, conciliation or mini-trial, the technique sets out to obtain a non-binding decision from an independent third party who is an expert in the relevant field. There is, therefore, no need for the neutral to have the requisite interpersonal skills or ability to assist parties in reaching an acceptable settlement. This ADR technique is all about – and only about – reaching a non-binding decision on the merits of the dispute so as to assist the parties in reaching a settlement outside of the confines of the process itself.

Since this is, in effect, a trial run of the parties' cases, there is no real need for any of the safeguards required by mediation/conciliation or mini-trial to prevent material used in the ENE from being relied on in any subsequent proceedings. That said, it is conventional for the parties to agree that their submissions and the results of the process shall not be relied on in subsequent proceedings.

Since the process is non-binding, the timescales for the production of a decision by the neutral can be kept short, so that the process can run in the same sort of timescales as an adjudication or dispute adjudication board for a small project. Of course, the key difference between the result of an ENE and that of an adjudication or Dispute Review and Adjudication Board is that the decision is not binding, even temporarily. This means that although the decision is expected to be indicative of a decision ultimately obtained in an arbitration, there is no need for the case to be worked up to quite the same degree as would be needed if the decision were binding, even if binding only unless and until overturned in arbitration. For this reason, it is not unusual for ENE procedures to be determined on paper only, or with very short oral submissions. It would be rare for witnesses to be called or for the evidence or credibility of factual or expert witnesses to be taken into account. For these reasons, there are many cases where there is no real value to be obtained by an ENE. The procedure is most suitable for questions of interpretation of documents or narrow technical disputes where the facts are not much in dispute. Where a neutral's decision in an ENE does depend on facts which are in contention the value of the decision

---

8   See http://www.cedr.co.uk and http://www.nai-nl.org/english/.

will be much reduced, since each side will then aim off from the decision obtained through the ENE by its anticipated chances of success on the disputed facts.

III. WHICH ADR TECHNIQUE SHOULD BE USED?

As readers will have appreciated from the description of the procedures set out above, the most commonly used form of ADR is mediation/conciliation. Of the two other types of ADR which have been identified the last, ENE, is clearly unsuitable for most construction disputes since these cases tend to be fact and document intensive. However, despite this, ENE may be suitable if there are key elements of larger disputes which are suitable for the procedure.

In particular, a good number of disputes in construction projects involve the applicability or non-applicability of clauses limiting or excluding liability. Of course, in many cases these issues are resolved by treating them as a preliminary issue in an arbitration, but this is often resisted on the basis that one party feels that the whole of the case needs to be deployed in order to put the clause in its correct commercial context. In addition, even a preliminary issue which is binding on the parties and which may be determinative of the outcome of the claim may involve considerable expense. Perhaps because of these factors, a number of such issues are not resolved in this sensible and relatively economical fashion, but are left to be determined by the tribunal in the course of a general decision on the merits. Where this happens, huge amounts of time and money can be wasted on proving the substantive merits of the case when what is really in issue is only, for instance, whether the claims are effectively unlimited or restricted to a small percentage of the claim. Once the answer to that threshold question is known, a suitable settlement can be swiftly negotiated, almost regardless of the arguments on the merits. ENE may be a suitable vehicle for getting the parties to a position where they are sufficiently confident of the answer to such questions to know what figure the negotiations should be around.

Leaving aside the possible use of ENE in the circumstances described above, the bulk of construction claims will generally be suitable for mediation or conciliation. Typical construction cases involve considerable disputes over the facts and large quantities of paper, both factors tending towards an ADR technique which focuses on harnessing the parties' knowledge of their cases rather than requiring the case to be distilled into a form that can be presented to a tribunal and then ruled upon. In addition, in significant parts of the industry (particularly in the process and power sectors, as well as in many public sectors), there is a pattern of repeat business, which lends itself to constructive solutions to problems on particular projects.

Whether or not mini-trial finds an increased role for itself is a matter of some debate. Quite obviously, the process lends itself to organizations with a strong central management where a decision maker can make – and is accustomed to making – decisions without first obtaining a consensus within a large part of the organization. Perhaps because of this it may be difficult to see much scope for mini-trial in the public sector or in cultures where it is important to obtain consensus before decisions are made.

It has to be recognized, however, that not every case is suitable for ADR. There are cases which can clearly only be determined by a binding decision made by a judge or arbitrator. Typical (though less likely to be heard by an arbitrator) are cases where fraud or other significant malpractice is alleged. Cases falling within this class are very difficult to settle on a voluntary basis because of the stigma which so often attaches to the person or organization against which the allegations are made. Even where the costs of fighting the case and proving innocence make a commercial settlement eminently sensible, in practice cases such as these prove remarkably difficult to settle. Also falling within this general category are claims against organizations where a decision to compromise a claim may be perceived as being 'soft' or, worse, involve some form of benefit to the person within the organization making the decision to settle. In these cases, it is easier for the organization and the individuals who comprise it to press the matter to a decision of an arbitrator or judge – even if the decision is adverse – than to reach a sensible commercial compromise before that time.

IV. ADR'S INTERFACE WITH ARBITRATION

There are two principal issues arising out of the interface between ADR and arbitration. The first is when is it appropriate to engage in an ADR procedure when the matter is to be referred to arbitration. The second is in what circumstances, if at all, is it appropriate for the arbitrator to engage in a form of ADR and the consequences of the arbitrator continuing in office if he does.

A. ADR AND ARBITRATION – A QUESTION OF TIMING

As already described, any form of ADR process leading to a consensual result (including ENE, to an a extent) requires that the parties to the process are fully aware of the facts surrounding the dispute and their chances of succeeding in resolving it in their favour if the matter is referred to arbitration. It is for this reason that it makes no sense for a party committed to mediation to try to 'bounce' the other party into a settlement by refraining from revealing part of its case before the mediation. Trial by ambush may work in an environment where it is permitted and where there is no opportunity to take time out to answer the point. But in the overwhelming majority of cases, such last minute disclosure in an ADR procedure merely renders the prospect of reaching any agreed settlement remote. Not only will the 'bounced' party want time to reconsider the impact of the new information on its own analysis of the situation, but it will also want the opportunity to educate the party responsible for making the late revelation why it is not the smoking gun that it was portrayed to be.

Accordingly, mediation cannot take place before the claim, its basis, its size and potential impact have been explained by the claimant and then not before the respondent has had an opportunity to analyse it with the benefit of its own advisers and on the basis of the facts as its own records and witnesses understand them to be. In practical terms, in a construction contract this probably means that where an engineer, architect or contract administrator is charged with forming an initial view on the validity of claims, mediation is unlikely to be successful before this process has been completed. However, with the benefit of this information and the respondent's response to the claim, both parties should be in possession of sufficient information to allow their respective organizations to participate in an ADR procedure and to make an informed decision about the circumstances of the dispute. In addition, it is hoped that the wider issues relevant to the benefits to both parties of settling rather than fighting the matter to the bitter end would be readily available at this time.

Where there is no such procedure, or the procedure has not been followed but the parties nevertheless decide to proceed with an ADR procedure regardless, it is probably fruitless to attempt to go straight to ADR without some form of formal claim being made and considered. While the position papers exchanged in mediations or mini-trials could be (ab)used for this purpose, their real function is not to argue the merits of the substantive dispute but to prepare for the mediation. In these circumstances it would make sense either for the standard contract procedures to be followed (if there are any) or for an analogous or shadow procedure to be adopted, perhaps on a truncated timetable.

So far the question of timing has been approached on the basis that the decision to try ADR is either written into the contract as one of the early steps in the dispute resolution process, or that the parties have simply decided (as sometimes happens) that the potential savings of ADR are worth going for even though the contract does not require them to. However, in the absence of guidance in the contract and any positive desire to resolve disputes in a cost efficient and effective manner, it may be that the first time that ADR is considered is after arbitration proceedings have been contemplated, or even started. At this point, it is likely that there would be ample information available to both parties to enable them to evaluate both their own and the other party's case – certainly sufficient to allow them to reach an informed decision to settle. The question is then not whether the parties are able to enter into an effective ADR procedure but whether it is in their interests to do so at that time.

If the question of attempting ADR has not been addressed up to this point, it is likely that a considerable amount of investment will already have been made by one or both parties and the opportunity to save a considerable proportion of the costs of a conventional arbitration will have been missed. From this belated starting point the relative merits of proceeding with ADR sooner rather than later need to be considered. In essence, this is a trade off between the additional information that the arbitration process will produce for each party against the costs of achieving additional certainty about each party's own case and its knowledge of the other's. There is, unfortunately, no clear answer to this question.

In fact, only two clear conclusions can be drawn. The first is that the sooner ADR is attempted, the greater the chances become of a lower overall spend on the dispute resolution process. What this does not, however, reveal is whether, from each party's individual perspective, it will achieve a better overall settlement (costs included) by participating in an ADR process at an early stage, rather than waiting for more information to become available. For instance, the availability of a wide scale disclosure process may offer significant actual or perceived opportunities to one or other party, which would suggest leaving an attempt at ADR until after the disclosure process is complete. Alternatively, it may be that it is important to see whether a particular person will be called as a witness for the other side and, if so, what he has to say. Or the case may revolve around a technical issue, the merits of which will not become clear until the exchange of experts' reports. The existence of any of these factors would suggest that commencement of an ADR procedure should be delayed.

The second conclusion is that there will come a point – before the costs of the hearing itself are incurred – when the parties will know as much about their own and the other party's case as they will ever know. The consequence of this is that in almost every case there is a window of opportunity for the parties to achieve both certainty of outcome and a reduction in the costs of resolving their dispute by attempting ADR. While in some ways this window is far too late in the process to be ideal, achieving both of these objectives certainly makes the exercise worth attempting.

B. ARBITRATORS AS MEDIATORS: MED-ARB

The one argument against attempting mediation at this late stage is that if it is unsuccessful, further money and effort will have been wasted in bringing a third party up to speed on the case. This inevitably leads one to question whether one or more of the arbitrators could not sensibly fulfil the function of a mediator, thereby eliminating or (at the worst) significantly reducing the possibility of ADR adding to (rather than reducing) the costs of resolving the dispute. The obvious advantages include the fact that the arbitrator(s) have already been brought up to speed or will have to be brought up to speed on the case, so the additional cost of educating him in the context of the mediation is either non-existent or one which will be incurred anyway. It is also probable that the arbitrator(s) have been selected for their suitability to determine the dispute, with the probable result that they have relevant background knowledge of the parties' businesses and/or of the subject matter of the dispute.

Whilst the arguments in favour of the arbitrator(s) acting as mediator(s) are superficially attractive, the following two matters need to be taken into account.

First, that the arbitrators have in all probability been selected for their abilities as arbitrators and not as mediators. While it is becoming increasingly common for arbitrators to train as mediators (as well as adjudicators, experts and members of all types of dispute boards), this does not necessarily mean that they are as good at the job of mediators as they are at the job they are best known for. Indeed, there is an inherent conflict when switching from carrying out a rights based analysis (judge/arbitrator) to contributing to an interests based solution (mediator). Some, but by no means all, judges/arbitrators successfully accomplish the transition. It therefore comes as no real surprise to find that some of the best mediators in the world act solely as mediators – whether this is explained by the way they approach matters or the amount of practice they get is immaterial.

Second, mediation involves a process of exchange of information between the parties and the mediator that is entirely contrary to the rules of natural justice found in arbitration (which effectively prohibit one party making private disclosure to the arbitrator). As a result, either the arbitration or the mediation will be compromised. If the arbitrator/mediator does accept private information from parties without disclosing it to the other there will be a real risk that that material may, consciously or unconsciously, inform that arbitrator's decision. This may then lead to reasonable grounds for concern and challenge of any arbitral award, both in the courts of the place of the arbitration and in the courts of the place of enforcement. If, on the other hand, the arbitrator/mediator either refuses to accept receipt of private information or accepts it only on terms that it may be used in reaching its decision in the arbitration and/or disclosed to the other party, then the process of mediation will be compromised. No right-thinking party in these circumstances would then voluntarily disclose anything to the mediator which would have the remotest possibility of impacting adversely on its case if the mediation failed.

For these reasons, the possibility of arbitrators acting as mediators and then continuing to act as arbitrators (Med-Arb) has been and remains the subject of considerable controversy. Notwithstanding this, a few countries, such as China, permit Med-Arb, at least where the parties expressly agree to the arbitrator/mediator reverting to the role of mediator after a failed mediation. This is also the position adopted in the UNCITRAL Model Law on International Commercial Conciliation. Whether or not this in practice happens – or whether it is a good thing for it to happen – is another question. The likelihood is that where it does happen mediation is not conducted in the same way that it is in England and the United States and that little additional (and no confidential) information is given by either party to the arbitrator/mediator while wearing his mediator hat.

Chapter 7
Commencement of an Arbitration

I. INTRODUCTION

This chapter outlines one of the main benefits of arbitration: the ability to tailor it to the needs of the parties. The parties are able to shape the process in various respects, the most important being their right to appoint the members of the tribunal. The fact that the parties can choose arbitrators who are specialized in particular technical areas and select the seat of the arbitration and the language to be adopted allows a unique flexibility, which means the process is particularly appropriate for international and complex technical disputes.

The decision as to whether arbitration is the most appropriate forum in which to resolve a dispute is usually taken long before any problem arises between the parties. The arbitration agreement that the parties draft at the time of entering into contractual relations will, to a great extent, dictate the process.[1] The parties must therefore think very carefully at an early stage, when the arbitration clause is being agreed, about issues such as the types of disputes which may arise, the appropriate seat and language of any arbitration and, where there is potential for a multi-party dispute, which parties are likely to have similar interests. It is often a difficult challenge to anticipate the way in which future disputes may arise, however this chapter will try to and address some of the considerations one should bear in mind when drafting an arbitration clause in order to ensure the proper procedures are in place, should it ultimately be necessary for the parties to arbitrate.

As discussed in the previous chapter, there are several arbitral institutions that will provide guidance and structure to the process, should the parties opt for an institutional arbitration. However, the parties should be aware that there is an additional cost, in that fees are payable to those institutions. The alternative is an

---

[1] The process will depend on whether (and if so which) institutional rules have been adopted and the national arbitral legislation of the arbitral seat.

ad hoc arbitration, which provides more flexibility and party control, but carries with it certain disadvantages. For example, an ad hoc arbitration may take more time, as the parties have to co-operate and agree the procedure and there are no institutional deadlines driving the process. It may also be necessary to have recourse to national courts for matters such as the appointment of arbitrators further slowing the process. There is, however, a middle road, whereby the parties elect to adopt the framework of a particular institution, but reserve the right to derogate from its rules by agreement.

Parties from different countries may favour different models, often depending on whether they are from civil or common law jurisdictions.[2] In international disputes, the procedural expectations of parties from different legal traditions can lead to difficulties, although as arbitration becomes more common these problems are diminishing. Nevertheless, it is vital that before selecting a particular structure, the parties consider the pros and cons of the different rules of the various arbitral institutions and assess the flexibility each provides.

II. SELECTION OF THE TRIBUNAL

The principle behind the parties' right to appoint the arbitral panel is that they should be free to have the dispute resolved by 'judges of their own choice'.[3] The quality of an arbitral tribunal is the most important factor in ensuring that the process is both efficient and effective. The process works best where the parties are able to weigh up all the circumstances of a particular dispute before making a decision as to the skills that are required of the particular tribunal. There is, accordingly, a tactical decision to be made at the time of drawing up the arbitration agreement, with regard to how much detail the parties want to include about the composition of the tribunal.[4] Where the arbitration clause is drafted before the dispute arises and includes restrictions as to the specific experience an arbitrator must possess or his nationality is specified in some detail, the choice of available arbitrators may be significantly limited. For this reason it is generally safer to provide a broad set of criteria and allow the parties (or failing them) the relevant institution, to make the selection based on this guidance. It is also generally preferable for the parties to be involved in the selection process even if agreement between them is reached less often than would be liked. Being involved in choosing their arbitrators gives the parties a sense that their case is being properly heard and may later assist them to accept the award.

---

2 The common law adversarial approach (as in the United Kingdom) is more party driven. The civil law approach (as in continental Europe) is more tribunal driven.
3 Convention for the Pacific Settlement of International Disputes (Hague I) 1907, commonly referred to as the Hague Convention 1907.
4 An example of this is where clauses have specified that the tribunal must be composed of 'commercial men'. The courts have held that this refers to 'practical commercial experience' – *Pando Compania Naviera SA v Filmo SAS* [1975] 2 All ER 515.

Ensuring equal treatment of the parties in the organization of proceedings, and in particular the selection and appointment of arbitrators, can be very difficult in a multi-party dispute. This situation may occur where there is a single contract between several parties (e.g. a joint venture) or where there is a web of interrelated contracts. Failure to ensure that all the parties have equal input may provide grounds for resisting enforcement of the award.

The parties should take time carefully to consider their options when carrying out the selection process. It is an opportunity to ensure that the panel is composed of the people with the requisite skill and knowledge to serve the needs of the parties and to reach the correct decision. This may be particularly important where the panel will be required to decide complex technical disputes, such as often arise out of construction contracts. The parties must, however, bear in mind any statutory, institutional or contractual time limits and ensure that in taking their time in considering the appointment, they do not fail to meet any deadline.

A.   NUMBER OF ARBITRATORS

This issue is usually decided at the time of drafting the arbitration clause. There is obvious merit in having an odd number of arbitrators; it is most common to have either three arbitrators or a sole arbitrator. A sole arbitrator will be cheaper and may deliver his decision more quickly, as he need not consult or discuss the issue with others. Co-ordinating meetings and hearings for three arbitrators will also be more difficult and may result in delays. Conversely, with a three-member tribunal there is less chance of an unusual result or a mistaken approach being adopted towards the case. However, the main disadvantage of proceeding with a sole arbitrator is that if the parties cannot agree on who to appoint, the decision will fall to an appointing authority and be taken out of the parties' hands. Where the arbitration agreement provides for three arbitrators, the usual practice is that the parties each appoint an arbitrator and the third is selected by the two party-appointed arbitrators or the arbitral institution. For the reasons outlined above, it is recommended that the parties retain control over the selection process by carefully drafting the procedure for the appointment of arbitrators in the arbitration clause.

As with many stages of the commencement process, much will depend on whether the parties have elected an institutional arbitration or an ad hoc arbitration. The different institutions have their own preferences. The LCIA[5] and ICC[6] Rules and the English Arbitration Act 1996[7] all provide that unless the parties agree otherwise, there will be a sole arbitrator. However both the LCIA and the ICC Rules make an exception where the dispute is such that in the circumstances, it warrants

---

5   Art. 5.4.
6   Art. 8(1),(2).
7   S. 15(2).

the appointment of three arbitrators. By contrast, the UNCITRAL Arbitration Rules[8] and the UNCITRAL Model Law[9] express a preference for three arbitrators unless, within 15 days, the parties have agreed otherwise. In China, CIETAC[10] is the most popular commission for international arbitrations and its Rules require that the tribunal is made up of three arbitrators unless the parties agree otherwise.[11] The pros and cons of the different options must be considered.

B.　　CONSIDERATIONS WHEN SELECTING A TRIBUNAL

The ability to select an arbitrator is one of the unique features of arbitration and is one which the parties should use to their best advantage. There are a wide range of potentially relevant qualities and skills that the parties should consider when tailoring the selection to their needs. These range from practical issues, such as language and experience, to the appropriateness of appointing a person with particular publicly-held views on an issue relevant to the dispute. The decision may also hinge on the number of arbitrators making up the tribunal. Ultimately, the more experienced the panel, the more efficient the process will be, thereby saving time and money for the parties.

As arbitration becomes increasingly popular in international disputes, the language of the arbitration becomes increasingly important to the parties. The languages of the parties may differ, documents in the case may be in a variety of languages (although of course these can be translated, albeit at a cost), witnesses may speak different languages and the seat of the arbitration may not be in the native country of one or more of the parties. Having a multi-lingual arbitrator or an arbitrator who speaks a party's own language may be a source of reassurance and comfort to the parties. Under the ICC Rules, if the parties cannot agree the language, it will be determined by the tribunal 'with due regard being given to all relevant circumstances including the language of the contract'.[12] Under the LCIA Rules, the language is based on that of the arbitration agreement, but the final decision is determined by the tribunal based on the LCIA recommendation, the parties' comments and any other relevant matters.[13] Under the UNCITRAL Arbitration Rules, the language is again determined by the tribunal.[14] It is also usual for sole arbitrators or chairpersons to be of a nationality other than the nationalities of the parties.

8　Art. 5.
9　Art. 10.
10　New rules were introduced on 1 May 2005, which try to bring CIETAC in line with its more modern international competitors.
11　Art. 20(2).
12　Art. 16.
13　Art. 17.
14　Art. 17.

At least one of the arbitrators on a panel should generally be an experienced lawyer, particularly where the arbitration involves issues of law – and there are few arbitrations that do not. The principal reason for this is that in most international commercial arbitrations the parties need to be able to entrust the process to someone who is capable of progressing it justly and effectively. Secondly, on a more substantive level, international commercial arbitrations frequently involve difficult questions of law. It is therefore sensible for at least the chairman of a three man tribunal to be a lawyer, so that he can deal with such legal issues as and when they arise.

In many jurisdictions, including the United Kingdom, it is both acceptable and desirable for some of the arbitrators on a panel to be non-lawyers in areas such as construction. It is clearly advantageous in a complex technical dispute for the arbitrators to have expertise in the relevant sector, although if the parties intend to use experts they might not deem it an essential requirement. The parties may agree that the arbitrator should be a person engaged in a particular trade.[15] The parties may even agree that the appointment should be made by a trade association. If an arbitrator is suitably qualified, he will be able to liaise competently with the experts (if they are also required) and help guide the tribunal in interpreting the experts' opinions and attaching weight to their findings. The Chartered Institute of Arbitrators runs courses and examinations for those who wish to qualify as arbitrators. Many arbitral institutions reserve the right not to confirm the appointment of a person they consider unsuitable.[16]

Arbitrators may also be selected because of their particular, publicly-held, views in relation to fundamental issues between the parties, for which reason the party nominating them feels that they will appreciate the party's case more readily. This is one area where the right to select arbitrators, one of the key advantages of arbitration over domestic courts, needs to be exercised with some care. While it is quite acceptable to nominate an arbitrator whose broad views – perhaps on legal theories or the roles of parties to construction contracts – align with those needed to win the case, care must be taken not to overstep the mark. Views held not just generally but on the particular subject matter of the dispute in question are likely to form the basis for a challenge for lack of neutrality. It is for these reasons that it is becoming increasingly common for the parties to interview arbitrators before nominating them. This is not just to assess whether the arbitrators have the appropriate qualifications or personality traits, but also to test their views on relevant issues. Clearly, any such conversation can be conducted in only the most general terms, since it would be inappropriate – and a ground for subsequent challenge of the award – for one party to have a private conversation with an arbitrator about the merits of the case he is appointed to determine.

---

15 *Myron (Owners) v Tradax Export S A Panama City R P* [1969] 1 Lloyd's Rep 411.
16 E.g. Art. 7.1 LCIA Rules.

C. METHODS OF APPOINTMENT

The method of appointment is usually agreed in the arbitration clause. There are two main ways in which the arbitral tribunal may be appointed. The process may be conducted by the parties or by the arbitral institution. The procedural law of the seat may also influence the process.

Where an institution is not involved, the most common method is that the parties seek to agree on the identity of a sole arbitrator or, where there is a panel of three, that each party appoints one arbitrator and the two party-appointed arbitrators then agree on the chairperson.[17] The appointment of the third member of the tribunal by the two party-appointed members is usually the most satisfactory method to the parties, as they will have faith in their appointed arbitrators' judgment in electing the chairperson. However, it is important to consider what is to happen in the event of any failure to agree on the identity of an arbitrator, or where one party simply fails to participate in any way in the appointment process. Whilst in most jurisdictions it will be possible to apply to the local courts to appoint an arbitrator in such circumstances, this can be a slow and uncertain process. For that reason, where the appointment of arbitrators is in the hands of the parties (as in an ad hoc arbitration) it is good practice to have an experienced body to act as an appointing authority – i.e. one whose principal function is to make appointments of arbitrators in default.

Where an institution is involved, the process tends to be under its control. While the parties may nominate arbitrators for appointment, the actual appointment is reserved to the institution itself. Thus under the ICC Rules[18] and the LCIA Rules[19] (where the parties have expressly agreed to be involved in the selection), the parties have the right to select one arbitrator of a three member tribunal each, however this is subject to institutional confirmation and the ICC or LCIA will appoint the chairperson. In addition, as described above, where any party fails to agree or to participate in the nomination/appointment process as required, the institution will fill the gap by acting as the appointing authority. In an ad hoc arbitration under the UNCITRAL Arbitration Rules, if the parties do not appoint the arbitrators, the Secretary–General of the Permanent Court of Arbitration at The Hague may, at the request of either party, designate an appointing authority.[20] This two-step process is cumbersome, but at least avoids the need to apply to a court to appoint the arbitrators.

There are, of course, other ways of achieving the same result. For instance, the new CIETAC Rules have altered the procedure for appointing the chairperson of a three member tribunal and now apply a modified list procedure, whereby each

---

17 E.g. Art. 7 UNCITRAL Arbitration Rules.
18 Arts 8 and 9.
19 Arts 5 and 7.
20 Arts 6 and 7 UNCITRAL Arbitration Rules.

side nominates one to three candidates and provides their names to CIETAC.[21] Where one candidate appears in common on both lists, such candidate will be appointed as the chairman of the tribunal. If more than one name is proposed by both sides, CIETAC decides the most suitable appointment. Where the parties do not nominate the same person or persons, CIETAC will appoint an individual who has not been nominated by either side.

The use of such lists may be 'an effective and happy compromise'.[22] It provides the opportunity for institutions to consult with the parties on the appointment of the arbitrators. The parties may be provided with a list of names and given an opportunity to veto anyone they do not think suitable or express a preference for others. Alternatively, the parties may draw up their own lists and exchange them simultaneously. This method helps to prevent the deadlock which often occurs when each party automatically rejects the other's nominations.

The UNCITRAL Arbitration Rules employ a system which is a hybrid of the party appointment and list systems.[23] The ICDR and Netherlands Arbitration Institute also employ this method. The appointing authority sends out the same list to each party, with at least three names on it. Each party grades the names on the list to show its preference and the tribunal appoints the arbitrators based on the parties' input. This method ensures that the parties retain the element of choice, but helps protect the arbitrators' independence. The parties are also more likely to be content with all three members of the tribunal if this method is employed. In the Netherlands, research shows that the preferences expressed by the parties are the same in 80 per cent of cases.[24] The use of a list also helps with adapting the process of selection to multi-party disputes.

D.   DUTIES OF ARBITRATORS

Once appointed, the tribunal has a duty to act fairly to both parties, even if the parties appointed their own arbitrators to the panel.[25] Arbitrators must meet the required standards of fairness, equality, independence and impartiality.[26] This has been described as a 'non-waivable mandatory principle'.[27] The presumption is that

---

21   Under the new Rules, rather than selecting from a very limited list of arbitrators, CIETAC now permit parties to nominate any person, subject to confirmation by CIETAC.
22   C Newmark and R Hill, *'The Appointment of Arbitrators in International Arbitration'*, Int. A.L.R. 7 (2004), 73–80.
23   Arts 6 and 7.
24   Newmark and Hill, *'Appointment of Arbitrators'*, 74.
25   The tribunal effectively enters into a mandate with both parties which enables it to bind the parties to the terms of any award.
26   Art. 15(2) ICC Rules, Art. 15(1) UNCITRAL Arbitration Rules, Art. 14(1) LCIA Rules, Art. 7(1) AAA Rules, and General Standard 1 IBA Guidelines on Conflicts of Interest in International Arbitration 2004.
27   Garnett, Richard, Judd Epstein, Jeff Waincymer and Henry Gabriel, *A Practical Guide to International Commercial Arbitration*, 2nd edition, (Oceana Publications Inc, 2000), p. 67.

the arbitrators should be independent and impartial even if they were appointed by one of the parties as a 'representative' of that party's views.[28] Because of this many (though not all) institutions require that, unless the parties agree otherwise, the nominated arbitrators must sign a statement of independence disclosing any previous relationship with the parties or the subject matter of the dispute.[29] One example where no formal statement of independence is required can be found in the AAA Rules, however, there is a clear duty to disclose any circumstances likely to give rise to justifiable doubts as to the arbitrator's independence and impartiality.[30] Another approach can be found in the UNCITRAL Arbitration Rules (where there is no institution) where any arbitrator may be challenged 'if circumstances exist that give rise to justifiable doubts as to the arbitrator's impartiality or independence'. The parties should try to anticipate any problems in order to try to ensure that their selection does not cause any unnecessary delay to the proceedings. The new CIETAC Rules impose a positive obligation on all arbitrators to treat the parties equally, fairly and independently. A requirement has also been introduced that arbitrators declare to CIETAC any matters which might raise reasonable doubts as to their independence and impartiality, whether such matters arise before or during the arbitral proceedings.

This duty may appear incongruous with the fact that the parties are able to appoint the arbitrators, especially considering the fact that the arbitrators may be selected because of their openly-held views on particular issues.[31] The arbitral institutions do not define the meaning of 'independent' and 'impartial'; instead they are defined on a case by case basis. The basic test is 'whether circumstances exist that give rise to justifiable doubts about the arbitrator's impartiality'.[32] A challenge cannot be based solely on grounds of lack of independence. The established test is whether there is a real possibility that the arbitrator is predisposed to, or prejudicial against, one of the parties in the arbitration.[33] There must be apparent bias.[34] Lord Woolf M.R. suggested in one case[35] that, if different standards of independence and integrity were to apply to judges and arbitrators, arbitrators should observe the higher standard. However, the court in fact held that the same standard should apply

---

28   Art. 9 ICC Rules and Art. 5.2 LCIA Rules.
29   Art. 7(2) ICC Rules and Art. 5.3 LCIA Rules.
30   Art. 7(1) AAA Rules.
31   The duties of arbitrators are different however from those of a Judge in civil proceedings as the arbitrator does not have a duty to the state. In contrast in some jurisdictions, such as New York State, partiality is positively encouraged.
32   English Arbitration Act 1996, s. 24(1).
33   *Porter v Magill, Weeks v Magill* [2001] UKHL 67; [2002] 1 All ER 465.
34   In *Porter v Magill supra*, the House of Lords held that the test in *R v Gough* [1993] 2 All ER 724 should be slightly modified. The test is now: whether from the viewpoint of a fair-minded and informed observer there is a 'real possibility of bias on the part of the relevant member of the tribunal in question in the sense that he might unfairly regard (or have unfairly regarded) with favour or disfavour the case of a party under consideration by him'.
35   *AT&T Corp and another v Saudi Cable Co* [2000] 2 All ER (Comm) 625, 638.

to both. Previous contact between the parties and the arbitrator should be considered; for example, if an arbitrator has any employment history with a party[36] or has any financial interest in the outcome of the arbitration,[37] there is a clear argument that there is a possibility of prejudice. Even a pattern of repeated appointment (of a particular arbitrator) by one of the parties could raise concern. However, 'the more pragmatic would not disqualify any person who may have a sympathy or a predilection towards a particular position as long as the arbitrator is able to listen and to deliberate at the end of the day on the basis of the evidence presented'.[38]

In the last few years there has been a distinct trend away from viewing 'independence' and 'impartiality' separately and towards 'viewing these two elements as the opposite side of the same coin'.[39] The traditional view[40] is that 'independence' can be judged objectively as it has nothing to do with the arbitrator's state of mind. It relates to his relationship to the parties. 'Impartiality', however, relates to bias and can only be assessed subjectively. The ICC Court judges 'independence' to be the appropriate test, whereas the LCIA believes that an arbitrator may lack independence but still act impartially. Therefore the LCIA Rules look to impartiality.[41]

Neutrality is a separate and far broader requirement, used mainly in the United States. However, the AAA Rules only provide that the chairman may be of a different nationality to the parties; it is not a requirement.[42] The requirement that the chairperson is neutral as to the legal and political systems of the parties is not always a practical option, as this must be balanced against the need to find a person who is suitably experienced.

As this is a very complicated and controversial area, the IBA published *Guidelines on the Conflicts of Interest in International Arbitration* (the Guidelines) in July 2004 to help parties and arbitrators determine what information should be disclosed. The IBA believes that the duty to be independent and impartial should not be a hindrance to international arbitration. The Guidelines are aimed at clarifying the scope of the duty so that there are fewer challenges to, and withdrawals of, arbitrators. The IBA provides specific examples of potential conflicts and determines the appropriate action based on a traffic lights system: they have produced (i) 'Non-waivable Red', (ii) 'Waivable Red', (iii) 'Orange', and (iv) 'Green' lists. The 'Non-waivable Red' scenarios are regarded by the IBA as giving rise to justifiable doubts as to the arbitrator's independence and impartiality, and disclosure of the

---

36 For example if he is an officer of an associate company or managing company: *Edinburgh Magistrates v Lownie* [1903] 5 F (Ct of Sess) 711.
37 For example, if he has a substantial shareholding: *Sellar v Highland Railway Company* [1919] SC (HL) 19.
38 Garnett *et al.*, *A Practical Guide*, p. 58.
39 Alan Redfern and Martin Hunter, *Law and Practice of International Commercial Arbitration*, 4th edition, (Sweet & Maxwell, 2004), 4–54.
40 As explained in Redfern and Hunter, *International Commercial Arbitration*, 4–55.
41 Art. 5.2.
42 Art. 6(4).

situation cannot cure the conflict. The Guidelines' examples of the 'Non-waivable Red' category include where the arbitrator is a director or legal advisor of one of the parties or has a significant financial interest in the outcome of the arbitration. 'Waivable Red' situations are where the conflict is serious, but if the parties expressly agree to the arbitrator's appointment, the conflict can be waived (for example, where a close family member of the arbitrator is a manager of one of the parties or has a financial interest in the outcome). The Guidelines' 'Orange' list covers situations where one of the parties has justifiable doubts about a conflict such as an arbitrator's historical involvement with one of the parties, but if no timely objection is made, the arbitrator may conduct the proceedings. The 'Green' list covers situations where, objectively, there is no conflict. This may be where the arbitrator has previously published a general opinion concerning a relevant issue or owns an insignificant number of shares in one of the parties, which is publicly listed.

Either party can challenge the preliminary decisions made at the time of commencement. In an institutional arbitration, the challenge is usually addressed to the institution, however a challenge may be made through the courts of the country in which the arbitration is to take place if the national law allows it. Challenges should be made as soon as is reasonably practical. A challenge to an arbitrator can be used to challenge an award. However, it is frequently too late to raise an objection by the time the award is granted, unless grounds are discovered late in the day. If the objection is not raised at the time the grounds for objection became apparent, the grounds for challenge will be deemed to have been waived.

III. THE PARTIES: JOINDER/CONSOLIDATION

The arbitration model is best suited to bilateral disputes. The reason for this is that the arbitral process is based entirely on the agreement of the parties in the arbitration clause of the relevant contract, generally before any dispute has arisen. It is a consensual process. The general approach in the United Kingdom[43] is that there is no power to join third parties (i.e. parties who are not a party to the arbitration agreement) without their consent or to consolidate two arbitrations[44] without the consent of all the parties, even when common issues of fact or law arise.

Unfortunately, construction disputes often involve multi-party relationships, whereby various contractors and sub-contractors are tied into one project together. In those circumstances it is very difficult to draft the 'perfect' arbitration clause because one cannot predict the different sets of facts that may give rise to disputes between the parties. This is further complicated where there is an international element. In addition, because of the different times at which the various agreements are concluded (and the differing negotiating strengths of the parties to those

---

43 Other jurisdictions allow for court ordered consolidations or joinder.
44 English Arbitration Act 1996, s. 35(1).

agreements) the parties may each have consented to arbitrate but under different procedures. For example, the employer and the contractor may have agreed to arbitrate under the ICC Rules in Switzerland, whereas the contractor and its various sub-contractors and suppliers may have agreed to arbitrate under the LCIA Rules in London. The two sets of agreements may also make different provisions for language or the number of arbitrators. In the absence of any agreement between the parties, there is no scope to consolidate the two sets of arbitrations. The procedures under the institutional rules are also not always tailored to multi-party disputes.[45]

A.   CONSOLIDATION BY CONSENT OF THE PARTIES

Where disputes arise in relation to the same project, it may make sense to consolidate the proceedings, or at least those proceedings with common elements. However, this is not generally provided for in standard form arbitration clauses, and is only rarely achieved in practice. It can, in fact, be very difficult to achieve a workable consolidation arrangement. Parties to an arbitration clause should give thought at the time they enter into the first contract to including consolidation provisions if they are not addressed in the institutional rules adopted by the parties. Even where a consolidation arrangement is included in the institutional rules,[46] the two sets of parties may have agreed to incompatible provisions, for example regarding the seat or language of the arbitration, which could have the effect of hindering or preventing consolidation.

Where there is no power in the relevant rules to consolidate cases and where it is obvious that consolidation of cases arising between different parties to different agreements is likely to be beneficial, the parties may try to make express provision for consolidation of proceedings in two ways.[47] First, they may enter into what might be called 'push' or 'pull' arbitration agreements, so that A enters into an agreement with B and B with C, D, E and so on, with each agreement containing an express consent to have disputes arising under that agreement consolidated with those arising under other agreements (i.e. pushed into a joined proceeding). As importantly, such an agreement must also allow disputes under other agreements to be heard together with disputes under the agreement in question (i.e. pulled into a joined proceeding). All related contracts should, of course, contain arbitration clauses with the same seat, language, governing law etc. Done properly, judicious use of the ability to push disputes into other arbitrations or to pull other disputes

---

45   However, the ICC does provide that where there are multiple parties, they will jointly nominate an arbitrator, and if they cannot agree on a nomination then the ICC Court of Arbitration will appoint all three arbitrators. The LCIA Rules go further and allow the tribunal to join third parties where they have consented to be joined.
46   Art. 22.1(h) LCIA Rules.
47   English Arbitration Act 1996, s. 48, allows for consolidation only if there is express agreement.

into the arbitration of the dispute under the agreement in question will allow effective consolidation. The advantage of this technique is that the consent of all the parties is not required simultaneously.

The second possibility is that all the parties involved in the project enter into a separate and stand-alone 'umbrella agreement', which allows for all related disputes to be heard together or, for similar disputes to be consolidated. In such an arrangement, the parties may reserve the right to decide which proceedings are related, or may refer that decision to the tribunal. However, it is inevitable that disputes will not always arise at the same time, so the parties should consider giving the tribunal a discretion not to consolidate, so that if one set of proceedings is quite advanced when the other arises, time and money is not lost in unnecessary consolidation. Although an umbrella agreement requires a separate set of negotiations (which may not be easy to complete), the advantage is that it allows a joinder/consolidation regime to be applied to agreements which have already been negotiated without the conclusion of founder provisions and also permits (since it provides for a stand-alone arbitration procedure) differences in the individual arbitration provisions in each of the underlying related agreements to be overcome.

B. CONSOLIDATION BY COURT ORDER

In some jurisdictions,[48] joinder and consolidation can be ordered by the courts, without the consent of the parties when it is necessary or convenient, however this is still considered exceptional. Court-ordered consolidation or joinder contradicts the principle that arbitration procedure is 'in accordance with the agreement of the parties'[49] and private. In England, there is no scope for consolidation without consent. The rationale was clearly explained by Legatt J in *The Eastern Saga*:[50]

> 'The concept of private arbitrations, derives simply from the fact that the parties have agreed to submit to arbitration particular disputes arising between them and only between them. It is implicit in this that strangers shall be excluded from the hearing and conduct of the arbitration and that neither the tribunal nor any of the parties can insist that the dispute shall be heard or determined concurrently with or even in consonance with another dispute, however convenient that course may be to the party seeking it and however closely associated the disputes in question may be'.

---

48 For example, Netherlands, Hong Kong and California.
49 Art. V.1(d) New York Convention.
50 *Oxford Shipping Co Ltd v Nippon Yusen Kaisha, The Eastern Saga* [1984] 3 All ER 835 at 841.

C. ADVANTAGES AND DISADVANTAGES OF JOINDER AND CONSOLIDATION

The main advantage of joinder or consolidation is that it can, at least when the costs of all the parties are aggregated, be cheaper for the parties and make the process quicker and more efficient. The cost of setting up a second or even third tribunal will be saved even if the one arbitration actually conducted will be rather more complex. It also helps to reduce the risk of inconsistent awards and therefore increases the parties' confidence in the process by delivering a more certain outcome, particularly for the party such as the main contractor who is the 'meat in the sandwich', having to deliver to the employer what, in practice, its sub-contractors will be providing.

All is not, however, rosy even where time and effort has been put into agreeing a regime which permits joinder and consolidation. For instance, one of the disadvantages of joinder or consolidation provisions is that unless they are invoked at the start of the initial dispute then, if another party is sought to be joined at a later stage in the proceedings and the disputes consolidated, that joined party or parties will not have had an equal input into the composition of the tribunal. This is one reason why joinder and consolidation clauses often provide for the appointment of all tribunal members by an independent body, thereby eliminating any disparity in influence between the original and joining parties.

However, even the most artful drafting cannot overcome the principal problem of the late joinder of parties, which is the impact on the timetable of the arbitration into which they have been joined (questions of fairness dictate that some delay will have to be tolerated, to allow the joining parties to catch up). Nor can it really overcome the problem of the wasted costs of the joining parties if they have already been seeking to resolve their disputes in some other forum. The only practical answer is to limit the operation of joinder/consolidation clauses to specific windows around the start of the first relevant disputes, leaving the provisions inoperative if not invoked more or less contemporaneously with the commencement of the first arbitration.

D. NO JOINDER: ALTERNATIVE SOLUTIONS

In many cases it is simply not possible to obtain effective joinder at the contract drafting stage. In many cases, an employer, for instance, will say that it wants single point responsibility for delivery of its project by its main or general contractor and therefore has no interest in becoming involved in disputes with sub-contractors. Or, perhaps, the employer is in a position to dictate dispute resolution provisions which are unattractive to both the main contractor and its sub-contractors, with the result that it is less unattractive to have separate dispute resolution mechanisms than it is to have all disputes resolved in an unfamiliar forum.

In these cases there are a number of drafting solutions which can be adopted to allow disputes with common themes to be resolved in one forum. A common technique to protect the main contractor is the use of pay-when-paid provisions which prevent the main contractor from becoming exposed to its sub-contractors until in receipt of funds from the employer. These clauses are, however, becoming less attractive since not only do they eliminate the risk of differing decisions on the same issue (i.e. is payment due?) but they also pass down to the sub-contractors the credit risk that the main contractor has taken on the employer. This last point is generally regarded as unfair and has led, in the United Kingdom and in some other countries, for pay-when-paid clauses to be either wholly or partially outlawed.[51]

In addition, unless the main contractor has passed the whole of the risk down to a sub-contractor in an identifiable way,[52] pay-when-paid clauses can also have the effect that the sins of the employer are visited without good reason on its sub-contractors.

For all these reasons the trend in major projects is to move away from simple pay-when-paid clauses to provide protection to the main contractor and (to some extent) the sub-contractors by a scheme which provides that the sub-contractors are entitled to receive whatever the main contractor is entitled to receive – and only what it is entitled to receive – whether it actually receives it or not. The corollary to this is that the sub-contractors are normally given a role in the pursuit and defence of claims between the main contractor and the employer. This may or may not be visible in the course of a main contract arbitration, where the effective party will generally be hidden behind layers of advisers and representatives. In some cases, however, the main contract expressly recognizes the possible role of sub-contractors in main contract arbitrations, though as part of the main contractor's representation and not (as in full joinder) as a separate party.

By these techniques the risks associated with lack of joinder can be largely overcome. Where a sub-contractor's rights and obligations are effectively defined by reference to the establishment of the main contractor's rights and obligations as against the employer, by applying the results of a main contract arbitration down the chain as a matter of contract, overall risk can be minimized. While disputes may still be brought at the sub-contract level – indeed it is nigh-on impossible to stop claims being brought – if the parties' respective entitlements under the sub-contract are defined by reference to the equivalent entitlements under the main contract, a sub-contract dispute will clearly be fruitless unless and until those main contract rights have been determined. This, and the possibility of an adverse costs order for bringing frivolous claims, operates as a powerful disincentive to the bringing of such claims before properly established at the main contract level.

---

51 Somewhat unusually in the UK Construction Act 1996-pay-when-paid clauses are ineffective *unless* the ultimate payer is insolvent, protecting the sub-contractor against late payment but not, ultimately, against the insolvency of the employer.
52 As happens at the 'concessionaire' level in PFI and limited recourse financings where all of the concessionaire's obligations are normally cleanly passed down to a few separate contractors.

IV. THE REQUEST FOR ARBITRATION

The Request for Arbitration (Request)[53] is the first step towards arbitration. The Request is served on the respondent by the claimant,[54] and has three key functions. First, it is a means of giving notice and calling the arbitration into existence. Failure to comply with the rules for giving notice may render the award unenforceable. Second, it establishes the basis of the dispute and will inform the arbitral tribunal of the issues to be decided. Third, it is important in trying to promote settlement in the case. If the claimant presents a strong case in its Request, the respondent may be more willing to settle.

Different jurisdictions have different approaches to the manner and extent in which this first written document should shape the conduct of the arbitration. In the United States, often all that is required is notice of the claim, an outline of the basis of jurisdiction, the cause and nature of the claim and the remedy. It is considered premature to provide the full facts at such an early stage. Parties frequently shift their positions, submit new evidence and make new assertions after the Request stage. In contrast, in the United Kingdom and Australia, the basic model is that the document should contain all the material facts, so that it is more than a 'skeletal outline'.[55] Conclusions of law and evidence are, however, left to a later stage. In civil jurisdictions, the request is likely to be in the form of a free-flowing narrative containing facts, law and evidence.

To a large extent, the procedure to be followed in any particular arbitration should have been detailed in the arbitration clause. If the parties have adopted the rules of one of the various arbitral institutions, those rules will dictate what should be included in the Request, unless (and to the extent permitted under the rules concerned) the parties have agreed otherwise. The requirements of the ICC and LCIA are similar to those of a modern English particulars of claim. The full names and a description of the parties and their full addresses must be set out. The main body of the Request should contain a description of the nature and circumstances of the dispute and the relief sought. Reference should also be made to the relevant agreements in dispute and the arbitration clause that is invoked (these are normally attached). The Request should also contain a statement of any matters on which the parties have already agreed, such as place of arbitration, language, the number of arbitrators, qualifications of those arbitrators or even their identities. If agreement has not been reached on these issues, the claimant should set out its proposals in the Request. The Request must be filed with the appropriate body. For the ICC, the appropriate body is the ICC Secretariat and for the LCIA it is the LCIA Registrar. The Request must also be sent to the other party or parties. The claimant should check with the institution as to whether there is a fee to be paid on filing the request.

---

53 It is called a 'notice' in the UNCITRAL Arbitration Rules.
54 The ICC takes responsibility for serving the Request and acts as an intermediary between the parties (Art. 4(5) ICC Rules), however the other institutions do not (e.g. Art. 1.1(g) LCIA Rules).
55 Garnett *et al., A Practical Guide*, p. 67.

In an ad hoc arbitration, if the parties have not made reference to the rules of any arbitral institution, it may fall to the law of the country of the seat of arbitration to determine the content of the Request and the manner in which it is to be served on the respondent.

The differences between the requirements of the different arbitral institutions reflect to a large extent the different jurisdictional approaches. In particular they differ in terms of the factual detail required and the inclusion of legal argument in the Request. The LCIA Rules effectively require service of an English-style particulars of claim, setting out in sufficient detail the facts and any law on which the party relies.[56] In contrast, the AAA Rules reflect the United States approach and require only a bare indication of the facts.[57]

Receipt of the Request marks the commencement of the arbitration. The claimant must ensure that the respondent receives the Request, and should request that receipt is acknowledged.

V. SELECTION OF THE PARTIES' REPRESENTATIVES

As with court litigation, lawyers or other professional advocates may be used to present the parties' cases to the arbitral tribunal. Although there is often no particular requirement that the parties be represented or that the parties' representatives are lawyers or lawyers qualified in the place of the arbitration,[58] this is often the case.

Each party will normally appoint counsel (i.e. a lawyer or lawyers) to run their case. The qualities that the parties should look for in their legal team are not covered here. This section addresses only the question of whether the assistance of local lawyers is required. This is often an important consideration in international arbitrations. Take the example of an ICC arbitration with its seat in Paris, where the documents on which the dispute centres are governed by German law. French lawyers may be required to advise on certain local, mandatory, procedural laws which the parties cannot contract out of. That does not mean, however, that the arbitration needs actually to be conducted by French lawyers, and there is nothing in either French law or the ICC Rules that requires this. Indeed, the arbitration might most logically be conducted by German lawyers, though if the project documents were in another language such as Russian, even this apparently sensible solution might not be the most appropriate.

What it is safe to say is that, while the local laws of the place of arbitration may be relevant to the conduct of the arbitration and the parties may (probably should) require local law advice to identify any pitfalls peculiar to the conduct of arbitration in that country, the presentation of the case is separate from, and not

---

56 Art. 4(3)(b).
57 Art. 2(3)(e).
58 Occasionally a country's professional conduct laws for the legal profession reserve the conduct of arbitrations to lawyers admitted in the country which is the seat of the arbitration, though the trend is to remove such bars on the practice of international arbitration.

dependent on, the law of the place of the arbitration. In almost every case, factors such as familiarity with the substantive law of the agreement, the project, the business sector, the language of the arbitration, the arbitration process or even the arbitrators themselves is more likely to be relevant in choosing the most effective representation than familiarity with the law of the seat.

VI. CONCLUSION

There are many important decisions to be made by the parties when commencing an arbitration. They should be fully aware of the impact of these choices on the process. In particular, time and care should be taken over selection of the tribunal. The parties' ability to input their views and nominate members of the panel is unique to arbitration. The parties need to have confidence in the arbitrators' ability to conduct the proceedings and they will usually find it easier to accept an award if they know that they appointed the right arbitrator.

The level of detail to include in any Request is another key tactical decision. The parties may not want to reveal their hand at such an early stage, especially if they anticipate new facts coming to light in the future, however setting out a strong case in the Request may lead to early settlement.

Whether to consent to joinder or consolidation is another important practical decision. Each case must be considered on its own merits. Consolidation may save costs in some situations but create delays in others.

All these issues should be discussed with the parties' lawyers as soon as the dispute arises. In fact, the parties should think about them at the time of drafting the arbitration clause. If the clause makes provision for the relevant procedures, the process should then run more smoothly if and when it is necessary to commence an arbitration. More importantly, the parties, rather than the arbitral institutions or the national courts, will be in control.

Chapter 8
Control of the Arbitration

I.   INTRODUCTION

This chapter considers the arbitral tribunal's powers and duties to control the proceedings, with reference to institutional and international rules of arbitration and the arbitration laws of selected jurisdictions. As a starting-point, the source of the tribunal's powers and duties is examined, as well as the principle of party autonomy and the extent to which this underlies or restricts the tribunal's ability to conduct the proceedings. A discussion of the tribunal's procedural powers in respect of each of the key stages of the proceedings follows, starting with the preliminary steps following the tribunal's appointment, all the way through to post-hearing matters. Finally, the tribunal's 'case management' duties, encompassing the need to ensure procedural fairness and the duty to act 'expeditiously' are considered in the light of arbitral rules and case law, in order to assess what, in practice, is required from the tribunal.

II.   SOURCE OF THE ARBITRAL TRIBUNAL'S POWERS

The powers of an arbitral tribunal are 'those conferred upon it by the parties themselves within the limits allowed by the applicable law, together with any additional powers that may be conferred by operation of law'.[1]

Accordingly, the powers of the tribunal originate primarily from the parties, directly or indirectly: directly, where the parties specify in the arbitration agreement, arbitrators' terms of appointment or another written agreement the powers which they wish the arbitrators to exercise; indirectly, where the parties stipulate that the

---

1   Redfern and Hunter, *The Law and Practice of International Commercial Arbitration*, para. 5–03.

arbitration is to be conducted according to rules of arbitration, whether institutional or *ad hoc*, which confer express powers (general and specific) upon the tribunal. For example, the UNCITRAL Arbitration Rules give an arbitral tribunal general powers to conduct an arbitration 'in such manner as it considers appropriate' (Article 15(1)), while also giving it specific powers, such as the powers, absent any agreement of the parties, to determine the place of the arbitration (Article 16(1)) or the applicable law (Article 33(1)).

However, as stated above, the 'applicable law', constituted by the law governing the arbitration agreement and the law of the place of the arbitration (to the extent that these are distinct), will also have a role to play in determining the tribunal's powers. It may operate so as to supplement such powers: for example, the English Arbitration Act[2] confers general and specific procedural powers on the arbitral tribunal, subject to any agreement to the contrary by the parties.[3] Thus, an arbitral tribunal sitting in England may (unless the parties otherwise agree) appoint experts or other advisers, order a claimant to provide security for costs, give directions for the inspection, preservation etc. of any property which is the subject of the proceedings, administer oaths to witnesses and give directions to a party for the preservation of any evidence in that party's custody or control.[4]

The applicable law may also operate to restrict the powers of the arbitral tribunal. For instance, in almost any civil law country, the arbitral tribunal will not be entitled to administer oaths. According to academic commentary, the reason for this is the limited scope of the arbitration agreement, which does not extend to third parties such as witnesses.[5] In common law countries, however, arbitration laws quite often include provisions that allow the tribunal to administer oaths or to order the production of documents by third parties.

In practice, therefore, in order to determine the scope and nature of a tribunal's powers, regard should be had, in the following order, to four sources:

(a) the arbitration agreement;
(b) any arbitral rules to which the parties have subjected the arbitration;
(c) the law governing the arbitration agreement; and
(d) (to the extent that this is distinct from (c)) the law of the place or 'seat' of the arbitration (the *lex arbitri*).

---

2   See s. 2(1) English Arbitration Act 1996 according to which the Act applies to any arbitration with a seat in England and Wales.
3   See s. 34(1) English Arbitration Act: 'It shall be for the tribunal to decide all procedural and evidential matters, subject to the right of the parties to agree any matter'.
4   See ss 37, 38(3), 38(4), 38(5) and 38(6) English Arbitration Act.
5   See Hans-Patrick Schroeder, *'Die lex mercatoria arbitralis'*, unpublished PhD thesis, Diss. Hanover (2004), at 239 *et seq*. (Abschnitt 3.A.II.2), (due for publication by Sellier Legal Publishers in late 2005/early 2006). For a discussion of German law on this issue, see Joachim Münch, *Münchener Kommentar zur Zivilprozessordnung*, 3rd edition, (Beck, 2002), §1049 paras 27 and 35.

The law governing the arbitration agreement ((c) above) may be expressly chosen by the parties, though in fact it rarely is. In the absence of such express choice, there is a presumption that the law governing the arbitration agreement is the same as that governing the underlying contract and the substance of the agreement.[6] However, this is not an invariable rule and the presumption has been disregarded in some cases, in favour of the law of the place of arbitration.[7]

III.   GENERAL PRINCIPLES

A.   PARTY AUTONOMY

Since the arbitration agreement between the parties is the basis of the tribunal's jurisdiction, the guiding principle is that the parties are free to dictate the procedure to be followed in an arbitration, within certain limits and subject to the requirement to observe basic standards of procedural fairness, as discussed below. Rules of arbitration and national arbitration laws strongly affirm the principle of party autonomy. For example, both the UNCITRAL Model Law and the ICC Rules state that, subject to the relevant institutional rules, the parties are free to agree on the procedure to be followed by the arbitral tribunal in conducting the proceedings. It is only failing such determination by the parties that the tribunal may itself determine rules of procedure.[8]

The LCIA Rules also promote the parties' freedom to determine the conduct of the proceedings, stating that 'the parties may agree on the conduct of their arbitral proceedings and they are encouraged to do so' (Article 14.1). Again, under the LCIA Rules, the tribunal's discretion to determine the rules under which it should discharge its duties is subject to any agreement by the parties to the contrary (Article 14.2). The English Arbitration Act also states that the tribunal's ability to decide all procedural and evidential matters is 'subject to the right of the parties to agree any matter' (s.34(1)). Arbitration legislation in both Switzerland[9] and Germany[10] provides that the parties may determine the arbitral procedure themselves or by reference to arbitration rules, and, again, it is only failing such determination that

---

6   See *Sonatrach Petroleum Corp v Ferrell International Ltd* [2002] 1 All ER (Comm) 627.
7   See, e.g., *Bulgarian Foreign Trade Bank Ltd v Al Trade Finance Inc*, Swedish Supreme Court, 27 October 2000, Case No. T 1881–99, in Yearbook of Commercial Arbitration XXVI (2001), pp. 291–298.
8   See Art. 19 of the UNCITRAL Model Law and Art. 15(1) of the ICC Rules.
9   The Swiss Private International Law Act (AS 1987, 1779–1831 SR 291) (the Swiss PILA), which governs international arbitrations with a Swiss seat. Domestic arbitrations are governed by the Concordat of 27 March 1969 on Arbitration (the Concordat), a uniform cantonal legislation.
10 The German Arbitration Law 1998, in force 1 January 1998 (*Act on the Reform of the Law relating to Arbitral Proceedings of 22 December 1997*), Bundesgesetzblatt (Federal Law Gazette) 1997 Part I, p. 3224. This law is contained in the 10th Book of the Zivilprozeßordnung, or German Code of Civil Procedure (the ZPO).

the tribunal may decide how the proceedings should be conducted.[11] Finally, it should be noted that failure to respect the principle of party autonomy can be a bar to recognition and enforcement: under Article V(1)(d) of the New York Convention, one ground for challenging the recognition and enforcement of an award is where 'the arbitral procedure was not in accordance with the agreement of the parties'.

On the other hand, some arbitral rules envisage a transfer of control of the proceedings from the parties to the tribunal early on. Both the UNCITRAL Arbitration Rules and the AAA Rules, in almost identical provisions, state that the tribunal may conduct the arbitration in such manner as it considers appropriate, subject to the relevant rules and minimum requirements of procedural fairness (discussed further below).[12] This is supported by paragraph 4 of the UNCITRAL Notes on Organizing Arbitral Proceedings[13] which states that:

'Laws governing the arbitral procedure and arbitration rules that parties may agree upon typically allow the arbitral tribunal broad discretion and flexibility in the conduct of arbitral proceedings'

as well as by paragraph 7 of the Notes which states that procedural decisions by the arbitral tribunal may be taken with or without previous consultations with the parties, depending on whether the tribunal considers such consultations necessary or whether:

'hearing the views of the parties would be beneficial for increasing the predictability of the proceedings or improving the procedural atmosphere'.

In this context it should, of course, be noted that (unlike sets of arbitration rules), the UNCITRAL Notes on Organizing Arbitral Proceedings are generally not something that would be chosen by agreement of the parties or binding upon them. Rather, as the Notes explain, they are designed 'to assist arbitration practitioners by listing and briefly describing questions on which appropriately timed decisions on organizing arbitral proceedings may be useful' (Note 1). It is clear that reference to, or the use of, the Notes 'cannot imply any modification of the arbitration rules that the parties may have agreed upon' (Note 3).

---

11 See Arts. 182(1) and (2) of the Swiss PILA. The Concordat contains a provision (Art. 24) to similar effect for domestic arbitrations. See also ZPO, §1042, especially §1042(3) and §1042(4).
12 Art. 15(1) UNCITRAL Arbitration Rules and Art. 16(1) AAA Rules.
13 available at :
    http://www.uncitral.org/uncitral/en/uncitral_texts/arbitration/1996Notes_proceedings.html
    (accessed 28 July 2005)

B. RESTRICTIONS ON PARTY AUTONOMY

The ability of the parties to determine the conduct of the proceedings is not completely unfettered, even under those arbitral rules that go the furthest in espousing the principle of party autonomy. Types of restrictions on the parties' freedom include:

- 'due process' principles (in particular the independence and impartiality of the arbitral tribunal, equal treatment and the parties' right to be heard);[14]
- public policy/mandatory rules of the forum;
- institutional rules; and
- the principle that arbitral proceedings cannot have an impact on third parties.

These restrictions are examined in turn below.

Due process principles such as the ones outlined above are fundamental requirements of arbitral procedure. They are intended to protect the parties, but also operate to limit the parties' freedom to determine how the arbitral proceedings should be conducted: the parties cannot, for instance, agree that the proceedings be conducted in a way that treats one party differently from the other.

A prominent example of the impact of this principle is the *Dutco* case,[15] in which arbitration proceedings were commenced by a claimant against two other parties, the claims against each of the parties being different. In accordance with the agreement to arbitrate, the claimant appointed its own party-nominated arbitrator whilst the two respondents were required, under protest, jointly to appoint a second party–nominated arbitrator, as the arbitration agreement provided for arbitration by a tribunal composed of two party-nominated arbitrators and a chairman. The tribunal's partial award affirming that it had been validly constituted was set aside by the French *Cour de Cassation*, which stated that

> 'the principle of the equality of the parties in the appointment of arbitrators is a matter of public policy (*ordre public*) which can be waived only after a dispute has arisen'.

These basic standards of procedural fairness are reflected in the above-mentioned rules of arbitration and national laws, which all state that the tribunal must both (1) act fairly and impartially or otherwise treat the parties with equality; and (2) ensure that each party has a reasonable opportunity to present its case or is given the right

---

14 Fair notice (of the appointment of the arbitrators and the conduct of the proceedings) is sometimes raised as an independent due process principle, although arguably it follows from the principles of equal treatment and the parties' right to be heard.
15 *Société BKMI et Siemens c/ Société Dutco,* Cour de Cassation, 7 January 1992, reported in Yearbook of Commercial Arbitration XVIII (1993), pp. 140–142.

to be heard.[16] They are also part of international public policy, as embodied in the European Convention of Human Rights. Article 6(1) of that convention states:

> 'In the determination of his civil rights and obligations or of any criminal charge against him, everyone is entitled to a fair and public hearing within a reasonable time by an independent and impartial tribunal established by law'.

Although an agreement to arbitrate disputes may constitute a waiver by the parties of some of their Article 6(1) rights (e.g. rights to access to court, public hearing and public judgment), it does not necessarily waive all of the rights guaranteed by Article 6(1).[17]

As a result, failure to observe these standards will be a ground for challenging the recognition and enforcement of the award. A challenge may be launched under Article V(1)(b) of the New York Convention according to which recognition and enforcement of the award may be refused if there is proof that:

> 'the party against whom the award is invoked was not given proper notice of the appointment of the arbitrator or the arbitration proceedings or was otherwise unable to present his case'.

Refusal may also occur under Article V(1)(d) (non compliance of the arbitral authority or arbitral procedure with the agreement of the parties or the law of the seat of the arbitration) or Article V(2)(b) (public policy).

The parties cannot determine that the proceedings be conducted in a manner that is contrary to the mandatory rules or public policy of the forum state. These encompass the principles of procedural fairness discussed above. More generally, however, any agreement by the parties to perform an act that is contrary to the forum state's mandatory rules or public policy, or an act that is not capable of being performed under the law of the forum state (or the law governing the arbitration agreement) would be unenforceable in that country. Further, such an agreement could open any resulting arbitral award to challenge under Article V of the New York Convention, specifically under Article V(1)(a), which provides for challenge where the arbitration agreement is not valid under its governing law or the law of the country where the award was made and, depending on where recognition and

---

16   See Art. 15.1 UNCITRAL Rules, Art. 18 UNCITRAL Model Law, Art. 14.1(i) LCIA Rules, Art. 15.2 ICC Rules, Art. 16.1 AAA Rules, s. 33(1)(a) English Arbitration Act, Art. 182(3) Swiss PILA and the ZPO, §1042(1).

17   See William Robinson and Boris Kasolowsky, *'Will the United Kingdom's Human Rights Act Further Protect Parties to Arbitration Proceedings?'*, Arbitration International 18 (2002) 453–66. See also Neil McDonald, who states that 'Article 6(1) guarantees are certainly present within ... international public policy' (Neil McDonald, *'More Harm than Good? Human Rights Considerations in International Commercial Arbitration'*, Journal of International Arbitration 20 (2003), 523–38).

enforcement are sought, under Article V(2), which provides for challenge where recognition and enforcement of the award would be contrary to the public policy of the forum where they are sought.[18]

The arbitral rules chosen by the parties may also restrict the parties' freedom to determine the conduct of their proceedings, as illustrated by the UNCITRAL Model Law and the ICC Rules, which both state that the parties' ability to determine the rules of procedure is subject to the provisions of the relevant rules.[19] However, rules of arbitration generally contain few mandatory provisions over and above the basic requirements of procedural fairness.[20]

Finally, in light of the principle that an arbitration agreement cannot bind third parties, the parties' latitude to dictate the conduct of the arbitral proceedings does not generally extend to requesting procedural measures which might affect third parties, such as compelling third parties to produce documents.

Although party autonomy is the guiding principle underlying arbitral proceedings, in practice, from the moment that the arbitral tribunal is appointed, there is a gradual transfer of control from the parties to the arbitral tribunal. This is due to the fact that once a dispute has been referred to arbitration, the parties will rarely succeed in agreeing any aspects of the procedure on which they have not agreed in advance. Indeed, according to one commentator, the control of the procedure is, as a practical matter, handed over to the tribunal after the issue of the first procedural order.[21]

Accordingly, what follows is a discussion of the arbitral tribunal's power to control the proceedings at each of the key stages of an international arbitration, as reflected in rules of arbitration and national laws.

---

18   See, e.g., *Soleimany v Soleimany* [1999] 3 All ER 847, in which the English Court of Appeal held that an arbitral award purporting to enforce an illegal contract was unenforceable in England. The Court stated: 'There may be illegal or immoral dealings which are from an English law perspective incapable of being arbitrated because an agreement to arbitrate them would itself be illegal or contrary to public policy under English law'.
19   Art. 15.1 ICC Rules, and Art. 19(1) UNCITRAL Model Law.
20   Redfern and Hunter, *International Commercial Arbitration*, para. 6–08. Examples of mandatory requirements imposed by institutional rules of arbitration include, *inter alia*: the obligation of the arbitral tribunal to draw up Terms of Reference in an ICC arbitration (Art. 18 ICC Rules); and requirements concerning the consecutive exchange of written submissions under UNCITRAL Arbitration Rules (Arts 18 and 19).
21   Thomas Webster, *'Party Control in International Arbitration'*, Arbitration International 19 (2003), 119–142, 135, which states: 'After the signing of the terms of reference in an ICC arbitration or the issue of the first procedural order in many other arbitrations the control of the procedure is, as a practical matter, handed over to the tribunal. The subsequent procedural orders are unilateral acts issued by the tribunal'.

C.   PRELIMINARY STEPS

Preliminary steps taken by the parties and the arbitral tribunal at the outset of the proceedings will include some or all of the following:

(a) a preliminary meeting to discuss the organization of the proceedings;
(b) a determination of any preliminary issues, such as jurisdiction, the law(s) applicable to the substance of the dispute and the arbitral proceedings and whether any of the issues in the proceedings (e.g. quantum) should be heard separately (see Chapter 11); and
(c) the appropriateness of expedited remedies, such as the pre-arbitral referee procedure and 'fast-track' arbitrations.

As a general rule, neither institutional rules of arbitration nor national laws grant the arbitral tribunal any specific powers in respect of these preliminary steps. However, as discussed later in this Chapter, the tribunal has a duty to conduct the proceedings diligently and expeditiously, and such preliminary steps will be a key means of resolving procedural issues.

D.   WRITTEN SUBMISSIONS

One of the many purposes of written submissions is, arguably, to give the arbitral tribunal enough information to enable it to manage the conduct of the proceedings in an efficient and effective manner. This is reflected by a number of arbitral rules, which set out fairly detailed provisions concerning written submissions. These provisions usually cover some or all of the following: the contents of the written submissions, including counter-claims and set-off; the periods for filing submissions; and the possibility of amending or supplementing submissions. For example, Article 15 of the LCIA Rules (probably the most perceptive of the common institutional rules) provides for service, within 30 days of notification of the formation of the arbitral tribunal and then at 30-day intervals, of written submissions and reply submissions. These submissions must set out 'in sufficient detail the facts and any contentions of law' on which they rely and/or which they admit or deny (and the grounds for admitting or denying such facts or legal contentions), and, in the case of the Statement of Case, the relief claimed against all other parties to the extent that such matters have not been set out in the Request for Arbitration.[22]

Similarly the ICC Rules, while dealing only with the parties' initial submissions, state that the Claimant's Request for Arbitration should contain details of, *inter alia*, the nature and circumstances of the dispute giving rise to the claim, the relief sought (including, if possible, any amounts claimed) and the relevant underlying agreements, whilst the Respondent's Answer should respond to or comment on the

---

22   Arts 15.2-15.5 LCIA Rules.

information provided in the Request.[23] Of course, the ICC Rules also provide for Terms of Reference which are agreed before the tribunal makes its first procedural order and provide a detailed framework from which the tribunal can manage the process.

Admittedly, national arbitration laws are less prescriptive as to the form, content and process of exchanging written submissions. For instance the Swiss PILA is completely silent on the issue of written submissions, whilst the English Arbitration Act merely states that, subject to any contrary agreement by the parties, the arbitral tribunal may decide 'whether any and if so what form of written statements of claim and defence are to be used, when these should be supplied and the extent to which such statements can be later amended'.[24] That said, it must, of course, be remembered that such laws are generally intended solely to provide the bare essentials necessary to allow an arbitration to be conducted and are not in themselves intended (in the normal run of things) to operate without input from the parties – whether by bespoke drafting or the adoption of published sets of rules.

And of course, in virtually every case the tribunal retains some modicum of control over written submissions, even where the rules of arbitration are detailed. For example, the LCIA Rules state that the Rules' provisions on written submissions are subject to any contrary determination by the parties or the arbitral tribunal.[25] Under both the UNCITRAL Arbitration Rules and the UNCITRAL Model Law, the period for filing written submissions is to be determined by the arbitral tribunal.[26] In relation to additional written submissions, the UNCITRAL Arbitration Rules further provide that the arbitral tribunal:

> 'shall decide which further written statements, in addition to the statement of claim and the statement of defence, shall be required from the parties or may be presented by them and shall fix the periods of time for communicating such statements'.[27]

Such powers confer on tribunals the ability to further refine the parties' cases to assist in proper case management.

E.   EVIDENCE

The arbitral tribunal's role in the process of the taking of evidence may vary to a certain degree according to the place of the arbitration and the legal background of the arbitrators. Arbitrators from a common law background tend to act more like

---

23   Arts 4 and 5 ICC Rules.
24   See 34(2)(c) English Arbitration Act.
25   Art. 15.1 LCIA Rules.
26   Art. 23(1) UNCITRAL Model Law and Arts 18(1) and 19(1) UNCITRAL Arbitration Rules.
27   Art. 22 UNCITRAL Arbitrtion Rules.

referees or umpires monitoring procedural fairness, whereas civil law arbitrators tend to favour more active case management.[28] Although styles of proceedings may be converging and an internationally accepted approach seems to be developing, some differences in the methods of taking evidence persist,[29] for example in relation to document production.[30] The choice of arbitrator may therefore have a fundamental effect on the style of the proceedings.

However, experienced international arbitrators, whether they come from a civil law or a common law background, do not usually allow themselves to be limited by existing rules of evidence, which might be purely technical and/or prevent them from establishing the facts necessary for determining the issues between some parties. In international commercial disputes, the strict and formal rules of evidence, which were originally devised to protect juries, are not deemed to be appropriate.[31] On the contrary, modern arbitration rules provide that the tribunal may conduct the fact-finding in any way it deems appropriate,[32] which includes the application or rejection of rules of evidence.

As a general principle, it has been stated that 'the methods of presenting evidence to an arbitral tribunal on disputed issues of fact derive from a synthesis of party autonomy, discretion of the arbitral tribunal and court control at the stage of enforcement'.[33] The tension between the first two principles, party autonomy and the discretion of the arbitral tribunal, is illustrated, in different ways, by the various sets of institutional rules:

– The ICC Rules give the tribunal a wide discretion to decide rules of evidence, stating that it 'shall proceed within as short a time as possible to establish the facts of the case by all appropriate means'.[34] The ICC Rules go on to provide specific instances of the tribunal's autonomy, *inter alia* in relation to hearing

---

28   See Schroeder, *Die lex mercatoria arbitralis*, 187 *et seq.* (Abschnitt 3. A.).
29   See Claude Reymond, *'Civil Law and Common Law Procedures: Which is the more Inquisitorial?'*, Arbitration International 5 (1989), 357–368.
30   The common law concept of disclosure is quite foreign to the civil law systems. See S Wilske and D Mack, *'Germany: Production of Documents under the Revised German Code of Civil Procedure'*, International Bar Association Committee D News (newsletter of the Arbitration and ADR Committee of the Section on Business Law) 8 (2003) 43. See also Chapter 12 below.
31   See Roger Ward, *'The Flexibility of Evidentiary Rules in International Trade Dispute Arbitration'*, Journal of International Arbitration 13 (1996), 5–20.
32   Art. 20 ICC Rules; Art. 92. IBA Rules; and Art. 27 DIS Arbitration Rules (Deutsche Institution für Schiedsgerichtsbarkeit e.V., *DIS-Schiedsgerichtsordnung* 1998, arbitration rules published by the German Institution of Arbitration and available in both German and English at: http://www.dis-arb.de/scho/schiedsvereinbarung98-e.html (accessed 3 August 2005)).
33   Redfern & Hunter, *International Commercial Arbitration*, para. 6–68.
34   Art. 20(1).

the parties[35] and witnesses or experts[36] and taking measures to protect trade secrets and confidentiality,[37] although one limitation is that the tribunal cannot decide the case solely on documents if one of the parties wishes to be heard;[38]
- The LCIA Rules contain very detailed provisions concerning evidence, conferring similarly wide powers on the tribunal, subject, in most instances, to giving the parties a 'reasonable opportunity to state their views'.[39] Under the LCIA Rules, the tribunal may (1) conduct such enquiries as are necessary and expedient to ascertain the relevant facts; (2) order any party to make any property, site or thing under its control and relating to the subject matter of the arbitration available for inspection; (3) order any party to produce any relevant documents or classes of documents for inspection; and (4) decide whether or not to apply strict rules of evidence as to admissibility, relevance or weight of any factual or expert material.[40] Specifically in relation to witnesses, the tribunal has a discretion to allow, refuse or limit the appearance of witnesses (whether they are factual or expert witnesses),[41] to control questioning by the parties/put its own questions to the witnesses[42] and to dictate the form/manner of exchange of witness testimony;[43]
- Both the UNCITRAL Arbitration Rules and the AAA Rules provide that the tribunal may require either party to deliver a summary of the documents and other evidence which that party intends to present in support of the facts in issue set out in its claim, counterclaim or defence. Furthermore, at any time during the arbitral proceedings, the tribunal may require the parties to produce documents, exhibits or other evidence.[44] Concerning witnesses, the tribunal is free to determine the manner in which these are to be examined.[45] Finally, the tribunal shall determine the admissibility, relevance, materiality and weight of the evidence offered;[46]
- The UNCITRAL Model Law, on the other hand (as perhaps might be expected of an instrument intended to provide the framework within which arbitrations are conducted, rather than a set of arbitration rules per se), is more brief on

---

35   Art. 20(2) (ability of the tribunal to hear the parties together in person on the request of either party or of its own motion).
36   Art. 20(3) (ability of the tribunal to hear witnesses and experts in the presence of the parties or in their absence, provided they have been duly summoned) and Art. 20(5) (ability of the tribunal to summon any party to submit additional evidence).
37   Art. 20(7).
38   Art. 20(6).
39   Art. 22.1.
40   Art. 22.1(c) – (f).
41   Art. 20.2.
42   Art. 20.5.
43   Arts 20.1, 20.2 and 20.3.
44   Arts 24.2 and 24.3 UNCITRAL Arbitration Rules and Arts 19.2 and 19.3 AAA Rules.
45   Art. 25.4 UNCITRAL Arbitration Rules and Art. 20.4 AAA Rules.
46   Art. 25.6 UNCITRAL Arbitration Rules and Art. 20.6 AAA Rules. Under the AAA Rules, in particular, the tribunal has the discretion to direct the order of proof and exclude cumulative or irrelevant evidence (see Art. 16.3).

the issue of evidence, and clearly affirms the pre-eminence of the parties' autonomy. It states that, subject to the provisions of the UNCITRAL Model Law and any contrary agreement by the parties, the arbitral tribunal may conduct the arbitration in such manner as it considers appropriate, which power includes determining the admissibility, relevance, materiality and weight of any evidence.[47] The UNCITRAL Model Law also provides that the tribunal may meet at any place which it considers appropriate for the purposes, *inter alia*, of hearing witnesses or experts or for the inspections of goods, other property or documents.[48]

National rules also affirm the principle of party autonomy. The German ZPO contains almost identical provisions to those of the UNCITRAL Model Law in this respect.[49] Under the English Arbitration Act, the tribunal's right to decide evidential matters (including, whether any documents or classes of documents should be produced by the parties, whether to apply strict rules of evidence as to admissibility etc and whether there should be oral or written evidence) is subject to 'the right of the parties to agree any matter'.[50] On the other hand, Swiss law states that 'the arbitral tribunal shall itself conduct the taking of evidence'.[51]

Finally, the effect of the IBA Rules on the Taking of Evidence in International Commercial Arbitration 1999 (the IBA Rules)[52] needs to be taken into account. Unlike the UNCITRAL Notes on Organizing Arbitral Proceedings, the IBA Rules 'are designed to be used in conjunction with, and adopted together with, institutional or ad hoc rules or procedures governing international commercial arbitrations' (see Foreword to IBA Rules). In practice, parties frequently agree to the application of the IBA Rules, either of their own volition or at the prompting of the tribunal. Where adopted, the IBA Rules balance party autonomy and the tribunal's discretion by referring to party autonomy as underlying the arbitral tribunal's powers, whilst giving the tribunal wide discretion in relation to a number of specific evidential issues.[53] For example, if a party asks the tribunal to take steps to obtain documents from a third party, the tribunal may use its discretion to determine whether the documents are relevant and material and, if so, it should take the necessary steps to obtain them.[54] At any time before the arbitration is concluded, the tribunal may

---

47 Art. 19.
48 Art. 20(2).
49 ZPO, §1042(3) and (4) and §1043(2).
50 See s. 34 English Arbitration Act, especially ss 34 (d)-(f) and (h).
51 See Art. 184(1) Swiss PILA.
52 International Bar Association, available at: http://www.ibanet.org/publications/IBA_Guides_Practical_Checklists_Precedents_and_Free_Materials.cfm (accessed 12 December 2005). The IBA Rules constitute a step towards establishing an international standard on the taking of evidence. A copy of the IBA Rules appears at Annex 9.
53 Art. 2(4) IBA Rules, which states that, insofar as the IBA Rules and any applicable institutional/ad hoc rules are silent and the parties have not agreed otherwise, the tribunal 'may conduct the taking of evidence as it deems appropriate'.
54 Art. 3(8) IBA Rules.

also request a party to produce any documents that it believes to be relevant and material to the case.[55] Further, the tribunal may order either party to ensure, or seek to ensure, the appearance for testimony of any person.[56]

In practice, the three main areas in which a tribunal in international arbitration is likely to exert control are:

– document production: in the majority of cases, the tribunal will seek to limit document production as far as possible, in order to focus on those documents or categories of documents likely to be relevant and necessary;
– witnesses: the tribunal will also, as far as practicable, attempt to shorten the oral stage of the proceedings, for example, by refusing to hear oral witness evidence or at least restricting the number of oral witnesses;[57] and
– the admissibility, weight and relevance of the evidence tendered: as stated above, an experienced arbitral tribunal will rarely exclude evidence on grounds of inadmissibility; however, it may use its discretion to attribute more or less weight to evidence (e.g. it may attribute less weight to uncorroborated witness testimony than to documentary evidence).

F.   EXPERTS

The parties may submit their own expert evidence to the arbitral tribunal, however, the tribunal usually has the power to appoint its own expert, if the tribunal considers it necessary and appropriate. This power will be expressly set out in the arbitration agreement or incorporated into the agreement by reference to institutional or international rules of arbitration, or implied under the law governing the arbitration agreement, if there is a provision in the governing law to this effect[58] or at least if there is no provision to the contrary.

The tribunal may require the parties to provide its expert with any relevant information or to produce for inspection any relevant documents, goods or other property that may be required by the expert.[59]

Generally, the parties have some control over the tribunal's expert. Some rules of arbitration provide that the arbitral tribunal should involve the parties in the process of appointing any such expert and in defining the scope of the expert's retainer. Both the ICC Rules and the IBA Rules, for example, state that the tribunal may appoint experts and define their terms of reference 'after having consulted the

---

55 Art. 3(9) IBA Rules. The relevant party may object to such a request on various grounds of inadmissibility, set out in Art. 9(2).
56 Art. 4(11) IBA Rules.
57 See s. 34(2)(h) English Arbitration Act, which states that the tribunal may decide whether and to what extent there should be oral or written evidence.
58 See, e.g. s. 37 English Arbitration Act 1996 and ZPO, §1049.
59 Art. 21.1(b) LCIA Rules; Art. 27.2 UNCITRAL Rules; Art. 26(1)(b) UNCITRAL Model Law; Art. 22(2) AAA Rules; and Art. 6(3) IBA Rules.

parties'.[60] Further, the parties may have the right to make objections to the tribunal if they have doubts as to the independence of its expert.[61] Finally, the parties must also be given the opportunity to question any tribunal-appointed expert at a hearing and to present their own expert witnesses to testify on the relevant issues (see Chapters 12 and 13).[62]

G.   INTERIM MEASURES

Subject to any agreement to the contrary by the parties, in most cases an arbitral tribunal will have the power to issue interim measures, usually for the purposes of taking or preserving evidence,[63] preserving the status quo and providing security for costs.[64] While the parties' right to apply to the relevant local courts for interim measures is preserved by national laws or institutional rules of arbitration,[65] once a tribunal has been established the parties are usually required first to seek relief from the arbitral tribunal, or at least to ask its permission before seeking interim remedies in the local courts. For example:

- unless the case is 'one of urgency', the English Arbitration Act provides that the court may make interim orders only if a party's application has been made with the permission of the tribunal or the agreement of the other parties (in all cases, the court may act only if or to the extent that the arbitral tribunal/institution does not have the relevant power or is unable to act effectively);[66]
- under German law,[67] for the purposes of taking evidence or in respect of the performance of other judicial acts which the tribunal is not empowered to carry

---

60 Art. 20(4) ICC Rules; Art. 6(1) IBA Rules.
61 Art. 6(2) IBA Rules.
62 Art. 20(4) ICC Rules; Art. 21.2 LCIA Rules; Art. 27.4 UNCITRAL Rules; Art. 22(4) AAA Rules; Art. 26(2) UNCITRAL Model Law; s. 37(1)(b) English Arbitration Act; and ZPO, §1049(2).
63 An arbitral tribunal does not usually have the power to compel the attendance of witnesses, as reflected by Art. 27 UNCITRAL Model Law, which states that 'the arbitral tribunal... may request from a competent court of this State assistance in taking evidence'.
64 Arts 25.1 and 25.2 LCIA Rules; Art. 26 UNCITRAL Rules; Art. 17 UNCITRAL Model Law; Art. 21 AAA Rules; and Art. 23 ICC Rules. See also s. 38 English Arbitration Act; Art. 183 Swiss PILA; and the ZPO, §1041.
65 See, e.g., Art. 26(3) UNCITRAL Rules; Art. 23(2) ICC Rules; and ZPO, §1033.
66 See ss. 44(3)–(5) English Arbitration Act. The recent case of *Hiscox Underwriting Ltd v Dickson Manchester & Co Ltd* [2004] EWHC 479 confirmed that the cases of urgency in which the court can issue interim injunctions are not limited to applications for orders to preserve assets or evidence, but effectively extend to any form of interim injunction.
67 See ZPO, §1050.

out, court assistance may be requested, but only by 'the arbitral tribunal or a party with the approval of the arbitral tribunal'; and
- although institutional rules of arbitration are generally not as clear as national laws on this issue, the LCIA Rules state that a party's right to apply to local courts for interim or conservatory measures shall only be exercised in 'exceptional cases' once the tribunal has been formed.[68]

Consequently, the parties should turn to the tribunal as their first port of call before pursuing remedies in the local courts, although they should bear in mind that the tribunal's ability to issue interim measures will be subject to limitations, as discussed in Chapter 11.

H. HEARINGS

Most rules of arbitration give the tribunal the right to determine whether or not oral hearings should be held, whilst preserving the right of the parties to request to be heard orally if they so wish, unless they have previously agreed otherwise.[69] Furthermore, the rules usually state that it is for the tribunal to determine the time and place of such hearings,[70] whether this is at the request of the parties or on its own initiative.

By the stage of oral hearings, control will usually have shifted fully from the parties to the tribunal. This is reflected by the ICC Rules, which state that the arbitral tribunal 'shall be in full charge of the hearings' (Article 21(3)). In practice, in an international arbitration, the tribunal tends to take an active role, more typical of the civil law tradition than of the common law one, at least for the purposes of establishing the facts.[71] For instance, the tribunal may give specific guidance to the parties concerning the presentation and content of their evidence,[72] as well as putting questions directly to the witnesses. A starker illustration of the tribunal's

---

68 Art. 25.3 LCIA Rules.
69 See Art. 15(2) UNCITRAL Rules; Art. 24(1) UNCITRAL Model Law; Art. 19 LCIA Rules; Art. 20(2) and (6) ICC Rules; s. 34(2)(h) English Arbitration Act; and ZPO, §1047(1).
70 See, e.g., s. 34(2)(a) English Arbitration Act, Art. 21(1) ICC Rules, and Art. 19.2 LCIA Rules.
71 See H. M. Holtzmann, *'Fact-finding by the Iran-United States Claims Tribunal'*, in Richard B. Lillich (eds.), *Fact-finding Before International Tribunals* (Transactional Publishers, 1991), p. 101, para. 6–10. Holtzmann states that 'it is wise for an arbitral tribunal to take an active role in augmenting the parties' presentation of the facts... Arbitration is more effective and efficient when the arbitrators actively seek to elucidate the facts, rather than merely evaluating what the parties chose to present'.
72 See UNCITRAL Notes on Organizing Arbitral Proceedings, para. 80: 'arbitration rules typically give broad latitude to the arbitral tribunal to determine the order of presentations at the hearings... it may foster efficiency of the proceedings if the arbitral tribunal clarifies to the parties, in advance of the hearings, the manner in which it will conduct the hearings, at least in broad lines'.

empowerment is the technique of witness conferencing (see above and Chapter 12), which consists of 'the simultaneous joint hearing of all fact witnesses, expert witnesses, and other experts involved in the arbitration',[73] although it should be noted that this practice is not yet widely used in the international arbitration community. Finally, in the event that one party refuses to appear at the hearing, the tribunal may proceed with the hearing and issue its award.[74]

I. POST-HEARING MATTERS

Once the hearing is over, the arbitral tribunal will declare the proceedings closed. Thereafter, the parties may not make any further submissions or produce any new evidence, unless requested or authorized to do so by the tribunal.[75] It is becoming increasingly common for tribunals to exercise their discretion to allow the parties to submit post-hearing briefs, which may be used to allow parties to address any new material or submissions made by the other party or answer any questions posed by the tribunal during the hearing to which they did not have the time to respond, or produce new evidence arising which came to light after the hearing (but before the tribunal has issued its award).[76]

IV. THE IMPORTANCE OF EFFECTIVE CASE MANAGEMENT

In exercising its various powers at each stage of the arbitral proceedings, the arbitral tribunal must respect the following fundamental principles: the parties' due process rights (as outlined above) in particular, the principle of equal treatment and the parties' right to be heard and the need to act efficiently and fairly, avoiding unnecessary delay or expense.[77]

A. PROCEDURAL FAIRNESS

As discussed in the first part of this chapter, the parties' due process rights (principally, the right to an impartial and independent tribunal, the principle of equal treatment and the right to be heard) are fundamental requirements of the arbitral procedure, restricting both the parties' and the arbitral tribunal's freedom to determine how proceedings should be conducted. If the parties' due process

---

73 See Wolfgang Peter, *'Witness "Conferencing"'*, Arbitration International 18 (2002), 47–58 and Chapter 12.
74 See, e.g. Art. 25 UNCITRAL Model Law.
75 See, e.g. Art. 22(1) ICC Rules.
76 See generally Redfern & Hunter, *International Commercial Arbitration*, paras 6–123 – 6–126.
77 E.g. s. 33(1) English Arbitration Act 1996 and Art. 14.1 LCIA Rules.

rights are not respected in the course of the proceedings, any ensuing award could be open to challenge under Article V.1(b) of the New York Convention. How, in practice, does this impact on the arbitral tribunal's conduct of the arbitral proceedings?

The principle of equal treatment will, in most cases, require the tribunal to treat the parties with plain equality, for instance, giving each party the same opportunity to participate and express its views on the evidence. However, it is worth noting that some rules of arbitration only impose a duty on the tribunal to act 'fairly'.[78] According to one commentator, in most cases this will mean treating the parties with equality, but in some circumstances it may mean that the tribunal has to treat the parties on less than equal terms, to preserve substantive equality.[79]

Concerning the right to be heard, there is some debate in the case law as to the extent to which the tribunal must adhere to the arguments pleaded by the parties.[80] For example, a tribunal may award relief of a different nature from that requested by the claimant, provided this is available under the applicable law, within the limits of the claim and (therefore) within the parties' reasonable contemplation.

One example of this is the Swiss case of *Bank Saint Petersburg PLC v ATA Insaat Sanayi ve Ticaret Ltd*,[81] concerning an arbitral award in which the tribunal had granted damages rather than specific performance as requested by the claimant. The Swiss Federal Tribunal stated that there was no breach of the principle *ne eat iudex ultra petita partium* (or the judge must not award more than what is claimed by the parties) 'when the tribunal adjudicates in law and within the limits of the claim, and yet does not base its legal considerations, or does so only partially, on the legal arguments set forth by the parties'. The Court, added, however, that the parties would have the right to be heard again in circumstances where the tribunal wished to base its award on legal arguments which the parties had not presented and the relevancy of which they could not have reasonably anticipated.

Another example of the tribunal's freedom in this regard is that it is not necessarily bound to rely exclusively upon the authorities supplied by the parties. For instance, in a recent English case concerning an application to set aside an arbitral award on the grounds that the award contained references to several authorities not referred to by either the parties or the arbitrator in the course of the proceedings, it was held that the applicant had failed to point to any injustice caused

---

78   E.g. s. 33(1) English Arbitration Act 1996 and Art. 14.1 LCIA Rules.
79   Georgios Petrochilos, *Procedural Law in International Arbitration*, (Oxford University Press, 2004), para. 4.87. Petrochilos cites as an example the possibility that the arbitral tribunal may have to allocate costs in unequal terms, in the event of one of the parties being impecunious, in order to maintain a level playing field between the parties.
80   According to Petrochilos, *Procedural Law in International Arbitration*, para. 4.89, it seems to be 'a question of degree, rather than anything else, to what extent a tribunal may stray, as it were, from the parties' submissions in general'.
81   *Bank Saint Petersburg PLC v ATA Insaat Sanayi ve Ticaret Ltd*, 2 March 2001, ASA 3/2001 531.

by the arbitrator's failure to cite the authorities earlier in the proceedings.[82] It is, however, undoubtedly the case that good practice would suggest that an arbitrator wishing to rely on additional authorities in an award would be well advised to let the parties know of this intention at an appropriately early stage and to permit submissions on them. At the very least, this would avoid the possibility – even if it were ultimately unsuccessful – of a challenge on the basis that no prior mention of the authorities had been made.

On the other hand, the authors are aware of circumstances in which departure by the tribunal from the matters strictly pleaded by the parties might be held to have infringed their right to be heard. For example:

- where the tribunal introduces issues which are not in dispute between the parties. In one English case concerning payments for the sale of sugar, the issue in dispute related to quantum only, as the respondent did not challenge the claimant's contention that it was in default. The arbitrators, deciding the dispute as a documents-only arbitration, put a number of questions to the parties in writing, most of which related to quantum. In their award, they dismissed the claimant's claim in full, concluding that the respondent had not been in breach of its contractual obligations. The claimant's application to set aside the award was granted by the court, which stated that the arbitrators had not given the parties sufficient notice that they were going to reopen the issue of liability, which had been very firmly closed by way of pleading between the parties and that in those circumstances, an award had been made 'on a basis which the claimants never had a reasonable opportunity of making the subject of their submissions or the subject of evidence';[83]
- where the tribunal uses particular items of evidence/fails to consider other evidence without allowing the parties to address this. In another English case, a landlord claimed that the award determining the level of rent should be set aside for serious irregularity because the arbitrator had determined the rent on the assumption that the upper floors of the relevant premises would remain vacant and that a rent-free period would be required in lieu of contribution to the cost of works, and had also chosen to ignore a substantial premium in a comparable transaction on the assumption that this was attributable to fixtures and fittings. The court set aside the award, stating that the arbitrator should not have made such assumptions without affording the parties the opportunity to comment on them.[84]
- where the tribunal fails to give the parties the opportunity to address it in relation to issues affecting its determination of an important point (e.g. costs) which are not advanced by either party. In an English construction case, for

---

82 *Sanghi Polyesters Ltd (India) v International Investor KCSC (Kuwait)* [2000] All ER (D) 93.
83 *Pacol Ltd v Joint Stock Co Rossakhar* [1999] 2 All ER (Comm) 778.
84 *Guardcliffe Properties Ltd v City & St James Property Holdings* [2003] EWHC 215 (Ch).

example, an arbitrator initially issued an interim award in favour of the claimants, but then made a final award in favour of the claimants on some issues and in favour of the respondent contractors on other issues. In the final award the arbitrator awarded costs in favour of the respondents, relying on two matters which had not been raised by either party. The court stated, *inter alia,* that the arbitrator's power to award costs was subject to the general duty to act fairly and impartially between the parties under section 33(1) of the English Arbitration Act and that 'a tribunal does not act fairly and impartially if it does not give a party an opportunity of dealing with arguments which have not been advanced by either party';[85] and

– where the tribunal proceeds to decide a case on a legal basis different from that pleaded by the parties, such as principles of *lex mercatoria*, without seeking argument thereon from the parties.[86]

Overall, the courts are reluctant to interfere with arbitral awards on grounds of procedural defects, provided that the parties' rights to due process have been respected.[87] For example, in one United States case, the court rejected one of the parties' argument that the arbitrators had acted improperly in limiting its ability to cross-examine the witness of the other party, stating that the arbitrators were charged with the duty of determining the relevance of evidence and that 'barring a clear showing of abuse of discretion, the court [would] not vacate an award based on improper evidence or the lack of proper evidence'.[88] The English case *Egmatra v Marco*[89] concerned the award rendered in an arbitration under the auspices of the London Metal Exchange, which the respondent challenged on the basis that the arbitrators had rejected its request to hear expert evidence on certain issues. The court upheld the award, stating that the arbitrators, who were experts in the metal industry, were entitled to decide that they did not require expert evidence to assist them. Finally, in a Swiss case, the Swiss Federal Tribunal held that an interim award on costs preventing the party against whom it was issued from offsetting the payment of the costs against any possible claim for compensation that might be granted in

---

85   *Gbangbola and another v Smith & Sherriff Ltd* [1998] 3 All ER 730.
86   See Petrochilos, *Procedural Law in International Arbitration*, para. 4.90, stating that 'in no case should an arbitrator proceed, without seeking argument from the parties, to decide a case on principles of lex mercatoria or the like, whose content is on any view indeterminate'.
87   See Rufus von Thülen Rhoades, Daniel M. Kolkey and Richard Chernick (eds.), *Practitioner's Handbook on International Arbitration and Mediation* (Juris Publishing, 2002), para. §8.05.
88   *Laminoirs-Tréfileries-Cableries de Lens, SA v Southwire Company and Southwire International Corporation*, United States District Court, N.D. Georgia, Newnan Division, 18 January 1980, 484F Supp 1063, 1067.
89   *Egmatra A.G. v Marco Trading Corporation* [1999] 1 Lloyd's Rep 862. This case emphasized that, 'substantial injustice' under s. 68 English Arbitration Act 1996 would only arise in 'extreme cases where the tribunal has gone so wrong in its conduct of the arbitration that justice calls out for it to be corrected'.

a final award did not violate the principle of equal treatment or Swiss public policy.[90] Overall, therefore, it has been commented that 'the trend of the day is definitely non-interventionist; courts of law do not intervene if they do not have to ... during the proceedings arbitrators enjoy large freedom to render procedural decisions knowing that local courts of law will not interfere'.[91]

B. THE ARBITRAL TRIBUNAL'S DUTY TO ACT EXPEDITIOUSLY

For the purposes of complying with its duty to act fairly, as well as ensuring procedural fairness, the tribunal must conduct the arbitral proceedings diligently, efficiently and within the shortest time-span in order to avoid unnecessary costs. According to one commentator, 'the diligent exercise of the arbitration procedure is ... the best antidote against dilatory tactics in arbitration procedures'.[92]

A duty to manage the proceedings so as to avoid undue delay and expense is expressly imposed on the arbitral tribunal by some institutional rules of arbitration. For instance, the LCIA Rules state that the tribunal has a duty to 'adopt procedures suitable to the circumstances of the arbitration, avoiding unnecessary delay or expense, so as to provide a fair and efficient means for the final resolution of the parties' dispute'.[93] The English Arbitration Act contains an almost identical formulation,[94] while the AAA Rules state that the tribunal 'shall conduct the proceedings with a view to expediting the resolution of the dispute'.[95] As an alternative, in some jurisdictions, as well as under the ICC Rules, a specific time-limit within which to render the award is imposed on the arbitral tribunal,[96] although this can usually be extended.[97]

Failure by the arbitral tribunal to comply with its duty to act expeditiously may give rise to sanctions such as the following: termination of the tribunal's mandate;[98]

---

90 *A, B and C v D*, Schweizerisches Bundesgericht, 17 December 2002, 4P.196/2002.
91 See Sigvard Jarvin, *'To what extent are procedural decisions of the arbitrators subject to court review?'*, ICCA Congress Paris No. 9 (1998).
92 See Juan Eduardo Figueroa, *'Ethics in International Arbitration'*, Mealy's International Arbitration Report 18 (July 2003), 41–50.
93 Art. 14.1(ii) LCIA Rules.
94 S. 33(1)(b) English Arbitration Act.
95 Art. 16(2) AAA Rules.
96 Art. 24(1) ICC Rules (six months), and Arts 813 and 820 of the Italian Code of Civil Procedure.
97 Art. 24(2) ICC Rules.
98 Art. 14(1) UNCITRAL Model Law, stating that if an arbitrator 'fails to act without undue delay, his mandate terminates if he withdraws from his office or if the parties agree on the termination'.

deprivation of remuneration;[99] or even removal by a competent court.[100] There is also the (slim) possibility that the members of the arbitral tribunal could be made liable for damages for undue delay, on the basis of a contractual duty of care.[101]

Concerning the sort of behaviour likely to constitute a failure to act expeditiously on the part of the arbitral tribunal, the English construction case of *Pillar v Edwards* is highly instructive.[102] In this case, a dispute concerning the final value of building works was referred to arbitration by the contractor and in the course of the arbitration proceedings the costs and fees incurred by both parties mounted to nearly four times the value of the original claim (GBP 100,000). The judge, stating that 'there [was] something inherently wrong with an arbitral process which involve[d] such large sums being spent in costs relative to the size of the sums in dispute', held that the arbitrator had breached his duty to adopt procedures avoiding unnecessary delay or expense under section 33(1)(b) of the English Arbitration Act, on the basis of the following: the arbitrator had failed to direct the parties to put right their pleadings, which were prolix and diffuse, as well as the schedules they had produced summarizing their respective cases, which, in breach of the arbitrator's original directions, were not Scott Schedules and failed to consolidate all aspects of each disputed work item; the arbitrator and the parties had not consolidated the two arbitrations into one so as to keep costs to a minimum; the hearing was disproportionately long (ten days), as a result of the number of expert witnesses, the failure of the parties and the experts to define and reduce the number of issues and the absence of time-limited cross-examination; the award, which took several months to be prepared, was diffuse, lacked substantial reasoning, incorporated verbatim the extensive pleadings and contained a significant number of errors and omissions, resulting in a large number of corrections being required.

The AAA's (new) Code of Ethics also provides some guidance on the types of behaviour likely to be sanctioned by a court, by indicating that for the purposes of acting fairly and efficiently, an arbitral tribunal must make reasonable efforts to prevent delaying tactics, harassment of the parties or other participants or other abuse or disruption of the arbitration process.[103]

---

99   E.g. in Colombia (see Mantilla-Serrano 'Colombia' in Nigel Blackaby, David Lindsey and Alessandro Spinillo (eds.), *International Arbitration in Latin America* (Aspen Publishers, 2003), p.121).

100   See s. 24(1)(d)(ii) English Arbitration Act, stating that a party may apply for the removal of an arbitrator on the ground that he 'has refused or failed – (1) properly to conduct the proceedings; or (2) to use all reasonable despatch in conducting the proceedings or making an award, and that substantial injustice has been or will be caused to the applicant'.

101   For a fuller discussion, see Redfern & Hunter, *International Commercial Arbitration*, paras 5-15–5-23.

102   *RC Pillar & Sons v Edwards and another* [2001] All ER (D) 232.

103   See Canon I (F) of the AAA Code of Ethics for Arbitrators in Commercial Disputes, in force since 1 March 2004, which states: 'An arbitrator should conduct the arbitration process so as to advance the fair and efficient resolution of the matters submitted for decision. An arbitrator should make all reasonable efforts to prevent delaying tactics, harassment of parties or other participants, or other abuse or disruption of the arbitration process'. Available at: http://www.adr.org/CodeOfEthics (accessed 28 July 2005).

On the other hand, a decision by an arbitral tribunal to make a preliminary determination of certain points, such as jurisdiction, notwithstanding that this may involve making determinations closely related to the substance of the dispute to be arbitrated, does not necessarily constitute a failure to act expeditiously. In an English case on this point, one of the parties sought to challenge an award on the grounds, *inter alia*, that the arbitrator's decision to determine his own jurisdiction as a preliminary point, notwithstanding the fact that it involved deciding an issue closely related to the substance of the dispute to be arbitrated, constituted a serious irregularity under section 68 of the English Arbitration Act. The judge rejected this contention, stating that it was commonplace in international arbitration for jurisdiction and liability both to be disputed and that the mere fact that the arbitrator chose a course which might involve the issue of his jurisdiction being first determined by him on a preliminary basis and then all over again could not constitute a basis for an allegation of breach of his duty under section 33(1)(b) English Arbitration Act.[104]

In the light of the above, whether or not an arbitral tribunal may be held to have breached its duty to act expeditiously appears to be a question of degree. Costs and delays that are grossly disproportionate to the complexity of the dispute and/or the relief claimed will, in any event, be a sure indicator that the tribunal has failed to manage the proceedings.

From a practical point of view, the main means by which an arbitral tribunal may manage the proceedings more effectively is the use of meetings with the parties where procedural issues (such as issues relating to document production) may be thrashed out. The AAA Rules make specific reference to this, stating that the tribunal 'may conduct a preparatory conference with the parties for the purpose of organizing, scheduling and agreeing to procedures to expedite the subsequent proceedings'.[105]

---

104 *Kalmneft JSC v Glencore International AG* [2002] 1 All ER 76.
105 See Art. 16(2) AAA Rules.

Chapter 9
Preparation and Collection of Evidence

I.   INTRODUCTION

The collection of evidence for the conduct of a construction arbitration will almost invariably involve the following tasks:

- reviewing documents relevant to a claim (including for instance, head office correspondence, site correspondence, site diaries, invoices and receipts);
- interviewing witnesses who can give an account of the facts relevant to the claim; and
- obtaining expert evidence as to the implications or relevance of certain facts (from, for instance, programming, soil mechanics, accounting or quantum experts).

When the bulk of this work is done is, ultimately, a question of judgment. Clearly, undertaking all this work prior to the commencement of an arbitration (whilst a counsel of perfection) will lead to delay, considerable expenditure which may ultimately be wasted if matters are ultimately accepted by the other party or the case takes a different turn in the light of the other party's presentation of its own case. However, drafting pleadings (or the request for arbitration where the request is required to contain a substantial part of the claimant's case) without having first thoroughly reviewed the evidence is risky. Asserting a patently unmeritorious claim may lead to an award for costs in the other party's favour to the extent that the other party wasted costs defending the unmeritorious claim.

Ideally, therefore, a party will at the very least make sure that the facts underpinning the legal arguments set out in its pleadings are supported by the evidence *before* filing that document with the tribunal. Time, however, does not always permit an exhaustive review of the evidence at the time of drafting pleadings, in which case a party may need to amend its pleadings once the process of collecting evidence has been completed. Whether – and the extent to which – a party may in fact amend its pleadings depends on the terms of the parties' agreement to arbitrate and the terms of reference.

Ultimately, the method by which documents are managed during the arbitral proceedings is a matter to be agreed between the parties or ordered by the tribunal (depending on the parties' agreement to arbitrate and/ or the terms of reference). This chapter sets out some of the issues surrounding the review, analysis and management of documentary evidence in international construction arbitrations.

II. THE VALUE OF A CHRONOLOGY AND OTHER AIDS

Particularly in large disputes, the volume of evidence relevant to the facts in issue may be formidable. Moreover, the evidence may well be contradictory in places; witnesses may differ as to the date on which events occurred and documents may contradict the witnesses' recollections. Reviewing and synthesizing such information in order to assess the merits of a claim can be a particularly daunting task.

Perhaps the best way to review evidence is by preparing a chronology. This is not to say that preparation of evidence by reference to specific issues is not a useful tool. On the contrary, once the structure of the case by reference to a timeline has been established, it is normal for particular issues to be identified and the evidence relevant to those specific issues separately collated. But the first step, in all but those cases with the most limited number of issues, is the establishment of that timeline or chronology in which the analysis of the individual issues sits.

There are no hard and fast rules about the format of a chronology, but typically it will be set out as shown in figure 1, below:

Figure 1

Date	Event	Source
2 May	Site instruction XXX issued requesting contractor to carry out Y work.	Site instruction XXX
3 – 5 May (approx)	Contractor commences Y work.	Witness statement
8 May	Site is shut down by strike.	Site diary Z and witness statement.

Invariably, the first column of a chronology will be the date on which the event relevant to a claim occurred. Sometimes it will be difficult to state the exact date on which the event in question occurred. In that case it may be useful to place the event approximately within the chronology until such time as the date can be pin-pointed by better evidence (as in the case of the second event in figure 1).

Typically, the second column is the event or fact relevant to the claim. There are a number of issues to consider when deciding what to include in the chronology. It is, for instance, important to be discerning in the selection of the facts to be included. A chronology which is littered with irrelevant facts may become meaningless and unwieldy. Having said this, certain facts or events, while not directly relevant, may need to be included because they put other relevant facts in context. Of more concern – at least if the chronology is to be presented to the other party to the arbitration – is the extent to which the selection of relevant facts provides information, by what is included and what is not included, of the internal thought processes which underlie the approach to the case.

Finally, it is useful to have a column which sets out the evidence proving the facts set out in the chronology. Linking the facts to the available evidence in this way allows a party to assess where the evidence is lacking, as the source column will be empty where facts remain to be substantiated on the evidence.

Additionally, the source column allows a party to see whether there are any possible inconsistencies in the evidence (see, for example, the third event in figure 1, which event is recorded as being evidenced both by a site diary and a witness statement). Review of both sources of evidence should obviously be undertaken for consistency and, if there is a major inconsistency which cannot be resolved, the correctness of the fact has to be regarded as being uncertain.

A good chronology will form the backbone not only of a party's pleadings, but also of its witness statement and any expert reports it requires. Programming experts in particular cannot assess a claim for delay or disruption unless the reasons for and the dates on which the claimant was delayed are known with precision.

However, this is not the end of the chronology's value. Arbitral tribunals and courts alike often require chronologies to be submitted by the parties. Even the most carefully drafted pleadings, statements and submissions can be difficult to read and digest. In this regard, the ICC Construction Arbitration Report[1] notes:

'If there are any claims for delay or disruption, a chronology of events will be required from the parties. This should be ideally included with the claimant's request for arbitration or the respondent's answers.... Since most construction arbitrations are about the performance of a relatively long-term contract, it is in our view highly desirable that the tribunal should scrutinize any chronology with care'.[2]

In some cases an arbitrator will require the parties to submit an agreed chronology setting out the undisputed facts relevant to the dispute. Experience unfortunately suggests that the compilation of an agreed chronology can be a complex and time consuming affair. Whilst most of the genuinely key facts are likely to be undisputed, it is the inclusion of other 'facts' which tend to cause the problems. In the nature of things, there will be disputes as to whether something did or did not happen as described, the date on which it did or did not happen, and the significance of the event. Accordingly, it is not unusual for there to be only the very barest of an agreed

---

[1] ICC, *'Final Report on Construction Industry Arbitrations'*, ICC International Court of Arbitration Bulletin 12/No. 2 (2001).
[2] Ibid., para. 21.

chronology, with each party supplementing this with its own, more detailed (and probably more partial) spin on what took place. Insofar as disputed facts are concerned, the ICC Construction Arbitration Report encourages the arbitral tribunal to:

> '…compile a composite chronology from the material provided and send it to the parties, asking them to clarify any discrepancies. The tribunal should thereafter maintain the chronology, amending it as the case develops, circulating any revisions, and asking the parties to complete any gaps in it'.[3]

There is a clear strategic advantage to having one's own chronology, where possible, form the primary basis of the tribunal's chronology. For this reason, it makes good sense to prepare a chronology which is accurate, reliable, relevant and user-friendly. In many cases, time will be far better spent in the preparation of this document than in attempting to argue the case through the process of agreeing a chronology.

The ICC Construction Arbitration Report also suggests that arbitral tribunals:

> '…request information to enable it to create organization charts, layouts and glossaries…'.[4]

Such aids are also of great benefit to the parties insofar as they serve to present complicated facts or difficult legal arguments and may be worth preparing in advance of preparing pleadings or submissions.

III. DOCUMENT MANAGEMENT

A. DOCUMENT REVIEW AND DISCLOSURE

During the course of an arbitration a party will review its own documents for the purposes of preparing its pleadings, providing the back-up documentation for witness statements or the giving of evidence in chief and in reply, briefing expert witnesses and preparing for disclosure to another party (where such disclosure is agreed between the parties or ordered by the tribunal).

In each case, the documents required to be reviewed may be voluminous. As the number of documents required to be reviewed grows, the importance of using an appropriate system for the management and analysis of those documents increases. If the system for managing documents is inadequate, then certain documents may go un-reviewed and/or certain analysis may go un-captured. Such inefficiencies inflate the cost of proceedings to the extent that additional time is required to be expended reviewing the same material on multiple occasions.

---

3   Ibid.
4   Ibid., para. 22.

1. The Scope of the Disclosure

The scope of the disclosure is determined by the parties' agreement or by the tribunal's direction. Generally, however, the scope of the disclosure is defined according to the 'issues' required to be resolved for the purposes of determining the dispute. In the context of a construction arbitration, the ICC Construction Arbitration Report notes:

> 'The tribunal should make it clear at the outset that the documents should be directly relevant to the issues as defined by the tribunal and should be confined to those which a party considers necessary to prove its case (or to dispose of the case of the other party) or which help to make the principle documents comprehensible'.[5]

This is welcome guidance on the process of the disclosure of documents and the preparation of materials to be used at the hearing, whether conducted in a common law jurisdiction or elsewhere. Of course, an order for disclosure for the purposes of an international arbitration is typically narrower in scope than an order for disclosure of a common law domestic court for the purposes of litigation. In this regard the ICC Construction Arbitration Report notes that:

> 'Few are still in favour of the wholesale and indiscriminate production of documents by means of the common law process of discovery…. In any event, such a process as practiced in domestic fora must be justified if applied to an international arbitration. Otherwise it has no place in ICC arbitrations'.[6]

Once the scope of the disclosure is known, the tribunal and the parties will be in a position to assess the volume of documents required to be managed for the duration of the proceedings. It is at this stage of the arbitral proceedings that the tribunal and the parties are best placed to decide the most appropriate way to manage documents.
See Chapter 11 for further discussion on disclosure.

2. Electronic Document Management

There is a growing trend in international construction arbitration for tribunals to require – or, if not require, to suggest – the use of electronic document management systems (*EDMSs*).[7] To the extent that the parties may be able to agree on whether or not such systems should be used by the terms of their agreement to arbitrate, the following advantages and disadvantages ought to be considered.

---

5   ICC Construction Arbitration Report, para. 52.
6   ICC Construction Arbitration Report, para. 52.
7   See ICC Construction Arbitration Report, para. 51, which paragraph presupposes the use of an electronic document management system.

EDMSs are generally less expensive to use[8] and more accurate than paper based methods of document management, for the following reasons:

- *Storage*: While it is unrealistic to assume that the need to bring original paper documents together in one place will be eliminated, electronic storage of documents reduces the cost involved in maintaining large document libraries in multiple locations, with a consequent saving in paper and related stationary. Additionally, to the extent that electronic documents are accompanied by meta-data (which would include, for instance, information as to where the document was created and to whom the document was sent), that information may be captured by certain EDMSs.
- *Coding*: typically documents stored on EDMSs are coded with objective and subjective data fields. 'Objective' data fields capture such information as the date, author and subject matter of the document, whereas 'subjective' fields capture information such as the relative relevance of a document, or the issue to which a document relates. Many EDMSs allow for controlled data-entry by the creation of 'pick-lists' from which the data-entry operator is required to select when coding a document thereby promoting greater consistency in data-entry. Additionally, certain EDMSs allow related documents to be 'data-linked', for ease of navigation.
- *Searching*: specific documents or ranges of documents may be retrieved quickly by using search terms. Typically, EDMSs permit searching according to both subjective and objective database fields. Certain EDMSs, however, also allow for full text searching of the stored document itself by means of optical character recognition processes.

Clearly a system which retrieves documents quickly and accurately is likely to reduce the amount of time required to be spent searching for relevant documents and, in doing so, reduce the cost of document management, generally. Additional advantages of EDMSs include:

- *Printing and reporting:* search results may be reported electronically and printed, where necessary. Large volumes of documents may be printed relatively quickly for the purposes of disclosure or for the creation of evidence or tender bundles later in the proceedings.
- *Portability and multiple users*: EDMSs may be operated by multiple users simultaneously and are usually accessible from any computer terminal installed with the relevant software connected either through a local network or, increasingly through the internet. Additionally, information from EDMSs may be distributed electronically and quickly to others (including, for instance, the

---

8   The cost of creation of the data for an EDMS where not already in a suitable state is another issue.

arbitral tribunal, clients, counsel or to other parties for the purposes of disclosure).

There are, of course, disadvantages associated with EDMSs, including the cost of purchasing the necessary hardware and software and the cost of training data-entry operators and users. To the extent that the equipment and users are likely to be the same, these costs may be spread across several disputes. However, the costs involved in setting up the EDMS (including the cost of scanning, loading and coding documents) will be specific to the arbitration at hand and these are likely to be considerable.

3. Market Trends

EDMSs are used increasingly by law firms in the United Kingdom to manage documents relevant to disputes. Most large law firms have bought, or have developed for themselves, large-scale EDMS software. Where such capabilities are not available in-house, it is possible to outsource document management requirements to litigation support agencies in possession of an EDMS, as and when the need arises.

In some cases, organizations are so frequently involved in disputes that it is cost effective for them to manage all of their documents by means of an EDMS, irrespective of whether a dispute is anticipated. Consequently, documents are stored and coded automatically upon their conception on the presumption that they will be required for the purposes of preparing or defending a claim in future. Even where not, the increasing trend towards the paperless office and electronic project communication should make EDMS usage very much more of the norm than it has been up to now. Situations where the backlog of documents to be loaded onto an EDMS is too great to allow its effective implementation should in the future become few and far between. This is supported by the results of a study by the University of California[9] which found that 92 per cent of new information in 2002 was stored electronically. Given the rate of take up of electronic storage and interchange of data since that time, it is likely that EDMS use will grow in the future.

4. EDMS Software and Web-based Packages

EDMS software and/or web-based packages are available from a variety of sources. Moreover, each particular software and web-based package has its own particular strengths and weaknesses. Insofar as product trends are concerned, there is an increasing tendency to use web or intranet based products.

---

9 University of California at Berkeley's School of Information Management and Systems, *How Much Information 2003?*, available at: http:// www. sims. berkeley. edu/ research/ projects/ how- much- info- 2003/ (accessed 29 July 2005)

B.   MANAGING DOCUMENTS DURING THE HEARING

In addition to requiring a method for managing the internal review and external disclosure of documents, a party may also wish to consider a method for managing documents during the hearing. Included in the documents required to be managed at this stage will be the contemporaneous documentary evidence, witness statements, expert reports, pleadings and submissions.

The ICC Construction Arbitration Report again suggests that these documents be managed electronically:

> 'The tribunal will need to ascertain at the outset whether it is practicable to work from printouts or whether it would be better if the material were accessed directly by the tribunal, in which case it will be necessary for the tribunal and every other party to be provided with the necessary software'.[10]

Clearly such a system is potentially expensive, requiring the presence of computers, software and the relevant support staff during the hearing. However, the ability for all involved in the arbitration to access the same set of documents has a very considerable benefit which is difficult to quantify. Providing all the users are comfortable with the operation of such a system (and this may still be a generational issue for certain users, including witnesses), common access to a unified database eliminates the time consuming and frustrating process of ensuring that all parties are (literally) on the same page.

To the extent that it is decided that the tribunal will work from printouts, there is nothing in any arbitration rules commonly used today that requires documents to be managed in a particular way. Consequently, the manner in which documents are to be managed is likely to be dictated by order of the tribunal.[11] The ICC Construction Arbitration Report suggests that documents be collated into agreed 'working' files (containing, for instance, files containing complete sets of site minutes, programmes or instructions)[12] and agreed 'issues' files (containing documents relevant to a particular issue).[13] In some cases the tribunal may request exhibits to witness statements to be presented in the same file, in chronological order. Much depends on the preference and collective prejudices of the tribunal, though here as elsewhere there is much to be gained by taking the initiative and persuading the tribunal (if necessary against its first inclinations) to adopt the way of working which will most suit the presentation of your own case and the ease and comfort of your advisers, witnesses and experts.

---

10   ICC Construction Arbitration Report, paras 51 and 52.
11   Paulsson, Jan, William W. Park and W. Laurence Craig, *International Chamber of Commerce Arbitration*, 3rd edition, (Oceana Publications, 2000), para. 24.02.
12   ICC Construction Arbitration Report, para. 53.
13   Ibid., para. 52.

In agreeing working or issues files, however, the parties are not agreeing the admissibility of those documents or their evidentiary value, rather it is only the authenticity of the documents which is in question. Documents whose authenticity is disputed (a rare event in international commercial arbitrations) may be kept separately until such dispute is resolved. Certainly, agreeing working and issues bundles serves to eliminate any unnecessary duplication in the documents and is a practice used widely by domestic courts.

Furthermore, where a paper based document management system is employed, the ICC stresses the importance of uniquely identifying each of the documents by an agreed number system[14] and highlighting the relevant sections of those documents for ease of reference.[15] Another simple but effective suggestion is to 'colour-code' files according to their content; for instance pleadings, exhibit bundles, working and issues files might each be assigned their own colours. The importance of ensuring that where multiple sets of documents are used (i.e. not a shared database on an electronic management and display system) that all documents are identical and there is a readily understandable system of unique identification for every page of every document, cannot be over-estimated. Presentation may not be everything but if the arbitrator cannot readily be directed to the material that is being considered, the impact of the presentation can be materially damaged.

IV. SCOTT SCHEDULES

During the course of a construction arbitration it would be typical for a large number of pleadings and amended pleadings, particulars and so on to be filed by the parties. The statement of case, for instance, will set out numerous claims; generally each claim for defective work, for instance, will be separately pleaded.[16] Given the large number of claims, it can be difficult for the tribunal and parties alike to quickly identify which of the claims are particularized, defended, the subject of a reply or the subject of factual or expert evidence.

For this reason it has become the practice for schedules to be prepared setting out the state of the pleadings on each individual issue. Known in the English courts as 'Scott Schedules' these summarize the pleadings (and sometimes the evidence) into a single document, so that the parties and tribunal may very quickly identify the status of a given claim. They might be described as 'ancillary' to the main pleadings.

As with chronologies, there are no hard and fast rules for the format of Scott Schedules. Usually their form is dictated by the tribunal's order. However, they are usually similar in the form to the spreadsheet set out at figure 2 below. Typically,

---

14   ICC Construction Arbitration Report, para. 53.
15   Ibid., para. 51.
16   See further below for discussion concerning delay claims.

claims of a similar nature (such as claims for delay, variations and/or defective work) are dealt with in the same schedule.

Figure 2

Delay Claim	Particulars	Defence	Particulars	Response	Particulars	Quantum	Particulars
20–22 June 2006 Breach of contract (clause 23) – failure to give access.	Truck blocking access to site from 06.00 20 June 2006 until 17.00 20 June 2006. [Witness statement of J Builder dated 2 December 2006 Para 54]	Denied: claimant's workforce was on strike.	Industrial action commenced 15.00 19 June 2006 and ended 17.00 24 June 2006. [Witness statement of J Owner dated 2 December 2006 Para 34]	Not all of the claimant's workforce was on strike during the period 19 – 24 June 2006.	J Thoms, P Harmer, and H Harris were on site and ready for duties. [Witness statements of J Thoms, P Harmer, and H Harris dated 2 December 2006 Para 14]	£50,000.	Wasted labour costs £30,000. Wasted site overheads £2,000. Wasted concrete mix £18,000. [Report of quantum prepared by E Expert dated 4 December 2006 at Para 32].

Typically, the Scott Schedule is completed by both the claimant(s) and respondent(s). First, the claimant sets out the basis of its claim, including particulars. The claimant also sets out the quantum claimed, including particulars. Then the respondent sets out its defence, including the basis on which the claim is denied or not admitted, including particulars. Next the claimant sets out its response and the basis on which it responds, including any particulars.[17] Clearly, this is best achieved by passing the schedule from party to party in electronic form. Often a column on the right is left blank for the use of the tribunal.

Parties should approach the task of completing the Scott Schedule constructively; the simple denial of a claim unsupported by reasons, for instance, does not assist the tribunal to distil the respondent's argument.[18] The use of Scott Schedules is endorsed by the ICC:

> 'In the right hands it is a useful tool. It defines the positions of the parties and ultimately it will or can be used by a tribunal to record its views and decisions. At the pre-hearing stage its main value is that, if properly compiled, it establishes the position of each party where the existing submissions or pleadings do not already do so adequately'.[19]

---

17   This approach is endorsed in the ICC Construction Arbitration Report (paras 36 and 37).
18   ICC Construction Arbitration Report, para. 36.
19   Ibid.

The ICC Construction Arbitration Report suggests that Scott Schedules be prepared before the parties' evidence is filed.[20] At this stage in the proceedings, a Scott Schedule assists in determining the facts required to be proven in support of the parties' claims. Once the evidence has been filed, however, it may be worthwhile requiring each party to cross-reference the claims set out in the Scott Schedule against their lay and expert evidence. In this way, claims which are unsupported by the evidence may be quickly and easily identified. Additionally, as is pointed out in the ICC Construction Arbitration Report:

> 'If fully and properly completed, these schedules show which points are not in dispute and thus irrelevant and which have to be decided'.[21]

Scott Schedules are of less use where the claimant seeks to bring a global claim. The problems raised with global claims are discussed in more detail below.

V. EVIDENCE REQUIRED FOR COMMON CONSTRUCTION CLAIMS

A. INTRODUCTION

Very broadly, claims in construction cases can be divided into two main classes. Global claims – i.e. claims where the claimant does not attempt to identify any direct correlation between individual events and their effects – and claims arising out of specific events. Each of these broad classes can be further divided into claims for additional time and for additional cost. This part of this chapter looks at the preparation of evidence for claims for additional costs, including those arising from delay. Before discussing the types of costs recoverable in construction claims, it is worthwhile touching upon the sorts of evidence conventionally required to prove that such costs were or will be incurred.

Depending on the nature of the claim and the facts of the case, a claimant may claim its actual, reasonable or estimated costs. For instance, claims made on a *quantum meruit* basis are for the contractor's reasonable costs, whereas claims for defective work are for the employer's estimated costs (in circumstances where the relevant rectification works remain to be carried out). Of course, if the works have been carried out, the claim then becomes one for the employer's actual costs.

In such a case, where the claimant seeks its actual costs then, as a general rule:

> '...a claimant ought to be required to produce the *primary documents* that confirm the amounts claimed,'[22] (emphasis added).

---

20 Ibid., para. 39.
21 Ibid., para. 38.
22 ICC Construction Arbitration Report, para. 45.

The ICC's reference to 'primary documents' in this regard is a reference to the *receipts* or *invoices* which evidence the claimant's costs, which will include its own direct costs and the costs of subcontractors, plant hire and the like. Collating the receipts and invoices underpinning a claim can be a laborious and tedious process, especially where the claimant did not have a system for managing those documents in the first place. So laborious and tedious is this exercise that in many cases the arbitral tribunal (being human) will wish to avoid the need to examine such documents themselves. Thus in many large cases directions will be given for some form of accounting expert to report on the accuracy of the sums claimed. Indeed, while one party's primary documents in a large case may (in theory at least) be made available to the other party, it is not unknown for the validity of this evidence to be demonstrated by examination of a statistically significant sample, rather than of all the documents. The importance of document management in this area has been discussed in greater detail above and is nowhere more important than in this area. For a while, in practice, short cuts of the type described above may be and often are adopted, the theory remains that losses of this type should be capable of being proved by the production of primary evidence. Woe betide any party who cannot, when challenged, satisfy the tribunal of its obligations in this regard.

Where, however, the claim is not for actual but for estimated or reasonable costs, the evidence cannot be the invoices for the work done. Instead, the evidence of loss is almost always a mixture of documentary and expert opinion. The documentary evidence tends to include quotes, estimates or tenders for the work to be done or costs and rates for the materials and trades required for the work in question. In every case expert evidence is required to demonstrate the 'likelihood' or 'reasonableness' of a cost claimed. The role of experts and expert evidence is discussed in greater detail below and in Chapters 12 and 13.

B.   Contractors' Claims for Delay or Disruption

Some of the most common claims found in construction arbitrations are claims for delay, both by the contractor for delays caused by its employer and by the employer (typically for delay liquidated damages) for late delivery of the project. Of course, once delay is established (a topic in its own right) the determination of the employer's claim requires no particular evidence – at least where the contract provides for liquidated damages. The contractor's claims, on the other hand, require considerably more effort and evidence to establish. The way these claims are presented often includes the matters set out below.

1. Claims for Overheads

a.   Site Overheads and Preliminaries

Where the contractor suffers delay for which the employer is responsible, then the contractor may claim the actual costs incurred while maintaining an idle presence on site during the period of delay. Where contracts provide a contractual right to these costs, it is not uncommon for the contractor to be required to take reasonable steps to avoid these costs. The same is effectively true if the claim for delay is not based on a contractual right for additional payment but for damages for breach, when the contractor generally becomes under an obligation to mitigate its losses. Subject to these obligations, the contractor's claim for delay will include:

- the cost of paying salaried or contracted site-staff during the period of delay;
- the cost of leasing site equipment or plant during the delay;
- the cost of maintaining owned site equipment or plant during the period of delay (perhaps including 'lost hire fees' if the contractor can show the equipment would have been hired out but for the delay keeping it on site, depreciation and finance charges);
- the cost of maintaining water and electricity to the site during the period of delay; and
- certain miscellaneous costs including the cost of postage, stationary and/or telephone calls associated with the delay event.

In each case, the contractor will need to show the costs incurred by reference to the relevant invoices or receipts for the costs claimed (see discussion above) including for instance, invoices evidencing hire-charges, stationary or labour costs.

In some cases, however, rather than proving each cost on an item-by-item basis the contractor may choose to claim these costs by extrapolating from the rates for preliminaries set out in its bill of quantities. This process, which is sometimes referred to as an 'extended preliminaries calculation', can be regarded as a mini global claim and is frequently viewed with some scepticism.[23]

b.   Head Office Overheads

In addition to costs at the site level, there may be costs incurred at the head office level which are recoverable to the extent that they could not have been avoided by the contractor. Again, such costs include the cost of salaried or contracted head office staff, the cost of idle office equipment and other miscellaneous costs such as postage, stationary and telephone calls incurred or wasted in connection with the delay. Depending on the terms of the contract, the contractor may also be able to claim certain financing costs incurred by him as a consequence of the delay

---

23   *Wraight Ltd v PH & T Holdings* [1980] 13 BLR 26.

(where, for instance, he must pay to maintain a bank guarantee or funding during the period of delay). Where the contractor chooses to add a claim for financing costs it needs to demonstrate not only the additional cost but also the causal connection to the delay.

In fact, demonstrating that the costs incurred are directly referable to the delay is the principal difficulty in claiming head-office overheads. Project specific, time-related costs (such as fees for a bank guarantee) present no particular problems, but general financing charges, such as the contractor's overdraft, are less easy to demonstrate as losses (or additional costs) caused by the delay to any particular project. Similarly it may be difficult or impossible to demonstrate what proportion of an employee's time was wasted as a consequence of the delay where the relevant individual worked on a number of different projects at the relevant time.

2. Net Lost Profits

Where the contract was profitable, the contractor may seek to recover profit lost as a consequence of the delay. Typically, such claims are estimated according to formulae which seek to determine the contractor's lost net profit; that is, gross profit minus site and head-office overheads.

The most common of these calculations are the Hudson, Emden and Eichleay formulae. The Hudson formula,[24] for instance, calculates net lost profit as follows:

$$\frac{\text{HO/Profit Percentage}}{100} \times \frac{\text{Contract Sum}}{\text{Contract Period}} \times \text{Period of Delay (in weeks)}$$

As is noted in Building Contract Disputes: Practice and Precedents:

> 'The effect of the formula is to calculate the rate at which the contractor would have earned gross profit had the contract run to time and to extrapolate that rate into the period of delay'.[25]

As with the extended preliminaries calculation, however, there has been some doubt expressed as to the correctness of relying on the Hudson formula where better evidence of the claimant's loss is available.[26]

In certain circumstances the contractor may wish to claim the net lost profit that would have been made on *other* contracts, had he not been delayed. Generally, the contractor will make this claim where the delayed contract was less profitable than other contracts being awarded at that time. Such a claim requires evidence that

---

24 I. N. Duncan Wallace, *Hudson's Building and Engineering Contracts*, 11th edition, 2 vols (Sweet & Maxwell, 1995), Vol. I, para. 8–182, as originally put forward in the 10th edition of that text, in 1970.

25 Robert Fenwick Elliott, *Building Contract Disputes: Practice and Precedents*, 13th Release (Sweet & Maxwell, 2004), para. 1–422.

26 *Tate & Lyle Food and Distribution Ltd v Greater London Council* [1982] 1 WLR 149.

other contracts would have been awarded to the contractor had he not been delayed and proof of the profit margins that would have been recovered on those other contracts had he carried them out. Obviously, evidence is required to show that other work was available, that it would have been profitable and that the contractor would have had won it. For this reason, evidence of tenders for other work with estimates of profit margins based on the tender price are unlikely to be sufficient to demonstrate a good claim for lost profits.

3. **Increased Costs in Executing Remaining Works Following Delay**

A contractor may claim for increased costs suffered as a consequence of delay including, for example, the increased costs involved in completing the works in a season or in a market other than that contemplated at the time of tendering for the work. Items recoverable might include the increased cost of materials (including handling costs), increased labour and plant costs. As with other claims, the evidence required is a mixture of documentary and expert opinion. Where the claim is for working in a season or market different from that anticipated at the time of the contract, the costs of both the work done in the original circumstances and those prevailing as a result of the delay will need to be proven, and the tribunal must be satisfied that any additional cost arises as a result of the change in circumstances and not for any other matter which the contractor may be responsible for.

C. DISRUPTION AND ACCELERATION

Where an employer is responsible for disruption to the progress of the works (i.e. where he has disturbed progress to items off the critical path so that the contractor is not delayed in the completion of the work but has suffered additional costs in completing the works), the contractor may claim the cost of wasted or increased overheads incurred as a consequence of the disruption. Likewise, where the contractor is required to accelerate (perhaps on the basis of a 'constructive' change order), he may claim the cost of increased overheads incurred by him as a consequence of the instruction to accelerate. As with other claims, the principal problem with providing evidence in support of such claims is not so much in identifying the actual cost incurred but in satisfying the tribunal that any additional cost claimed arises as a result of the event relied upon. In other words, the challenge is to satisfy the tribunal that, but for the disruptive event, the cost to the contractor would have been less than it actually turned out to be and that that difference arose as a result of the disruption.

D. CONTRACTORS' CLAIMS FOR REPUDIATION

There are occasionally extreme cases where a building contract is terminated in its entirety. Where the contract has been terminated in accordance with its terms, it is likely that the contract itself will provide a mechanism for determining what sums are to be paid to whom. Where, rarely, the contract is brought to an end by the employer for no good reason and outside the contract mechanisms, the contractor will have the opportunity of treating the contract as repudiated, that is as having been brought to an end by the employer's actions. In these circumstances, the contractor is generally under no further obligation to perform but may claim whatever loss it can establish as damages on general principles.

1. Repudiation Before the Date for Possession of the Site

Where a contract is improperly repudiated by the employer before work commences on site, the contractor may claim the net lost profit he would have otherwise made on the entirety of the work. Alternatively, the contractor may claim its wasted costs, including its preparation and mobilization costs so long as those costs were reasonably known to the employer.

Clearly, the evidence required in support of these two claims is completely different and a call ideally needs to be made at the time of preparing the claim determining which route the contractor will follow (though it is not impossible to claim in the alternative as a safety measure).

Where the claim is for loss of profit on the completed project, the contractor will need to show that he would have been able to complete the project as anticipated and that he would have made a profit in doing so. While this claim is a little easier to present than a claim for lost other work in cases of delay, it still requires satisfying the tribunal of the contractor's ability to complete the work in a profitable manner. Such a claim is also subject to an obligation to mitigate, in this case, perhaps, by obtaining other work. If other work is generally available in the market within which the contractor is working, and this work is at least as profitable as the cancelled work, the impact on the claim may be large. However, in this case, the boot is on the other foot and it is for the employer to demonstrate by its evidence that such other, profitable work existed and would have been available to the contractor. In both cases, the evidence will involve a significant amount of expert opinion in connection with the likely profit margins of the contractor and the availability of other work.

Where, perhaps because of the contractor's inability to show that he would have completed the cancelled project profitably, or perhaps because of an upturn in the market leading to other, profitable work being freely available, the claim is based on wasted costs, the evidence changes completely. In this case the claim becomes, in essence, a simple accounting exercise based on the primary documentation referred to above.

2. Repudiation Subsequent to the Date for Possession of the Site

Where work has been carried out by the contractor prior to the employer's improper repudiation of the contract then the contractor may claim for such work under the terms of the contract (and for which he has not been paid) as well as demobilization costs and the lost net profit on the remaining portion of the work.

In this case the evidence is an amalgam of straightforward documentary evidence to demonstrate the amount properly earned under the contract to date and, as described above, expert opinion on the amount that the contractor would have earned on the remaining part of the contract, had he been allowed to complete it.

E. GLOBAL CLAIMS

All of the examples outlined above are based on the premise that the claim is being presented in a traditional manner (in other words, that there is one event which, whether under the contract or as a result of a breach of contract, entitles the contractor to a claim for additional payment). Where a party makes such a claim he must demonstrate that the delay or extra costs complained of were *caused* by the event for which the other is responsible (an 'event'). Of course, in real life it is common for a number of events to occur in the same time period. Despite this, where there are multiple events the contractor (it is typically the contractor) must show how each of those events contributed to the delay or the increased costs complained of.

Occasionally, the contractor will be unable to prove the effect of each individual event on progress and costs because, given their complexity, it is impracticable or (rarely) impossible to disentangle the individual effects.[27] This may arise because the contractor has been on the receiving end of so much misfortune that it has become difficult to separate out cause and effect, or it may be that as a result of the failure to maintain adequate site and other records that at the time of the making of the claim this exercise has become too difficult (or too expensive) to undertake. Whichever is the correct reason, in such circumstances, contractors are generally inclined to make a claim for time and costs on the basis of those events collectively, requiring the arbitrator to infer that the employer's breaches caused the entirety of the delay and extra costs incurred. Such claims are referred to as 'global' or 'rolled-up' claims.

From the start it needs to be said that global claims are approached with a considerable degree of scepticism, because they effectively ignore the issue of causation in any proper sense.[28] Additionally, pleadings and evidence in support of

---

27 *Wharf Properties Ltd v Eric Cumine Associates (no 2)* [1991] 53 BLR 1.
28 *British Airways Pension Trustees Ltd v Sir Robert McAlpine & Sons* [1995] 72 BLR 26 and *Holland (John) Construction & Engineering Pty Ltd v Kvaerner RJ Brown Pty Ltd* [1996] 82 BLR 81.

global claims tend to lack precision and contain irrelevant or redundant material. This creates further work for the arbitral tribunal[29] and allows the contractor effectively to sidestep much of the evidential burden of proving its case. In many cases, the presentation of a global claim is seen as an attempt by the claimant to sweep its own failings under the carpet whilst laying full responsibility for those failings on the respondent.

Implicit in the concept of global claims (though not often highlighted in the claim itself) is the fundamental notion that the contractor is not, itself, responsible in any way for the delay or extra costs complained of. A typical global claim comprises a litany of matters, all of which are events falling within the employer's responsibility, followed by a simple assertion (normally easily demonstrated) that the actual costs exceeded the contractual price by the amount of the claim. Consequently, where a global claim is presented the employer will normally attempt to demonstrate that some or all of the delay or costs complained of were caused by events for which the contractor is responsible including, for instance, poor project management, technical failures or a failure properly to scope or price the work before the contract was entered into. Clearly, any such evidence tears the heart out of a global claim and, where the contractor has not presented any alternative evidence to support its claim, ought to result in the dismissal of the entire claim.

However, this is a risky strategy for an employer if the tribunal is minded to accept the principle of a global claim. This might arise as a result of genuine difficulties in disentangling the effects from numerous potential causes, the vast majority of which are probably ones within the employer's responsibility. In such cases, much depends on the identity of the member(s) making up the tribunal, their professional backgrounds, their collective prejudices and their attitudes to strict rules of evidence and of pleading. For these reasons, an employer faced with a global claim might be well advised not simply to seek to show that certain of the events set out in the global claim are matters within the contractor's responsibility, thereby hoping to pull the entire global claim down. The alternative, though requiring much more evidence and effort from the employer, is to undertake the task that (arguably) the contractor ought to have done, carrying out a blow by blow analysis of the events relied on and their probable effects.

There are, however, problems with this second approach. To start with, it may well be the case that the employer has insufficient records to carry out the exercise, though to some extent the availability of disclosure from the contractor ought to provide some of the material required. Of more significance is that, whilst the approach may guarantee the rejection of the global claim, it may well be at the price of establishing for the contractor (but at the employer's expense) what the contractor's entitlement actually is. For these reasons, an employer faced with a global claim needs to consider very carefully its best defence strategy in the light of all the circumstances, including the likely attitudes of the tribunal to the claim.

---

29 *Rugby Landscapes Ltd (Bernhard's) v Stockley Park Consortium Ltd* [1997] 82 BLR 39.

1. Preparing a Global Claim

Despite the criticisms of the global claim implicit in the preceding paragraphs, there are occasional instances where a global claim is legitimate. Indeed, it may be the only way to present a case where it is genuinely impossible to disentangle the effects of multiple causes. However, because of the potential vulnerability of the global claim in the face of a robust tribunal accustomed to strict rules of evidence and proof of causation, it should be regarded as the last resort and not the norm. Accordingly, and to the extent that it is possible and practical to do so, a contractor should start by establishing a causal link between each event complained of and the corresponding delay or disruption and costs.[30] Where, following this exercise, there remain delays or costs which cannot be attributed to specific events then the contractor might wish to make a global claim in respect of them.[31] In doing so it should take great care to remove from this residual global claim any effects which can positively be identified as flowing from matters within its responsibility. If a contractor follows this course it may still not recover its global claim in the face of a determined attack by the employer, but its chances of success on the global element of the claim will have improved and it will, of course, always be able to recover those elements of the claim for which evidence was produced showing cause and effect.

F. EMPLOYERS' DELAY CLAIMS

1. Loss of Profits

Just as contractor's claims are typically for additional costs (whether as a result of change or delay) in completing the project, employer's claims are typically for the cost to them of delayed delivery of the project. Where the works are delayed through the fault of the contractor then the employer may be unable to draw revenue from, or use, those works as from the time originally intended (that is, the date for practical completion as adjusted under the contract). In such circumstances, the employer may (in principle) claim the loss which arises fairly and reasonably as a natural consequence of the delay. This might include loss of rent or loss of production depending on the nature of the works. To claim these losses, the employer must prove that the relevant circumstances were such that he would have made a profit had he been able to use the works. In addition to these losses, which are potentially recoverable in all circumstances, the employer may also claim for special losses (that is, losses other than those which arise fairly and reasonably as a natural consequence of the delay) where he can show that the contractor knew of the special

---

30 *London Borough of Merton v Stanley Hugh Leach Ltd* [1985] 32 BLR 51.
31 *British Airways Pension Trustees Ltd v Sir Robert McAlpine & Sons supra.*

purpose to which the project would be put and the likely loss to the employer as a result of delay.

As has been discussed above, the evidential burden of proving loss of profits is considerable, though perhaps easier for employers in certain circumstances. Where, for instance, the property is to be let for commercial occupation, the availability of tenants and the likely rents can be readily established by a mixture of factual (comparator) and expert evidence. So too can the loss of profit for a process plant, based on evidence of the market prices for the feedstock and the product and expert evidence of the cost of processing.

Despite this, there are circumstances where evidence of actual loss would be difficult to provide, such as the cost to an organization of not being able to move into a new headquarters building. Perhaps for reasons such as these, it is commonplace for the cost of employer delay in the commercial sector to be pre-agreed as liquidated damages. In such cases, no evidence of loss ought to be required to prove the claim. However, in certain jurisdictions (such as England) liquidated damages may be recoverable only if shown by the evidence to be a genuine pre-estimate of costs. In other jurisdictions even more evidence of actual loss may be required.

2. Wasted Expenditure

As an alternative to a claim for loss of profits (and in the absence of agreed liquidated damages) an employer may claim for wasted or additional expenditure. A claim for wasted expenditure might arise out of cancelled contracts relating to the occupation of the building, whilst additional expenditure might arise in cases where additional short term accommodation has to be organized at increased cost to take account of the unavailability of the completed building. Similar claims might arise in cases where feedstock had been contracted for a process plant and had to be sold in the market at a loss, or additional product bought on the open market at higher cost where the plant was intended for domestic supply. Obviously, the evidence in support of wasted or additional expenditure costs will vary greatly, depending on the circumstances on each case.

3. Repudiation by the Contractor

Where the contractor improperly repudiates the contract and its repudiation is accepted, then the employer may recover the difference between the original contract price and the actual or estimated cost of carrying out the relevant works (depending on whether or not those works have been carried out). In each case, the employer must show that the works were, or will be, completed substantially as originally intended, in a reasonable manner and at the earliest reasonable opportunity.[32] Generally the nature of the evidence required will depend on whether the works

---

32   *Mertens v Home Freeholds Co [1921] 2 KB 526.*

have been carried out at the time of the claim or are yet to be carried out. Where the works have been carried out in accordance with the intention of the original contract, the evidence will in essence be purely factual, subject possibly to the need for some expert evidence as to the reasonableness of the manner in which the works had been completed. Where, however, the works have not been completed the question becomes one for expert evidence as to the likely future cost of the works.

4. Defective Works

Where the works are defective through the fault of the contractor, then the employer may claim either the diminution in property value or the actual or estimated cost of reinstating the works.[33]

Where the employer claims for reinstatement, he must provide evidence that the work will be (or has been) completed substantially as originally intended, in a reasonable manner and at the earliest reasonable opportunity. Where, however, the employer cannot or does not intend to reinstate the works, or where the tribunal considers that it would not be reasonable to require the contractor to reinstate the works, the employer may claim for diminution in property value. Generally, a claim for diminution in value will be no greater than a corresponding claim for reinstatement. To make this claim the employer must show that the value of the work as built is less than the likely value of the work had it been built as originally intended.

Additionally, where the employer has opted to reinstate the works, he may also claim for any residual diminution in value of the work following reinstatement. Again, to make this claim the employer must show that the value of the work as built is less than the likely value of the work had it been built as originally intended.[34]

The employer may also be able to claim for certain miscellaneous expenses incurred as a consequence of the defective works, including any increase in insurance premiums and/or the costs of inspections or reports carried out in identifying, assessing or managing the defective works.

In certain instances, the contractor may claim that the reinstatement of the works will cause the employer to get something better than he bargained for. This argument may be resisted on grounds that the employer had 'no reasonable choice' but to carry out the reinstatement works in the way that he did.

---

33 *Ruxley Electronics and Construction Ltd v Forsyth* [1996] AC 344.
34 *George Fischer Holding Ltd (formerly George Fischer (Great Britain) Ltd) v Multi Design Consultants Ltd (Roofdec Ltd and others, third parties)* [1998] 61 Con LR 85, 145.

VI. EXPERT EVIDENCE

A. INTRODUCTION

While it can be seen that much of a party's preparation for proceedings centres around the collection of factual evidence, a party to construction disputes will (as indicated above) almost invariably need to adduce expert evidence in support of its claim or defence. For instance, where a party's claim is that the defendant has not acted in accordance with accepted industry practice, that party will need to establish exactly what 'accepted industry practice' is. Generally, this is done by way of expert opinion based on the knowledge and experience of a person suitably acquainted with the relevant industry.

Expert evidence is also used to make sense of complicated facts. Where, for instance, evidence of a large number of delays is adduced, expert evidence is commonly – if not invariably – used to demonstrate whether and to what extent the events complained of caused delay to the claimant (by reference to the effect of those delays on the critical path, for instance).[35]

Although, in the context of international construction arbitration, one might expect members of the arbitral panel itself to possess a certain degree of expertise, it is still very common for several experts with knowledge and experience in quite separate fields to be required for the proper conduct of a complex construction claim. As the authors of the ICC Construction Arbitration Report found:

> 'construction disputes often raise a variety of technical issues, some of which may be highly specialized and lie beyond the competence of an ordinary expert and others may necessitate a decision between two different schools of thought, towards one of which a tribunal member may have a leaning, as a result of training or experience'.[36]

For example, programming experts may be required to give evidence in relation to delay claims, accounting (or quantity surveyor) experts may be required to give evidence on quantum, and expert-engineers may be required to give evidence in relation to defects. The result is that expert evidence in international construction arbitration is very much the norm.

This section addresses the role of experts in arbitrations, the factors to be considered when selecting and briefing an expert for an arbitration and the nature and format of expert evidence.

---

35 The critical path and programming are discussed in further detail in Chapter 10.
36 ICC Construction Arbitration Report, para. 58.

B. THE ROLE OF EXPERTS IN ARBITRATION

1. Appointing Experts

Where an arbitral tribunal does not possess the expertise necessary to determine the claim before it then either it, or the parties, may appoint an expert(s), depending on the terms of the parties' agreement to arbitrate or the terms of reference. Generally, the terms of reference or the agreement to arbitrate will expressly state whether it is the parties and/or the tribunal who may appoint an expert. This is certainly the case where the parties have agreed to arbitrate in accordance with one or other of the accepted sets of institutional rules.[37]

Where the matter is not dealt with expressly then the power to determine how, and in what manner, expert evidence is to be adduced will generally lie with the tribunal as an implied term in the agreement to arbitrate. It may, however, be necessary to refer to the law of the place where the arbitration is taking place to see what powers the arbitrators have to appoint experts or to admit expert evidence.

a.   Party-Appointed Experts

Parties from common law jurisdictions, who are accustomed to adversarial-style proceedings, typically prefer to appoint their own experts to give evidence on the technical aspects of the dispute. Such parties value their ability to consult with the expert as he prepares his evidence to ensure that all relevant facts and matters have been properly considered.

Where not already provided for in terms of reference, the number of experts each party may retain is usually agreed during a pre-hearing conference or direction hearing. The scope of each expert witness's remit and the manner in which his evidence must be presented is also commonly agreed by the parties or directed by the arbitral tribunal at this time.

b.   Tribunal-Appointed Experts

Occasionally, the terms of reference will provide (or the tribunal will direct) that it is the tribunal and not the parties who may appoint an expert. These tribunal-appointed experts are also commonly referred to as 'neutral assessors'. While certain parties (particularly those from civil law jurisdictions, who are accustomed to inquisitorial-style proceedings) may be comfortable with this arrangement, most parties from common law jurisdictions are not. Many parties are wary of the inclusion of a further, unendorsed member in their carefully selected arbitral panel whose opinion and views expressed in private conversations with the

---

[37] Art. 27 UNCITRAL Arbitration Rules; Art. 20.4 ICC Rules; Art. 12 LCIA Rules; Art. 22 AAA Rules; and Arts 5 and 6 IBA Rules.

tribunal members, although not binding, are likely to be highly influential on the decision-making process.

Again, whether or not the tribunal may impose a tribunal-appointed expert on the parties depends on the terms of reference and/or the agreement to arbitrate. Tribunal-appointed experts are certainly contemplated in most sets of institutional rules.[38]

Where an arbitral tribunal decides to appoint its own expert, the consensus is that it should involve the parties in the selection process (by requesting that the parties agree on an expert from a list, for instance). Indeed, consultation is a necessary prerequisite to the tribunal appointing an expert under the ICC Rules.[39] The parties should also have the opportunity to see any report provided by the expert to the tribunal, and to question the tribunal's expert at the hearing.

Whether a tribunal can prevent parties from appointing their own experts also depends on the terms of reference. Under the ICC Rules for instance, the arbitral tribunal is under no express obligation to hear party-appointed experts. However, the consensus is that:

'Even if they intend to appoint a neutral expert, ICC arbitrators are not well advised to reject the presentation of expert testimony by the parties'.[40]

In this regard it should be noted that an arbitral tribunal may rely on expert assistance only '…to obtain any technical information that might guide it in the search for truth'[41] and must not delegate any of its decision making powers to the expert. To this end, the tribunal-appointed expert should not participate in the tribunal's ultimate deliberations nor should his opinions be treated as binding by the tribunal members.

2. Tribunal and Party-Appointed Experts

In the case of party-appointed experts, it is for the arbitral tribunal to determine which expert opinion is to be preferred in the case of conflicting evidence. This may be difficult where the arbitral tribunal lacks the relevant expertise. In these circumstances, the arbitral tribunal may take the view that it requires its own expert adviser (in addition to those already appointed by the parties) to assist in distinguishing between the varying opinions.

Aside from any concerns they may have as to the influence such an expert advisor might have on the decision-making process, the parties may resent the cost

---

38 See, e.g., Art. 21.1 LCIA Rules. See also Art. 20(4) ICC Rules, Art. 27 UNCITRAL Arbitration Rules and Art. 22(1) AAA Rules.
39 Art. 20(4) ICC Rules.
40 Paulsson et al., *International Chamber of Commerce Arbitration*, p. 442.
41 See *Starrett Housing Corp. v The Government of the Islamic Republic of Iran*, Award No. 314-21-1, para. 264 (August 14, 1987) 16 Iran-U.S. C.T.R. 112, 196 quoting the ICJ in the *Corfu Channel Case* (1948) ICJ Reps 15.

of an additional expert. Nevertheless, in the case of highly technical claims, this arrangement may be unavoidable.

3. Selecting an Expert

a. *Finding Candidates*

Most lawyers practicing international construction arbitration keep lists of experts in the major fields of expertise (including for instance, quantum and delay, programming).

Occasionally, parties require expertise of a more obscure nature; a claim for defects in connection with steam turbines for instance, would require the expertise of a specialist mechanical engineer. In such circumstances and if the client is unable to suggest candidates, it may be necessary to request the assistance of an institution such as the ICC International Centre for Expertise, in order to locate an appropriate expert.[42]

b. *Qualifications and Expertise*

Ideally, an expert will possess relevant professional qualifications and current experience in the industry. Additionally, where the expert is likely to give oral evidence, or to be cross examined, he should be able to communicate his opinions in a credible and authoritative manner. There may, however, be a tension between the possession of up-to-date experience in the relevant field, and the ability to present fluently and persuasively in the somewhat artificial setting of an arbitration hearing. All too often it is the current day-to-day experience that is lacking, with the emphasis being placed on the forensic skills of 'professional' expert witnesses.

There is, in fact, nothing at law which requires expert witnesses to possess any professional qualifications. As Mr Justice Lloyd put it in *James Longley & Co Limited v South West Regional Health Authority* [1983] 25 BLR 56:

> '…an expert may be qualified by skill and experience, as well as by professional qualifications'.[43]

Nevertheless, such qualifications are of undoubted value insofar as they lend weight to the expert's evidence and credibility, generally.

On the other hand, recent industry experience is crucial.[44] Expert evidence may be entirely rejected if it can be shown that the supposed expert is not:

---

42 See http://www.iccwbo.org/drs/english/expertise/all_topics.asp (accessed 29 July 2005).
43 *Longley (James) & Co Ltd v South West Thames Regional Health Authority* [1983] 25 BLR 56, 62.
44 *Royal Brompton Hospital NHS Trust v Hammond (No 7)* [2001] EWCA Civ 206; 76 Con LR 148.

'... someone who had experience of the practices [in the industry] of the time relevant to the [claim]'[45]

irrespective of the expert's professional qualifications.[46] Likewise, expert evidence will be rejected where it can be shown that the expert, although possessing expertise, does not have expertise in the precise area which is the subject of dispute.[47]

The importance of ensuring that the expert possesses expertise in the appropriate area cannot be overstated. Opinions in relation to matters in which the expert has no experience are inadmissible on grounds of irrelevance (in the same way that the opinions of factual witnesses are generally inadmissible in that regard). As Auld J put it in the context of engineers, architects and surveyors:

> '... it is only the evidence of the surveyors, Mr Dyson and Mr Greenham, that may be of value in this issue. Mr Shaw (a civil engineer) and Mr Fairhurst (an architect), however confident they may be in their own professions cannot speak with authority on what is expected of an ordinary competent surveyor'.[48]

It is important to check the expert's curriculum vitae to ensure that he possesses qualifications and *recent* experience in the field relevant to the claim at hand. It is therefore important to note that experience as an expert witness does not amount to experience in the industry. Consequently, it is important to beware of professional expert witnesses who have lost touch with current industry practices.

c.   *Credibility*

In most common law courts, expert witnesses owe a duty to the court '...to offer dispassionate and disinterested assistance and advice to the court...' in the court's determination of the facts relevant to the claim. The courts take a dim view of experts who '...take upon their own shoulders the mantle of advocacy and themselves to seek to persuade the court the desired result',[49] or '...consider that it is [their] job to stand shoulder to shoulder through thick and thin with the side which is paying his bill'.[50]

While, strictly speaking, it is probably true that an expert owes no 'duty' to an arbitral tribunal, the greater the apparent independence of the expert, the more

---

45   Ibid., 166.
46   Ibid., 166.
47   *Whalley v Roberts & Roberts* [1990] 1 EGLR 164 per Auld J.
48   Ibid., 169.
49   Arab Bank plc v John D Wood (Commercial) Ltd [1998] EGCS 34.
50   *Cala Homes (South) Ltd v Alfred McAlpine Homes East Ltd* [1995] FSR 818 as per Laddie J. See also *London Underground Ltd v Kenchington Ford plc (Harris & Sutherland, third party)* [1998] All ER (D) 555 and *Munkenbeck & Marshall v The Kensington Hotel Ltd* [2000] All ER (D) 561

weight the tribunal will give to his evidence. To this extent, the case law surrounding the duty of experts in court proceedings is relevant to the assessment of their credibility in arbitral proceedings.

While an assessment of the candidate's likely credibility is largely a matter of judgment, the following ought to be considered when selecting an expert witness:

First, does the candidate have any 'connections' with the client? If the expert is any way connected with the party which retains him (say for instance, he is an employee or otherwise receives work from the client), his credibility will be called into question. It is therefore important to ensure that there is no connection, or as little connection as possible between the expert and the client.

Second, does the candidate communicate his opinions credibly? Experts are generally regarded as 'impressive' where they refrain from acting as an advocate for the party which retains them[51] and show a willingness to concede points where sensible and appropriate to do so.[52] The views of an expert that are communicated authoritatively but dispassionately are more likely to be adopted than those that are not.

Where possible (and in addition forming your own opinion on the general demeanour of the expert), where the expert has given evidence before, it is essential to review any relevant judicial or arbitral decisions for commentary on the candidate's performance as an expert witness. In particular, when appointing an expert it is important to establish that there has been no adverse criticism of the expert's performance, integrity or judgment as this will inevitably impact adversely on the weight given to his evidence.

d.   *Timing of Appointment*

In certain fields, there may be relatively few people who are sufficiently qualified to give an expert opinion. There may be fewer still, once those who are connected with the client have been eliminated from the list of candidates. In these circumstances, it may be wise to identify and retain an expert as soon as possible, to prevent any other party to the proceedings obtaining the services of the 'best' or (worst still) the 'only' expert in the field.

C.   BRIEFING AN EXPERT

1.   Briefing Party-Appointed Experts

The manner in which the expert is briefed is crucial to the case, especially where the expert has never given evidence before (in which case he will need to be carefully guided through the task of preparing a report and the process of giving oral

---

51   *Great Eastern Hotel Co Ltd v John Laing Construction Ltd* [2005] EWHC 181.
52   Ibid., para. 111.

evidence). The following are some of the essential points for a party and its advisers to consider when putting together a brief to instruct an expert to prepare a report:

a.   *Scope*

As mentioned above, often the scope of the expert evidence required to be adduced by the parties is agreed between the parties, or directed by the arbitral tribunal during a pre-hearing conference. Where this is the case, it is important that the expert be requested to prepare his report within the confines of the scope. Opinions that venture outside the mandated remit may be inadmissible and a waste of costs. For instance, in *Pozzolanic Lytag Ltd v Bryan Hobson Associates*,[53] (in proceeding brought under the English Arbitration Act) it was held that the expert evidence adduced (the experts' reports directed to issues with which they should not have been concerned) by both sides went well beyond what the official referee had authorized, everything possible should have been done to discourage this.

Where the scope of expert evidence has not been agreed or directed, then it will be for the party to define the matters to be addressed. Usually, the party will define the scope of the expert report by reference to a series of carefully formulated questions, which assist the expert in directing his attention to the relevant facts that need to be proven to make out the claim or defence. In formulating these questions take care not to request:

> '… evidence that really amounts to no more than an expression of an opinion by a particular practitioner of what he thinks he would have done had he been placed, hypothetically and without the benefit of hindsight, in the position of the defendants, [which] is of little assistance to the Court… that is the very question which it is the court's function to decide. It is often said that an expert may not be asked the very question which the court has to decide, for the reason that that would be to usurp the court's own function'.[54]

b.   *What to Give the Expert*

Depending on the stage of the proceedings, a brief should include the following documents:

- all portions of the claim and any corresponding defence;
- all documents and other evidence; and
- all portions of factual witness statements;

that are relevant to the facts which the expert is required to analyse. It is not wise to hold back documents or any witness statement (especially those from the other

---

53   *Pozzolanic Lytag Ltd v Bryan Hobson Associates* [1999] BLR 267.
54   *Longley (James) & Co Ltd v South West Thames Regional Health Authority supra.*

party) which might be prejudicial to the client's claim.[55] The other party will be likely to pick up on these omissions and require the expert to deal with them in cross-examination.[56]

Where certain facts are not known, it may be necessary to prepare a list of assumptions upon which the expert must prepare his opinion. In that case, it is important to ensure that the report is amended and updated as and when the correctness of the assumptions becomes clear.

c. *Instructions*

Any brief to an expert should include, among other things, instructions in the following terms:

– *The report should be prepared by you alone* (the courts have shown a distaste for expert reports which are primarily prepared by subordinates and with which the 'expert' is only vaguely familiar).[57]
– *Answer only those questions which fall within your area of expertise* (see above as to the importance of expertise).
– *Set out the assumptions upon which your opinion is based* (which should include any assumptions set out in the brief).[58]
– *Set out the documents considered by you in reaching your conclusions* (which should include all documents accompanying the brief and relevant codes and standards).[59]
– *Set out the methodology employed in reaching your conclusions.*[60]
– *Advise if you are of the view that more information is needed before you can reach a final decision. Place a caveat in your report in that regard, stating that your opinions are provisional.*[61]

Each of these instructions is aimed at improving the transparency of the expert's decision-making process so that it may be open to analysis and review by the other parties. It is this transparency which lends credibility to the expert's evidence.

---

[55] See criticism of an expert for failing to have had regard to all of the evidence at in *Great Eastern Hotel Co Ltd v John Laing Construction Ltd supra*, para. 117.
[56] As was the case in *Great Eastern Hotel Co Ltd v John Laing Construction Ltd supra*, para. 128.
[57] *Skanska Construction UK Ltd (Formerly Kvaerner Construction Ltd) v Egger (Barony) Ltd* [2004] EWHC 1748, para. 415.
[58] *National Justice Cia Naviera SA v Prudential Assurance Co Ltd, The Ikarian Reefer* [1995] 1 Lloyd's Rep 455
[59] *National Justice Cia Naviera SA v Prudential Assurance Co Ltd, The Ikarian Reefer supra.*
[60] *National Justice Cia Naviera SA v Prudential Assurance Co Ltd, The Ikarian Reefer supra.*
[61] *National Justice Cia Naviera SA v Prudential Assurance Co Ltd, The Ikarian Reefer supra*, and Newman, 'Expert Witnesses'.

2. Directing Tribunal-Appointed Experts

Where the expert is appointed by the tribunal the usual practice is for the tribunal to prepare specific questions for the expert to address on the basis of the evidence filed by the parties in the proceedings. The consensus is that arbitrators should invite the parties to comment on the questions prior to their submission to the expert. Likewise, it is common practice to allow parties to comment on the answers provided by the tribunal-appointed expert and to reply with expert evidence of their own. Allowing the parties to be involved in determining the tribunal-appointed expert's remit, and analysing any report prepared by him, will reduce later objections to that evidence.

D. WHAT FORM DOES EXPERT EVIDENCE TAKE?

As with factual evidence, expert evidence can be written or oral. Generally, the rules of evidence applied by the arbitral tribunal in connection with the presentation of expert evidence will be identical to those applied in the case of factual evidence. While the rules of evidence may vary across proceedings, expert evidence which strays beyond the agreed or directed scope is technically inadmissible. In such circumstances the parties may agree, or the tribunal may direct, that the original remit should be expanded on the condition that each other party be given the opportunity to respond to the additional expert evidence.

1. Reports

Written expert evidence is usually presented to the arbitral tribunal in the form of a report. These reports are generally filed with the tribunal after the factual witness statements, since in most cases the expert reports depend on matters covered by the factual witness evidence. Occasionally, though rarely, there is a case for expert reports to precede the delivery of the factual witness statements as it can be helpful for the parties to have a full understanding of a technical matter (for instance the cause of a specific failure) in order to establish what factual evidence is relevant to the question of responsibility for the failure. Whichever sequence is adopted, party-appointed experts may normally prepare additional reports during the arbitral proceedings, setting out criticisms of the expert evidence presented by an opposing party or, indeed, defending criticisms launched by others. Likewise, tribunal-appointed experts may need to prepare additional reports addressing points of criticism raised by the parties.

In keeping with the instructions discussed above, the expert report should include, among other things, the following:

- a curriculum vitae, setting out relevant professional qualifications and experience;

- a list of the documents reviewed by the expert;
- a list of the assumptions relied upon by the expert; and
- a statement of the methodology underpinning his analysis of the facts.

2. Oral Evidence

An expert will normally be required to submit to cross-examination on the contents of his report. He may also be required to face questions from the arbitral tribunal. It is at this point that his general demeanour as a witness becomes particularly important.

An increasingly common method of dealing with expert evidence is to arrange expert conclaves, which are attended by the experts alone and in which the experts are required to identify the points upon which they agree or disagree. In such circumstances the experts will prepare a joint report, for submission to the tribunal (and the parties) setting out the outcome of the conclave. In this way the technical matters in dispute can be refined quickly. Such conclaves present a number of issues for the parties and their advisers. Given the fact that these conclaves are generally not attended by the parties or their advisers, there is a general (if not in many cases a genuine) concern that the expert may 'give away' a part of the case of the party employing him.

There is no easy way round all aspects of this problem. Issues of fact can be dealt with by specifying in advance that the experts are not permitted to express views on the existence or validity of facts on which their opinions depend. This is, after all, a matter for the factual evidence of the parties and the determination of the tribunal. It is most definitely not for experts to purport to reach a determination of the relevant facts. The experts' views, on the other hand, are not something which the parties can demand are not expressed in these conclaves. Nor would it be appropriate for the parties' representatives to be present to guide the expert in the views he expresses in the course of the conclave. For this reason there is (from the parties' perspective) a considerable risk inherent in the course of expert conclaves which can be managed but not eliminated. In particular, the expert new to the process needs to understand the uses to which the product of the conclave (typically a list of agreed and not agreed issues) may be put. A newcomer needs to be well briefed as to the legitimate boundaries of the process and must understand that it is quite proper both to maintain divergent views and to resist expressing any views where the factual matrix necessary to form that view is in dispute between the parties.

In addition (or, perhaps, as an alternative) to expert conclaving, some tribunals require the experts to give oral evidence during the hearing (see Chapter 12 for further discussion). In these circumstances, the experts are required to answer questions raised by the tribunal and/or the parties and/or the other expert simultaneously. This method of oral presentation is useful insofar as it allows the tribunal to consider both parties' evidence at the same time. As with expert conclusions, this approach may eliminate the need for excessive written evidence in reply.

Chapter 10
Programme Analysis

I.   INTRODUCTION

One of the most common features of any construction project is that the project takes longer to complete than anticipated. This may arise as a result of changes in the scope of the work required by the employer, failure by the employer to provide access or to deliver information required by the contractor, failure by the contractor to perform its own obligations as expeditiously as it intended to or (occasionally) just plain bad luck. Whatever the reason, the delay almost invariably gives rise to a claim by the contractor for additional payment and/or an extension of time. At the same time it will also generate a claim by the employer for the costs to it of delayed access to the completed project, most often in the form of delay liquidated damages.

   The performance of the construction industry in achieving on-time (and on-budget) delivery of projects is constantly improving. Despite this, the nature of the construction process, involving change brought about as a result of factors both within and outside the parties' control and the inevitable discovery or occurrence of the unexpected, means that there will always be delay to some projects. As a result, disputes involving the analysis of delays will continue to be a feature of construction arbitrations for the foreseeable future.

   For this reason this chapter sets out to identify the particular issues which delay claims raise in the context of a construction arbitration. In particular this chapter discusses the role of the programme, the significance and meaning of critical path analysis and the legal and practical issues surrounding the thorny questions of concurrent delay and ownership of 'float'. To some extent these topics cannot be examined independently of a system of law or of contract terms and conditions which regulate how the parties intended to deal with specific matters. However, neither this book or this chapter is intended to be an authoritative text on such issues, with the result that the matters will be approached, where possible, from a relatively theoretical standpoint. This is not to deny the benefits of, for instance, review of the considerable body of case law that can be found, e.g. in the United States as a result of construction cases being more commonly determined in the courts than in private arbitrations. However, the extent to which cases on specific forms of contract in the light of specific previous decisions will be recognized as relevant in another jurisdiction remains uncertain.

II.   DATE FOR COMPLETION AND THE ROLE OF THE PROGRAMME

The starting point for any party to a construction project that has taken, or is likely to take, longer than intended is to establish both when delivery of the completed project was to take place and how that was to be achieved. This involves the determination of at least three separate issues, namely, what state was the project to be in when delivered to the employer, when was the delivery date and how was the necessary work intended to be – and actually – carried out.

The first and second of these three issues might seem to be readily determinable. So far as the state the project has to be in to be ready for delivery, it is generally recognized that the process of construction means that a construction project need not be 100 per cent finished in order to be complete and ready for delivery to the employer. Most standard forms of contract contain a concept equivalent to 'practical completion'. Though rarely defined in entirely adequate terms (and in some cases not defined at all) the concept is immediately recognisable to anyone in the industry. In short, a project will be complete when everything essential for its use has been completed and shown to work. In addition (and this is important), what remains to be finished off must also not affect the use or operation of the project by the employer, either because of the fact that it has not been done or because of the necessary disruption to the employer's use and occupation of the project while the work is completed. Unsurprisingly this concept often gives rise to heated disputes, particularly where loosely defined and where (for instance in process plants) the execution of the remaining work has to be carried out in an operational plant subject to a health and safety regime the construction industry has yet to come close to. In such circumstances completion of even minor omitted matters can acquire considerable significance.

Recognizing that in some circumstances the definition of completion for the purposes of delivery may be problematic, it ought at least to be possible to determine when that point should be (as opposed to has been) reached. After all, virtually every standard form (and most non-standard forms) of construction contract will specify the date when the project is to be completed by. That, however, is only the starting point since it is almost equally true that almost every form of construction contract will contain some form of mechanism for changing the date for completion of the project. As has already been discussed in Chapter 2, the original reason for the granting of extensions of time was in order to protect the employer's right to claim liquidated damages in circumstances where the employer had himself delayed the contractor in completion of the project. Nowadays, extensions of time may be given for any number of reasons, including matters which might well be regarded as within the contractor's risk (such as the availability of labour or materials) as well as for force majeure and similar events which are generally regarded as relieving both contracting parties of their obligations to the extent affected.

As a result, at the time of any arbitration it is likely that there will in the first instance be a factual question of whether any extensions of time have already been awarded and, if so, what their effect is. Fortunately, the effect on the completion date of issued awards for the project is likely to be relatively simple, amounting to no more than a simple calculation of the additional time the contractor has to complete the works subject, of course, to there being no live dispute about the amount of time awarded.

In addition to determination of this simple element of the dispute, the contractor will also be looking for a further extension of time to cover the period between the contractual completion date (as already extended) and the actual or anticipated completion date for the project. This statement may be a little unfair, for there may be instances where a contractor and its advisers recognize, as does the employer on equally rare occasions, that it might have had something to do with why the project is late. However, experience indicates that in this area unless and until a determination of the parties' relative rights and obligations has been made, the debate tends to be almost completely polarized.

From the employer's perspective the delay is all down to the contractor and the issue is only about when it gets its delay liquidated damages (and/or calls on the performance bond for payment). At the same time, the issue for the contractor is not about how much the employer has contributed to the delay (normally at least all of it) but how much additional payment the contractor is entitled to upon the issue of the award for an extension of time. As a result, based on the same factual matrix, it is by no means uncommon for each party to be claiming tens or even hundreds of millions of dollars on a project of any substance (such as large power station or petrochemical facility).

The key elements in the factual matrix relied on by each party are what the contractor intended to do and what in fact happened which affected its progress in accordance with that plan. Central to the first issue is the programme. Unfortunately, despite the huge advances in the preparation, presentation and analysis of programmes which has been made possible by the development of computer based systems, the contractual position in relation to programmes is normally unsatisfactory. It is, of course, understandable that in most cases the contractor's actual programme should be a matter for it to decide in the light of its assessment of how best to complete the works on time and in the most efficient and effective manner. That said, where the contractor's progress relies on input or actions from third parties, it is appropriate for it to have to declare its hand to let the employer know when the relevant inputs into the process adopted by the contractor have to be provided. It is not by any means acceptable for the employer to learn of these requirements as part of its review of a claim for delay to the project based on late or inadequate supply of information.

This does not, however, mean that a contractor should be obliged to prepare a detailed programme and stick to the letter of it, since specific key interfaces between the contractor's work and matters within the control of the employer, or those it employs or contracts with, can be specified. Where this is impractical (as it probably is for matters such as non-critical design approvals or routine information supply with little or no lead time involved) the interface can be managed by a process which requires the contractor to give notice of the need for such inputs on the fly, with the employer responding within specified periods. If necessary, differing periods for the giving of notices and responding to those notices can be specified.

For these reasons it is generally not appropriate to make compliance with a contractor's programme – and certainly not a detailed programme – mandatory. If this were done there would be no flexibility in the management of the project and both the contractor and the employer would be perpetually in breach of their obligations, either to perform in accordance with the contract or not to interfere with the other's due progress. And in practice this is the case. While most contracts contain some form of contractually binding programme it is often only a rudimentary bar chart giving not much more than a broad indication of what is to happen and the sequence of activities. Where there are key interfaces between the employer and the contractor, such as access dates, dates for decisions on long lead time items, dates for delivery of possession of parts of the works to the employer or a need for the employer to provide staff, feedstock or energy (in a process plant, for instance) on specific dates, these may be dealt with as activities on the bar chart or may be identified in text form in a schedule separate from the programme itself.

Despite the fact that there is not normally a detailed 'contractual' programme, the need for a detailed programme which is kept up-to-date throughout the project is absolutely essential to the analysis of delay claims. Again, sadly, this information is rarely available, or if it is available it is not made available to all participants in a construction project.

The reason why this information is essential is that the first two questions that an arbitral tribunal has to answer in claims based on delay are (i) what actually happened in the course of construction of the project and (ii) how did this impact on its progress? The first question can generally be answered by reference to contemporaneous records. Traditionally this would have been the site logs and the diaries that every engineer (and some others) seems to be trained from birth to keep. Where they exist, these sources of information are extremely valuable, since rarely is there anything as contemporaneous or as untainted by subsequent ex post facto re-analysis of what happened as a manuscript diary or notepad. However, the likelihood is, on most projects being carried out today, that as much or more information will be found in analysis of the e–mail traffic which surrounds any significant project. Indeed, since e-mails:

- are already in a convenient electronic form facilitating collection, review, analysis and presentation of their contents;
- are quite difficult to erase completely from any significant corporation's records; and
- tend to be composed contemporaneously and used in a way which makes them inherently more likely to be free from spin;

it is e-mails which are likely to be the richest seam to mine when determining what happened in a project on any particular day and what the immediate impact of that event was.

The second question – what was the impact of the event – can only be answered by a full understanding of where the project was at the time of the event, what the plan was at that time for the completion of the project and what in fact happened after the event, preferably with an understanding of not just what happened, but why. Of course, this understanding can, laboriously and expensively, be re-created to some degree by a retrospective delay analysis undertaken in the course of the arbitration.

Generally, re-creation of a project's history is unsatisfactory for two reasons. The first is the one just mentioned. It is far easier, cheaper and more accurate for data to be collected and recorded at the time than it is to re-create the data from miscellaneous records at a later stage. And the second point derives from the first. It is inevitable that any retrospective analysis will suffer from a lack of some information, both hard data and more subjective data such as the intentions of the parties in response to the event in question. This in turn leads to a need for each party (or its expert) to make a series of assumptions or subjective decisions in the course of the retrospective re-creation of events and their analysis.

Where original data is missing and re-creation takes place, the result is that there is no single, objectively determinable set of facts available to the parties from the moment the project is first delayed to the conclusion of any subsequent dispute. What instead happens is that two selective sets of facts supporting each party's expert programming (scheduling) evidence are identified and used by each party's expert to prepare his expert opinion on the extent to which each party is responsible for the delay. Inevitably the expert evidence is prepared before any primary finding of the facts on which the analysis is based has been made, occasionally leading to academically correct analyses being rendered useless as a result of the facts relied upon not being accepted by the arbitrator or judge. This, coupled with the lively debate on the correct method of analysing delay claims, means that the outcome of the whole process becomes incapable of accurate assessment in advance of any hearing. In turn this results in unnecessary time and money being spent in the resolution of disputes which, if accurate site data had been available from the start, could have been avoided.

The answer to this begins with the preparation of an appropriately detailed programme at the start of the project. What is 'appropriate' for a project will depend on the nature of the project, its complexity, the sophistication of the sub-contract supply base and the amounts in issue. It is, for instance, not necessary or appropriate for a contractor on a simple project involving nothing more than sequential site clearance, foundation installation, envelope construction and fitting out to be required to have a fully functioning CPM[1] programme (though there would be no argument against having one if the contractor concerned was equipped to produce and operate it). However, even where a fully detailed programme is not produced, all key interfaces of the type identified above need to be recognized in the programme or associated documentation, so that it is clearly understood what constraints there are on the contractor's ability to carry out the work.

Such constraints would not only include matters required by the contractor to progress the work as-planned but also constraints on its ability to do so; for instance having only limited access to the site or needing to give early possession of part or parts of the works to the employer. Traditionally a contractor would only normally be expected to give early possession irrevocably, in accordance with sectional completion (see Chapter 2). Nowadays a contractor's life is made more difficult by sometimes being required to give early possession to the employer on a temporary basis. For instance, in more complex schools PFI projects, temporary possession of part-completed works may need to be given to the employer in order to allow an operational school to continue to function around construction activities without significant impairment of the services offered to pupils. This raises additional grounds for dispute over both the condition of the works at handover to the employer and on return to the contractor.

As well as being 'appropriately' detailed, the programme must also be accurate. This doesn't mean that the programme must be the programme that will actually be carried out by the contractor whatever the circumstances. On the contrary, change is a fact of life in construction contracts and the management of change is one of the main contractor's principal tasks. Rather, the programme must be accurate in the sense that it must properly reflect the contractor's genuine proposed scheme for the carrying out of the works. There is absolutely no purpose or benefit in the contractor presenting a programme which bears little if any resemblance to what is proposed to be carried out. Because of this need for accuracy it is commonplace to find that provisions in construction contracts dealing with the preparation and supply of detailed programmes frequently refer to the programme being an 'approved' programme. The rationale for this is obvious. Where the programme is not supplied at the time the contract is executed (and a fully detailed programme rarely is), it obviously makes sense from a project management perspective for the programme offered by the contractor to be accepted by the employer as the basis on which the work will be carried out. In other words, one which the employer and its professional advisers feel properly reflects the contractor's obligations and

---

1 Critical Path Method.

expectations and also one which probably represents the likely sequence of construction, all things going to plan.

Despite the obvious common sense of this approach, such clauses often themselves give rise to further disputes, namely whether the programme offered by the contractor is or is not an approved one for the purposes of the contract. Alternatively, whether, despite the employer's refusal to 'approve' it, the programme should nevertheless be regarded as one which should be treated as an approved programme under the contract. Such issues occasionally find their way into construction arbitrations by way of an element of the claim – the contractor arguing that the employer's refusal to approve the programme gave rise to delay. Where this happens the number of issues to be determined by the tribunal is increased, rather than, as intended, reduced.

For this reason, among others, it makes little sense to make employer 'consent' to or 'approval' of a detailed programme a contractual requirement. Instead the contract should spell out the objective requirements of the detailed programme. These requirements should include recognition of the key contractual interfaces, proper reflection of the logic the contractor is working to and compliance with the level of detail set out in the contract specification. These are ordinary contractual obligations of the contractor, breach of which would carry the usual sanctions. This would possibly include a deduction from sums otherwise due or a limitation on the contractor's remedies under the contract to the extent that the employer would have incurred less cost had the programme been properly prepared and the correct information made available to the employer. Given such obligations, the employer or the employer's representative will probably have sufficient ammunition to obtain contemporaneous programming information. Assuming that these powers are exercised and a sound initial programme obtained, the degree of uncertainty in the subsequent analysis of the effects of delay should be very much reduced.

Obtaining a sound initial detailed programme was described as the beginning of the solution. Such a programme is helpful, though not in itself sufficient. Building on that initial programme, the contractor needs to supply regular updates to the detailed programme. These updates should reflect not only actual progress to date but also changes to the proposed sequence for completion of the works, whether changed for the contractor's own convenience or as a reaction to external events, including matters for which the employer is responsible. The need to have such a programme seems obvious and, in reality, in most projects one will probably exist somewhere, if only as a working tool to allow the contractor to manage the day to day work being carried on by its sub-contractors.

After receipt of the first detailed programme, the issue tends to revolve not on whether an updated programme is actually prepared but on how effectively it is prepared and whether that information is openly shared with the employer. All too often where the project is delayed the programmes produced resolutely show completion on the contractual date and retain the existing logic despite it being evident that the project is in delay and that something has to give. What then tends to happen in such cases is that activity durations become shortened or (in some cases) activities are omitted completely, leading to an increasing disparity between what is happening at site and what is happening at the 'contractual' level.

The answer to this looking-glass situation is to build on the production of the initial programme and to require the delivery of regular updates which demonstrably show where the project actually is and how the contractor proposes to complete the works from whatever the current position happens to be. Logically, amended programmes should in all circumstances identify the actual anticipated completion date, with any identified delay being accompanied by an application for an extension of time, to the extent thought appropriate. To encourage this it would have to be accepted (in the interests of obtaining accurate information and avoiding inflated claims) that any failure to match anticipated delay with an extension of time would not amount to an admission by the contractor that the balance of the delay was for the contractor's account.

As with the original programme, it probably doesn't make a great deal of sense to require updated programmes to be approved by or on behalf of the employer, since this will only give rise to as many disputes as it resolves. The solution is, as before, to provide constructive guidance to the parties to participate in good project management practices, at the same time providing the parties and any construction arbitrator with the raw data from which a proper analysis of any delay to the project can be undertaken. If both enlightened drafting and professional project management fail to produce a programme which is properly updated, the only option – and one which should certainly be undertaken where a project is in delay in these circumstances – is for the employer to take steps to procure for the project and its own purposes the programme data necessary to analyse the causes for delay. While it is true that such data will not have the benefit of the contractor's no doubt extensive programming/scheduling resources and will not be able to reflect the contractor's future intentions (save to the extent obvious from objective scrutiny), the collection of contemporaneous 'as-built' data will prove invaluable if a matter has to be referred to arbitration. There is little doubt that someone who can speak authoritatively from a personal and intimate knowledge of the events affecting progress, supported by accurate data, will be a much more compelling witness than an expert subsequently employed who is remote from the action and working from a set of hypotheses and unproven facts.

III. CRITICAL PATH AND DELAY ANALYSIS

Given the relevant data, the next issue is how it is to be analysed. Invariably this will involve one or more experts and it is likely that the experts from the various parties will approach the task armed with a different set of facts and approach the analysis from different – possibly widely different – theoretical bases. Quite what the task is will depend on the extension of time and other delay provisions of the particular contract. Very broadly, however, the question is likely to resolve itself

Programme Analysis

into what caused the delay and whether the cause of the delay was a matter within the risk of the employer or the contractor.

Leaving on one side both the question of the method of analysing the effects of one or more events whose effects on the project overlap, and the thorny and much debated issue of who owns 'float' in the project, both of which are discussed below, the issue is in essence one of causation. In other words, looking at the regular process of the works, what is it that has caused the project not to be completed by the anticipated date?

Of course, one answer to this is that nothing has happened to disrupt the regular progress of the work and that it was always going to take longer than anticipated because of an error in the determination of the original programme duration. It is not uncommon in analysing tenders for the conclusion to be reached that on the basis of the information supplied there is a possibility, perhaps a significant possibility, that the contractor, with the resources and technology identified at the tender stage will not achieve completion within the contracted timescale. While it would be a reckless employer which went ahead with a contract which it had analysed as having little or no hope of being completed anywhere near on time, it is normal for projects to proceed on the basis that there is a probability that there will be some delay. Naturally, from the employer's risk perspective this will be covered in the contract provisions for compensation for delay. In addition, the employer will also have made an assessment of the ability of the contractor not just to complete the work but also to make good the employer's potential financial losses. Where such a risk is identified, it will be all the more important that appropriate steps are taken to provide for proper management of delay and for the information and data identified above to be collected and retained: as already observed, the chances are that a delay to completion will give rise to a claim.

In addition to there being a distinct possibility that the programme as originally conceived was overly ambitious, with a risk of delay even without any material adverse problems arising, it is also a distinct probability that the contractor's (or its sub-contractors') own execution of the works will give rise to some delay. This might arise either as a result of matters taking longer to complete than might reasonably have been expected or as a result of re-work of defective parts of the works. Any form of analysis must take account of these possibilities since these are at least as likely to be the causes for any delay to project completion as the types of external matters typically covered by standard forms of extension of time provisions.

A.   THE CRITICAL PATH

In practice, expert evidence will be submitted which relies on identifying the 'critical path' to completion and the effect that identified events have on activities falling on the critical path by one of a number of recognized forms of analysis. A brief description of those more commonly found are set out below. It is, however, important to understand the nature of the critical path and some elements of programming (or scheduling as it is often called) to appreciate the strengths and weaknesses of the various forms of analysis.

The critical path in a project can be defined as that series of connected activities commencing at the start of the project and finishing at the end which determines the overall project duration. This is determined by reference to a programme which identifies the nature of the activities, their durations and the logical connection with each other. For instance, in a simple project for the fabrication of a house (excluding fit-out) the critical path would lie through (1) site clearance (2) foundation construction (3) erection of walls and (4) completion of the roof. Delay to any part of these activities is likely to impact on the total time for completion since each of the four suggested activities relies on completion of the preceding one before it can start. Each of these activities is, therefore, on the critical path or a critical path activity.

However, some other matters which do need to be done before the house shell can be complete will not necessarily lie on the critical path. For instance, glazing of the windows and hanging of doors may be scheduled to take place as part of the construction of the walls, but delay to these activities will not affect the overall completion date as they can be completed either as part of erection of the walls or while the roof is being put on. Of course, if these activities are very much delayed and cannot be completed before the roof is complete, then the project will be delayed and these items, not formerly on the critical path, will now be critical path items on whose completion project completion will depend. This simple example demonstrates not only the concepts of activities and the critical path but also that the critical path, even on a simple project such as this, is a flexible and variable creature. Of course, most projects are considerably more complex than this and the identification of one single critical path becomes more difficult. In practice, more than one critical path may exist as a result of there being two parallel work streams of the same overall duration. The disentangling of the various strands and the assessment of the impact of various events may be carried out by any of the following forms of expert analysis.

B.   AS-PLANNED VS. AS-BUILT

Perhaps the simplest method of analysis (not actually involving determination of a critical path) is the direct comparison between the original plan and what has actually happened on site. Apart from a knowledge of the original programme and an accurate as-built programme, little additional information apart from the nature of the events alleged to have delayed the project is required. This makes the process simple but restricts its ability to take account of more complex activities or multiple events causing delays. Given a simple project such as the construction of the shell of the house referred to above, an *as-planned vs. as-built* analysis might be appropriate (bearing in mind the likely size of the sums in issue) and sufficient (as a result of the simple logic flow through the limited number of activities). However, even here, this approach would only be appropriate where there is a simple and accepted logic in the as-planned programme which flows through unchanged into the as-built programme. If this were not the case this would involve comparing apples with pears.

In addition, in order not to cover up the possible optimism of the as-planned programme and possible inefficiencies in the methods of working adopted by the contractor, the as-planned programme must be not only logically acceptable but also be based on reasonable durations and resources. Finally it would only be appropriate to use this form of direct comparison where there were relatively few identified events impacting on the project duration with discrete, identifiable results. For these reasons, this technique would not be appropriate for more complex situations, other than, perhaps, as a starting point for a more sophisticated analysis. Such an analysis would normally be required because a simple comparison between planned and actual activities cannot properly take account of:

- the consequences of delay (such as re-sequencing of the works) which takes place in a more complex project; or
- the myriad of factors which in practice account for the difference between what was expected at contract inception and what in fact happens before project completion.

C. AS-PLANNED, IMPACTED

A more complex analysis, and one used commonly within the construction industry, is the *as-planned, impacted* method. As its name suggests, this method of analysis takes the as-planned programme from immediately before the event(s) which is (are) the subject of the investigation and adds to it, as discrete new activities, the effects of the events which entitle the contractor to an extension of time. These events, their durations and their correct categorization as 'excusable events' are assessed by the expert from the material available to him.

As with all methods of analysis, a considerable degree of subjective or expert assessment is required in order to undertake the process. To start with (similar to the process undertaken in relation to the *as-planned vs. as-built* method described above) the available as-planned programme has to be investigated to determine whether it is a proper programme to use as the basis for the further analysis. Assuming that it is, delay events within the employer's risk (i.e. those entitling the contractor to an extension of time) are identified and durations of activities representing the effect of these events on the programme are determined by the expert. In doing so the expert assumes the use of similar resources planned by the contractor to be available to it – in other words using the same assumptions made by the contractor in preparing the original programme. The new activities are then inserted in the programme current at the start of the delaying events with the appropriate logic and the programme re-calculated using the existing logic modified only as necessary to accommodate the new events. The new end date obtained then represents the theoretical impact of the delaying events on the programme. Since, by selection and definition, the added activities represent events which entitle the contractor to an extension of time, the difference between the new completion date and the previously planned completion date represents the extension of time that the contractor is entitled to.

This process, though often used, is subject to two principal criticisms over and above the considerable degree of subjective input the expert is required to make. The first criticism is that its use depends on the existence of a reliable programme at the start of the effect of the events being analysed. If the programme is not reliable, either because its logic is not being followed or the duration of its activities are not properly representative of the work to be undertaken, the analysis has no sound foundations on which it may proceed. In these circumstances, the expert may feel that his first task is to reconstruct a reliable as-planned programme which can be used as a base-line. While this may be a more reliable place to start from than the available documentation, nevertheless this introduces yet further additional subjective elements into the process and with it more grounds for concern that the result is not reliable.

The second criticism is that the result is entirely theoretical. The product of the *as-planned, impacted* analysis relies on the assumption that the contractor would proceed with the then current plan subject only to the effects of the events identified by the expert and imposed on the plan in the course of the analysis. The result does not, for instance, take account of the contractor's failure to proceed as fast as anticipated, or, perhaps, to bring additional resources into play, thereby accelerating its progress.

For these reasons, *as-planned, impacted* analysis is only suitable for relatively simple projects and (perhaps) more complex projects where the events causing delay are capable of clear identification and their effects can be determined with some certainty. Overall the technique is regarded as being more theoretical than real, more suited to determining extensions of time where the work has yet to be completed than where the project has been completed and only rarely appropriate for use in determining, retrospectively, a contractor's entitlement to an extension of time. As a result, particularly where there are records to show what actually happened in the course of the project, other forms of analysis are likely to find favour with tribunals.

D.   As-Built, But For

At one level this might be regarded as the opposite process to that adopted in the *as-planned, impacted* method of analysis. As set out above, where the *as-planned, impacted* method is adopted, the contractor's programme is taken as the starting point and activities are *added* to reflect matters which entitle the contractor to an extension of time. By way of contrast, the *as-built, but for* method takes what factually happened (i.e. the as-built programme) and *deductions* are made from it to reflect matters which are the employer's responsibility, producing the earliest date at which the contractor could have completed the work without the effect of matters within the employer's risk.

The actual method of carrying out an *as-built, but for* analysis is complicated and outside the scope of this book. The short point is that the method is very definitely not a theoretical exercise since it is rooted in what actually happened over the duration of the project. For that reason it is generally regarded as being an appropriate tool by which to compare what actually happened with a verifiable alternative programme which maps out what would have happened had the contractor's progress not been delayed by events which entitle it to an extension of time.

Of course, there are weaknesses in this method of analysis. These start with the fact that both the preparation of the as-built programme and the identification and classification of events impacting the proper performance of the work will take a considerable amount of time, effort and money. In addition, the as-built programme necessarily contains within it the critical path to completion determined as at the completion date. In undertaking and reviewing an *as-built, but for* analysis it has to be recognized that the critical path and the construction logic at various points in time throughout the progress of the works would in all probability not have been the same as they were at the end of the project. This creates the potential for there to be different consequences to events occurring at early stages in the completion of the works from those indicated in the analysis.

E.　　　Time Impact

The final technique discussed here is the *time impact* analysis. This involves an assessment, as of the start of the impact of each event which delays the progress of the works, of the consequences of the delay caused by that event. The key factor which distinguishes this approach from the others is that the analysis starts from an accurate as-built position determined from contemporaneous records, taking account of all delays to the project and changes in programme logic and critical path activities which have taken place prior to that date. From that solid foundation, the analysis proceeds by applying the then current as-planned programme for the incomplete part of the works to the starting position to determine the duration of that part of the works still to be completed. Importantly, the analysis identifies the consequences of events falling within both the employer's and the contractor's responsibility, providing a complete analysis of what happened in the course of the project. The process is repeated as often as is necessary to deal with all delay events. Once the consequences of the events have been properly analysed, the parties' relative responsibilities can then be determined.

The principal problem with *time impact* analysis is that if there are multiple causes of delay each of which needs to be analysed independently, the process can become extremely complex. Despite this, if as-planned programmes are maintained on a frequent basis and take account not only of major changes to the project (such as variations) but also the state of progress at regular intervals, the process of *time impact* analysis can be less complicated than it might otherwise appear.

F. CHOOSING AN APPROPRIATE METHOD OF ANALYSIS

As can be seen from the very brief descriptions set out above, there are radically different approaches to the analyses which may be taken. Of course, not all of them will be appropriate for any given project, though technically all could probably be applied, given enough time, effort and ingenuity on the part of the expert. Given this, there are three factors which are relevant in determining which method should be chosen.

The first of these is whether the exercise that is being undertaken is an ex post facto determination of what entitlements should have been or a 'real time' determination of what the entitlements should be. In practice, in most arbitrations the exercise being undertaken will be the former, since only rarely will applications for an extension of time be brought before an arbitral tribunal before the completion of the project renders the exercise an historic one. However, if the arbitration is one of those rare ones where the determination is to be prospective, it is likely that the *as-planned vs. as-built* and *as-built, but for* methods of analysis will not be appropriate. Either of the two remaining methods is clearly better suited for a prospective analysis. The corollary is that where a retrospective delay analysis is required, it is more likely that the *as-built, but for* and the *time impact* methods of analysis will prove to be appropriate and acceptable. Of the two, *time impact* analysis was identified[2] as the preferred means of analysing delay and compensation for that delay by the authors of the Society of Construction Law's Delay and Disruption Protocol[3] (the *Protocol*).

The second factor relevant to the decision is the availability of records to support the analysis. In extreme cases the availability of records – more accurately the lack of available records – virtually dictates the method of analysis. Thus, for example, where there has been no attempt to produce a networked or any other form of programme, the only form of analysis which can sensibly be conducted is the *as-built, but for*, since this relies only on an as-built programme and such of the project's records as are in fact available to identify what events were impacting on the progress of the works during the period to completion. That said, one can only

---

2 Protocol (below), para. 4.8.
3 The Delay and Disruption Protocol published by the Society of Construction Law, October 2002, reprinted October 2004.

speculate about the quality and reliability of the records that would in fact be available in a project conducted without a networked or other programme.

At the other end of the spectrum if there was no as-built programme (and it was felt unnecessarily expensive to re-create one from the available records) the parties and the tribunal would probably conclude that the only appropriate method of proceeding would be by way of an *as-planned, impacted* procedure. Where more information is available, the choice is wider, with both *as-planned vs. as-built* and *time impact* analysis being capable of use, provided both planned programmes (networked in the case of a *time impact* analysis) and as-built records are available. As indicated above, a *time impact* analysis is the preferred method of the two in virtually all cases where the relevant information is available, save where the cost of the exercise would be disproportionate to the sums involved.

The third and final matter to be taken into account in reaching a conclusion on the appropriate method of analysis is the cost of the exercise. Not only must the cost be reasonable but also proportionate to the sums in issue. The result of this is that the use of a *time impact* analysis, though generally recognized as producing the most reliable results when applied properly, will be a less obvious candidate (because of its cost) than others in small projects. For this reason, although criticized on other grounds, *as-planned vs. as-built* and *as-planned, impacted* are more likely to be regarded as appropriate methods for analysis of delay in small projects. *As-built, but for* falls somewhere in-between the two extremes, since although relatively simple to operate, the emphasis on the factual investigation into the logic of the as-built programme will add to the costs in all but the simplest of cases.

The conclusion that has to be drawn from the wide range of methods of analysis (the summary above is only of the most common types) is that it would assist both the parties and the tribunal for there to be a single agreed approach to the methodology of delay analysis. Once agreed, the parties could then adduce appropriate evidence to establish one common set of facts required by the agreed methodology, leaving the experts to debate the application of the agreed methodology to the one set of facts. Where this happens the possibility of the tribunal being presented not only with a choice of methodologies but also an unnecessarily extended set of facts on which to rule is eliminated. In fact, the Protocol[4] recommends the agreement of a single methodology and goes on to conclude that at the same time the parties should agree on one expert to conduct the agreed method of analysis. Given that in the real world there are likely to be some significant disagreements not only about the proper methodology to be adopted but also the identity of the expert to carry out the analysis, the Protocol's suggestion that the parties will be encouraged to follow this course by the imposition of an appropriate costs sanction for failure to do so is unlikely to prove sufficient in practice.

---

4 Protocol, paras 4.17 and 4.18.

IV.   CONCURRENT DELAY

One issue which is likely to arise no matter which form of analysis is used is the question of 'concurrent delay'. The term is commonly used to cover a fairly wide variety of situations. As noted in Chapter 2, the term is commonly used to describe circumstances where two events, one within the responsibility of the employer and one within the responsibility of the contractor, might be regarded as causing or contributing to delay to the project. As will be appreciated, the resolution of which event has in fact caused the delay to the project is crucial. If the matter is within the responsibility of the employer then the contractor will be entitled to an extension of time and possibly also to the costs associated with the delay, depending on the terms of the contract. If, on the other hand, the matter is within the responsibility of the contractor, then no extension of time will be granted and the contractor will not only have to bear the additional costs associated with the delay itself, but will also probably be required to pay the employer liquidated damages for delay. As has been noted earlier, the difference between these two situations in any significant project could easily amount to tens or even hundreds of millions of dollars, making resolution of this question, when it arises, a very hotly contested debate.

To put the matter in context it may help to consider an example of how the point may arise. In the construction of a power station, for example, the employer is commonly responsible for the connection of the power station to the national grid of the country concerned. This is necessary to export the power produced by the plant (essentially a matter for the employer) but also, and critically, also required to allow the contractor to carry out the performance tests on which its performance guarantees and in some cases completion of the plant is dependent. Assume, then, that completion of the plant is dependent on satisfactory completion of the performance tests and that at the time specified for completion of the plant (the time from which liquidated damages for delay commence):

- the contractor has only completed 80 per cent of the plant for reasons entirely within its own responsibility; and
- the employer has failed to secure a connection to the national grid with sufficient capacity to allow the carrying out of the performance tests.

In these circumstances the employer will no doubt claim that the project is delayed for reasons attributable to the contractor (the contractor is in 'culpable delay') and that the employer is entitled to liquidated damages. After all, the plant is only four fifths completed for reasons entirely unconnected with anything for which the employer is responsible.

Naturally, the contractor sees things a little differently. From its perspective it matters not one jot whether the plant is 80 per cent complete, 'practically complete' or has been completely finished and all snags sorted because, it will say, it still could not have achieved completion and relieved itself of the obligation to pay liquidated damages. This, it will say, is because however hard it had tried it would have been unable to carry out the performance tests for reasons which were both outside its control and within the employer's risk. On this basis, the contractor will say that it is unfair for it to be penalized where, had it been on time, it:

- would not have been able to complete the plant for reasons attributable to the employer;
- would not have been exposed to liquidated damages; and
- might have been able to recover the costs of any delay before the grid connection was established and the performance tests carried out.

Failing that, the contractor would commonly argue that since both the employer and it are in breach, the responsibility for the delay should be apportioned or shared in some way. One common version of this is that the contractor should not get additional payment for the delay period (after all, it is late with the construction), but equally the employer should not be entitled to claim liquidated damages because it is unable to run the plant because of the inability to complete the necessary tests and, for that matter, an inability to export power in commercial quantities to the grid. This latter point may in fact have some significance in jurisdictions where recovery of liquidated damages may be limited by reference to the loss actually suffered at the time of breach.

The first thing to observe about this situation is that although it is very common for arguments to be raised about concurrent delay, there are very few contracts which attempt to deal with the position. Nor are there many jurisdictions with authorities dealing with the issue squarely. For this reason the question tends to be addressed from first principles as a matter of causation, which at least provides a mechanism for the determination of an answer even if it does not address the inequity that one party suffers as a result of the good fortune of the other. Of course, if the matter were felt to be sufficiently iniquitous or important the question could be addressed at the contract drafting stage. The fact, as noted above, that this rarely happens indicates that while the point generates much heat when projects are delayed, the issue may be seen more as a means of avoiding a liability than a genuinely serious issue.

Where the matter is dealt with as one of causation, the question boils down to what has 'caused' the delay. Where there is only one event which impacts the regular progress of the work, there can be only one of two answers. Either the work was going to be late in any event (and this possibility, though awkward for contractors, must never be forgotten) or the project has been delayed by the event in question. Even where there is more than one event affecting the progress of the project, there is no real issue of causation if the consequences of the events all fall to be treated in the same way. In other words, no difficult questions of causation arise if all the events fall within one class of events, each of which, as a matter of contract, produces the same consequences. This is, in essence, the basis of the global claim discussed in Chapter 9 and where this occurs it becomes unnecessary to separate out and track through cause and effect.

The problem only arises, as in the example described above, where the events in question produce different consequences. So, returning to that example, is it possible to determine in a satisfactory manner which event has caused the delay? Does it, for instance, depend on the order in which the events arose. So, for instance, if it was apparent from the very start of the project that the employer would or might have difficulty in procuring the completion of the grid connection in time, does this (if in fact the grid connection is delayed) effectively relieve the contractor of all responsibility for any delays to construction which fall within its responsibility and deprive the employer of any remedy for the contractor's failings, no matter how serious and prolonged?

While the sequence of events may have some significance, there is certainly no hard and fast rule which requires that the first event to arise should be treated as the cause of the delay. In a most unsatisfactory manner the answer is that it all depends and that causation is not a simple matter of chronology. In practice, judges and arbitrators have a reasonably keen sense of what 'caused' an event, rejecting unlikely hypotheses which do not fit in with the commercial realities of the situation. Thus it is never going to be an answer, when there is an obvious failure by one party to perform what might be regarded as one of its primary obligations, for that party to try to rely on the other's failure to excuse the consequences of its own failings.

Thus, in the example given, it is likely (but put no higher) that a tribunal would regard the cause of the delay as being the fact that the plant was, at the contractual completion date, only 80 per cent complete and would reject the argument that the delay was the responsibility of the employer for not having procured something which the contractor did not, at that point, require. Obviously the analysis changes when the plant has been completed to the point where it is safe and ready to undertake the completion and performance tests. At that point if the grid connection is not in place, alternative solutions cannot be found or the relevant part of the tests are not waived, it seems clear that the employer would then be responsible for all ensuing delays.

What in fact has happened, is that this situation of superficial 'concurrency' has been capable of analysis into separate events with separate consequences. There is, therefore no true concurrency and no difficult questions arise. And it is this analysis which leads to the conclusion that there will in fact be few instances where there is genuine concurrency. In other words while there is considerable discussion about the need to resolve the problems of 'concurrent delays', this turns out to be a largely academic and theoretical discussion since tribunals, applying a degree of common sense derived from experience of the construction industry, in practice do not find it too difficult to separate out the effects of differing events, thereby allowing an allocation of responsibility to be made on traditional principles of proof of cause and effect. In this they are supported by doctrines such as the 'dominant' or 'effective' cause, which permit a finding of cause and effect even if there is a small contribution, or a possibility of a small contribution, to the consequence from another matter.

Yet despite this, there will be a small number of cases where it is impossible to separate out the effects of two events, where one is within the employer's responsibility and the other within the contractor's. In these cases it is generally accepted that there is a true 'concurrent delay' and that the solution is that in such cases the contractor is entitled to claim a full extension of time for the combined (but inextricably linked) consequences of the two delays. It is said that there is authority for this proposition in English law[5] but on careful reading it appears that this is the common view of the parties rather than the expressed view or decision of the judge himself. It is, however the clear and unambiguous position of the authors of the Protocol whose conclusion is that in genuine (albeit rare) cases of true concurrency, the contractor's entitlement to an extension of time should not be reduced by the fact that matters within its own responsibility have contributed to that delay. Interestingly this approach runs counter (albeit at the 'tie breaker' level) to the general solution to the presentation of global claims which (absent proof of each entitlement) only permits recovery where all the events not separately identified result in the same consequences.

V. OWNERSHIP OF FLOAT

Another general issue that parties to construction arbitrations and tribunals will have to deal with is the issue of who 'owns float' in a programme. By this it is meant those periods in a programme where activity durations are in fact shorter than the time available to carry out the work without delaying the overall completion. This might be regarded, in programming terms, as the presence of a time contingency equivalent to an unallocated amount of money in the contract price to cover the unexpected over and above the cost of all the foreseen work content. The rationale of its inclusion is often the same, namely to provide some protection for the contractor against the possibility of not everything going as smoothly as possible, thereby minimizing its potential exposure to liquidated damages. Arguably, if the programme contained less float then there would have to be a larger monetary contingency to guard against this possibility, at least if the contractor were to be left with the same level of risk. Float may be concealed within an activity's duration (i.e. one which is simply longer than it perhaps needs to be) or shown as free time between the end of one activity and the next following activity.

The issue of ownership of this extra time arises because the concept of entitlements, particularly the contractor's entitlement to extensions of time, is linked to the concept of delay to specified events. Traditional provisions providing for extensions of time, for instance, do not provide relief to the contractor for increased risk to it in completion of the project. Thus where there is identifiable delay to activities on the critical path of a project, the provisions will operate and, if their

---

5   *Henry Boot Construction (UK) Ltd v Malmaison Hotel (Manchester) Ltd* [1999] 70 Con LR 32.

conditions are satisfied, the contractor will become entitled to an extension of time to compensate it for the delay caused to it. The contractor may also become entitled to the additional costs associated with that delay, as well as relief from the imposition of liquidated damages.

However, where the activity delayed is not on the critical path to completion because of the existence of float in that activity, delay to that activity will not at first show up as delay to completion but merely as a reduction of the amount of float. Once the float has been completely consumed (but only then), the activity will lie on the critical path to completion so any further delay will be reflected in a delay to completion. In the first phase of this process, while float is being consumed, there is no delay to the project so typical extension of time provisions will not respond, though the risk to the contractor is becoming greater because it is now increasingly exposed to the consequences of its own failure to complete in time the activity that was 'in float'. Only when the process enters the second phase – when all the float has been consumed and the activity is on the critical path – do standard clauses offer the contractor any relief. In other words, where float exists, and particularly where it has been deliberately introduced to cover a contractor risk, the employer stands to benefit from that provision where it causes delay to the activity in float.

Not unnaturally this point is a matter of considerable contention since, as observed above, it may be that float has been deliberately introduced as a means of risk management and as a means of avoiding the inclusion of a monetary contingency to cover the risk of late delivery by the contractor of that element of the works. As a result, it is often argued that the existence of float should be recognized in delay analysis and preserved in the determination of extensions of time. The purpose of this is not only to protect the contractor against delay to the project caused by matters falling within the employer's control and responsibility, but also to leave the contractor with the same level of risk in carrying out the work.

The problems with this argument arise in two separate areas. The first is in the fact that typical standard form extension of time provisions are phrased in terms that require actual or, if in the future, anticipated delays to completion of the whole or a specified part of the works. As a matter of pure semantics, these clauses do not protect the contractor against increase in risk but only against delay to critical path activities. For this reason, many attempts at arguing that the contractor should be given the benefit of float fail, because for so long as float exists then, whatever the merits of the case, there will be no delay to completion and as a result there will be no entitlement to an extension of time. The second reason why the problem arises is that not all should necessarily be protected since float can arise for a number of reasons. For instance, as a matter of construction logic it will normally be the case that while there are activities which inevitably lie on the critical path to completion of the project, there will be others, probably many others, whose timely completion is not critical to project completion. Not because they are especially sensitive operations deliberately protected by float but only because they are matters which can be carried out within the duration of one of several activities on the critical path without affecting the execution of critical path activities or impacting on the completion date for the project. These activities are just as much in float as any other but it is by no means unreasonable to deny a contractor relief from an event which consumes float relative to these activities as there is no delay to the overall completion date and the increase in risk, if any, is (at least at the outset) marginal.

The answer, probably, lies in specific drafting to protect float which does not arise by accident of programme logic but whose deliberate insertion in the programme is for the purpose of risk management. This needs to be coupled with specific identification of this 'protected float' in a programme produced at the time of contract execution. Until then, the parties, their advisers and tribunals will have to continue grappling with the concept of 'ownership' of float, with detailed forensic and semantic analysis of contract provisions providing an unsatisfactory answer on a case by case basis.

Chapter 11
Procedural Issues

I. INTRODUCTION

This chapter looks at procedural issues. In essence, arbitration procedure is the set of rules that governs the contest between the parties. They bear on the entire arbitration process from beginning to end. For example, procedural rules govern the manner in which the arbitration is commenced, the appointment of the arbitrator(s), the role of written submissions (if any), the production of documents, the role of witnesses (if any), hearings (if any) and the process by which the arbitrator's award is to be delivered.

II. OBJECTIVE OF PROCEDURAL RULES

Procedural rules serve a number of purposes. Key amongst them are foreseeability and equality of treatment.

Whatever their derivation, procedural rules are generally set out in advance of a dispute.[1] As a minimum, the parties should have agreed at least the process by which the rules are to be determined – even if this is no more than granting the arbitrator(s) the discretion to make the rules so that specific rules can then be set at the outset of the dispute. This enables parties to know ahead of commencing the substantive arbitration process the rules by which the arbitration will be conducted and the rules by which they and other parties will be expected to conduct themselves in the arbitration process. Amongst other benefits, foreseeability affords a level of awareness regarding the approximate cost and length of the arbitration process, the number and identity of the personnel who may be required to be involved and the

---

1   Although, on occasions, parties may agree to amend certain procedural rules during the course of a dispute.

arrangements that need to be made for the various stages in the arbitration process, including any hearings.

Procedural rules also ensure equality of treatment and due process between the parties. There may be an express protection of such principles in arbitration rules.[2]

III. DERIVATION OF PROCEDURAL RULES

Arbitration is by its nature a consensual process and it is therefore unsurprising that parties are generally free to agree the procedural rules to be followed in the arbitration of any dispute between them subject to any applicable mandatory requirements.[3] Parties may agree some rules 'up front' in the arbitration agreement with others (often the more detailed rules) agreed between them at the commencement of an individual arbitration process once the nature, complexity and quantum of the dispute are known. In practice, procedural rules derive from a combination of sources.

First, certain procedural rules may be determined by the parties' choice of arbitral seat as in some jurisdictions there are mandatory procedural requirements that apply to all arbitrations conducted in that jurisdiction. For example, the provisions set out in Schedule 1 to the English Arbitration Act 1996 apply to any arbitration where the seat of the arbitration is England, Wales or Northern Ireland and have effect notwithstanding any agreement to the contrary. The provisions deal with matters such as stay of any legal proceedings regarding the same dispute, the power of the court to remove an arbitrator, determination of the preliminary point of jurisdiction, securing the attendance of witnesses, enforcement of arbitral awards and challenge to arbitral awards.

Secondly, procedural rules may be stipulated and set out in full detail in the parties' arbitration agreement. This is most common in the case of 'ad hoc' arbitrations where the arbitration process is to be run without the aid of an institution such as the ICC or LCIA. Alternatively, parties may agree in their arbitration agreement to adopt an established set of rules such as the ICC Rules, UNCITRAL Arbitration or LCIA Rules amongst others. Parties may either agree to adopt the established set of rules in their unamended entirety or to set out agreed amendments to the chosen rules in their arbitration agreement (see Chapter 3).[4]

---

2   For example, Art. 15(2) ICC Rules states: 'In all cases, the Arbitral Tribunal shall act fairly and impartially and ensure that each party has a reasonable opportunity to present its case'.

3   As discussed below, the parties' choice of some arbitration seats (for example, England) will attract some mandatory procedural requirements that apply as matter of law to any arbitrations conducted in that jurisdiction.

4   As a note of warning, amending certain fundamental features of institutional rules may not be permitted by the institution – such as the production of terms of reference and the review of the award by the ICC court. Material amendments will undermine the value of using institutional rules, which lies in part in the confidence that such tried and tested rules will assist enforcement of the ultimate award.

Thirdly, parties may agree certain procedural rules at the time of a dispute. This is normally the case regarding detailed procedural rules such as the all-important timetable for exchange of submissions and witness statements, hearings and other procedural steps.[5] Arrangements for communicating document exchanges (for example, by electronic transmission) and translation of documents, if necessary, are other practical issues typically addressed at this stage. The value of a site visit might also be considered at this stage. In ICC arbitrations, some such rules may be set out at the time of agreeing the terms of reference.

Lastly, procedural rules may be laid down by the arbitrator(s) – with or without consultation with or agreement of the parties. Arbitration rules may expressly reserve power to the arbitrator(s) to determine any procedural rules that have not been agreed by the parties.[6] Consultation with parties, where the arbitrator(s) deem appropriate, sometimes takes place in a preliminary conference attended by the arbitrator(s) and parties devoted solely to determination of procedural matters. According to the ICC Construction Arbitration Report this type of initial procedural meeting is considered vital in construction arbitrations.[7]

A useful reference for both parties and arbitrator(s) in agreeing or otherwise determining procedural matters is the UNCITRAL Notes on Organizing Arbitral Proceedings. As previously noted, the UNCITRAL Notes on Organizing Arbitral Proceedings are not arbitration rules but aim to provide a checklist of matters which arbitrators and parties may find useful to consider and decide upon when involved in an international arbitration. Some matters address significant procedural issues such as whether hearings will be held, but most are eminently practical matters such as arrangements for exchange of information (electronic or otherwise) and management of deposits in respect of costs. The UNCITRAL Notes on Organizing Arbitral Proceedings serve as a useful indication of the range of possible procedural options for each matter. More detail regarding specific aspects of the UNCITRAL Notes on Organizing Arbitral Proceedings is given below under the heading 'Administrative Issues'.

As the majority of arbitration rules cover mechanisms for taking evidence very broadly, another useful reference is the IBA Rules 'as a resource to parties and to arbitrators in order to enable them to conduct the evidence phase of international arbitration proceedings in an efficient and economical manner',[8] Parties may choose to adopt the IBA Rules in their entirety to supplement their chosen arbitration rules (institutional or ad hoc) either in their arbitration agreement or at the commencement

---

5 ICC Construction Arbitration Report, para. 48 provides guidance on timetables for international construction arbitrations.
6 For example Art. 15(1) ICC Rules states: 'The proceedings before the Arbitral Tribunal shall be governed by these Rules, and, where these Rules are silent, by any rules which the parties or, failing them, the Arbitral Tribunal may settle on, whether or not reference is thereby made to the rules of procedure of a national law to be applied to the arbitration'. Art. 2(4) IBA Rules is of similar effect regarding procedure governing the taking of evidence.
7 ICC Construction Arbitration Report, para. 29.
8 IBA Rules, p. 1.

of a dispute. As will be demonstrated below, the IBA Rules largely represent current international arbitration practice on procedural matters relevant to the evidence phase, and a blend of common law and civil law traditions. Alternatively, the IBA Rules usefully canvass the issues (and one approach to dealing with such issues) that typically require consideration ahead of the evidence phase of an international arbitration. Such rules can be used by parties and/or arbitrators to develop their own procedural rules on these issues.

The ICC Construction Arbitration Report is of particular assistance to the arbitration of construction disputes. The report is the result of a study commissioned by the ICC Commission on International Arbitration with a remit including the investigation of techniques used successfully to control construction arbitrations and the production of guidance for arbitrations in construction disputes. It sets out various recommendations for efficient management of large and complex construction arbitrations, with a particular emphasis on cost-effectiveness. The recommendations are mainly directed at arbitrators, but are also of use to parties involved in construction arbitrations to the extent that they play a role in designing the arbitral procedure. Although focussed on the ICC Rules, the report's recommendations are general enough to be applicable to non-ICC arbitrations. The conclusions are comparable in many respects to the UNCITRAL Notes on Organizing Arbitral Proceedings.[9]

IV. CONTRASTING CIVIL AND COMMON LAW PROCEDURAL APPROACHES

A. INTRODUCTION

An essential feature of international arbitration is the meeting of parties from different legal backgrounds with different philosophies and experiences regarding the procedural rules that should govern a dispute resolution process. Because arbitration is a flexible forum for dispute resolution and capable of being moulded by the people involved, an individual arbitration may bear characteristics of the legal tradition associated with the national origins of any of the parties, their counsel, the arbitrator(s) and/or the seat of arbitration. For example, where each of the parties (or, often more crucially, their legal counsel) and the arbitrator(s) is from a common law tradition, the arbitral procedure may bear more hallmarks of the common law tradition than would a similar dispute featuring civil law parties (or legal counsel) and arbitrator(s). A mix of legal backgrounds may create tensions, or at least differing expectations between parties of how an arbitration will be conducted.

Procedural rules differ in a number of respects between common law and civil law traditions. The differences can be encapsulated by the differing philosophies underlying an adversarial process (common law) as opposed to an inquisitorial

---

9 ICC Construction Arbitration Report, para. 8.5.

process (civil law). In common law jurisdictions (such as the United Kingdom, Commonwealth countries and the United States), the dispute resolution process is characterized by two adversaries pitted against each other with the task of presenting their respective cases to the decision-maker who has the task of deciding between them. In civil law jurisdictions (including Continental Europe), the decision-maker plays a more active role in investigating the case before him or her, with the role of the parties' counsel being to aid the decision-maker, in this process. This basic point of distinction between common law and civil law procedure goes a considerable way to explaining the key areas of procedural difference between the two.

B.   EXTENT OF DISCLOSURE

A key point of difference between the common law and civil law traditions is the role of disclosure or 'discovery' of documents. In the common law tradition, parties are generally required to disclose to each other all documents in their control or possession that are relevant to the issues in dispute between them, including documents that may be damaging to their cases.[10] This requirement is foreign – often unfathomable – to counsel from civil law jurisdictions, who are used to producing to the other party (and the decision-making tribunal) only those documents on which their clients need to rely to support their cases. Most civil law tribunals do have varying degrees of power to compel parties to produce documents but these extend only to specifically identified documents or possibly a narrowly described category of documents – a long way short of the common law notion of full discovery.

The extent to which a party's documents may be required to be disclosed to other parties is also subject to rules and norms regarding legal privilege and, in some jurisdictions, without prejudice communications. Concepts of legal privilege or professional secrecy, which prevent documents and other communications from disclosure in proceedings, are widely recognized. However, the extent and scope of such concepts vary between jurisdictions. In common law jurisdictions, privilege is a rule of evidence and is therefore procedural. The most relevant privilege for arbitration purposes is 'litigation privilege' which, generally speaking, protects documents created for the purpose[11] of proceedings (existing or contemplated) from being disclosed in the subject proceedings. The other primary head of privilege is 'legal advice privilege' which applies to communications between lawyers and clients for the purpose of requesting or providing legal advice. In some civil law jurisdictions, on the other hand, the notion that certain documents are protected from disclosure arises from the ethical obligations of practitioners rather than

---

10   Subject to legal privileges such as lawyer client, legal advice or litigation privilege which protect the documents against disclosure.
11   Either the sole purpose or the dominant purpose, depending on the jurisdiction.

procedural rules. For example, in Germany, a lawyer is obliged to keep confidential everything relating to knowledge gained in the exercise of his profession. In France, correspondence between an *avocat* and his client or between an *avocat* and the other party's *avocat* is privileged and cannot be disclosed.[12]

In common law jurisdictions, the concept of without prejudice communications may also limit parties' disclosure obligations. It operates to keep confidential and protect from disclosure in proceedings, settlement communications. The concept is not widely recognized in other jurisdictions. For example, in Austria, it is permissible to show settlement offers to a tribunal, and this is commonly done to indicate attempts made at settlement. Whilst it is possible, in principle, to agree that settlement communications ought to remain confidential, this is not automatic, and must be specifically agreed between the parties. Similarly, there is no equivalent to the common law 'without prejudice communication' concept in other civil law jurisdictions such as France, Spain and Germany.

C. WRITTEN SUBMISSIONS

In contrast to the common law focus on oral argument or presentation of witness evidence, civil law proceedings are characterized by written submissions. Although there may be short oral arguments, the principal method by which civil law cases are presented is exchange of detailed written submissions ahead of witness evidence (if any). These tell the whole story – including allegations and facts. In addition, written submissions typically attach copies of all documents on which the parties rely. See further Chapter 12.

D. PREPARATION OF EVIDENCE

In contrast to the civil law emphasis on written submissions, the common law emphasis is upon witness evidence. Witness evidence is a crucial mechanism by which the parties present their version of the facts to the decision-maker and by which they seek to disprove the case presented by the opposing party/ies. Witnesses therefore play much more of a central role in common law than they do in civil law proceedings. Consequently, the rules governing witnesses and their evidence differ in a number of respects.

Craig, Park and Paulsson have neatly summarized the contrast between common law and civil law systems of introducing evidence into proceedings as follows:

---

12   This does not apply in the case of in-house lawyers in France, hence communications between a company's lawyer and other members of the company are not privileged. The same is true in the Netherlands.

'The continental civil-law system of proof is dominated by the exchange of documents between the parties. Hearings serve principally as an occasion for arguments based on facts revealed in written evidence already submitted. The common-law system, on the other hand, uses hearings to develop facts and to introduce documents into evidence'.[13]

In the common law there is no bar on who may be called as a witness. The only requirements are that the person is competent and that his evidence is relevant to issues in the proceedings. On the other hand, civil law procedure typically limits the category of persons who can be called as witnesses to third parties. A party's own officers and employees are not permitted to give evidence on the presumption that they would be partisan. The common law's response to this risk is to allow extensive cross examination of witnesses (addressed further below) by the opposing party with the aim of testing witnesses' credibility, amongst other things. The common law tribunal is thereupon called to assess a witness' credibility and the weight to be given to his evidence.

Preparation of witnesses prior to their giving evidence at a hearing is another area of tension between the common law and civil law traditions. In civil law jurisdictions, such as the French-speaking Brussels bar, ethical rules prohibit contact between counsel and potential witnesses on the basis that it would be improper and because it is for the judge to question the witness. However, witness preparation is the norm in common law proceedings. Indeed, it would be negligent for common law counsel not to investigate witnesses' responses to issues in the case before a hearing, although witnesses may not be coached in the sense of providing answers, and contact with a witness must usually cease once his evidence has begun. This approach reflects the rigour and scope of common law cross examination, which should be capable of uncovering any improper preparation. Questioning in civil proceedings is more limited, given the greater emphasis on documentary evidence, and the fact that there is no cross examination in the civil law tradition.

In the context of international arbitration, preparation of witnesses is generally acceptable, and widely practiced.

See Chapter 12 for more information on witnesses.

V. TYPICAL PROCEDURAL DIRECTIONS – A FUSION BETWEEN COMMON AND CIVIL LAW

Whilst arbitral procedure varies greatly, there has gradually evolved a body of international procedural norms that have come to be adopted by parties to international arbitrations with reasonable regularity. As a result there is now a largely settled body of typical procedural directions that represents a compromise

---

13 Paulsson *et al., International Chamber of Commerce Arbitration*, pp. 427–429.

between common and civil law procedure.[14] From civil law, international arbitration practice has adopted the reliance upon written submissions and relatively brief oral submissions. From common law, it has adopted practices on production of documents and witness evidence.

A.   EXCHANGE OF WRITTEN STATEMENTS OF CASE

International arbitrations typically commence with the parties exchanging written statements of case. These are usually in narrative form rather than in common law pleading-style, and are provided in succession, with the claimant being required to provide its statement of case first and the respondent to follow. This is the process adopted, for example, under the UNCITRAL Arbitration Rules, where the claimant (within a period of time determined by the arbitral tribunal) is required to provide its statement of claim including particulars of the parties, a statement of the facts supporting the claim, the points of issue, the relief or remedy sought and annexing a copy of the relevant contract giving rise to the dispute and the arbitration agreement (Article 18). The statement of claim may also annex or reference documents relevant to the claim (Article 18). The respondent is required to respond with a statement of defence replying to the particulars set out in the statement of claim, and again may annex or reference documents that are relevant to the claim (Article 19). Article 15 of the LCIA Rules is similar. The authors of the ICC Construction Arbitration Report give a salutary reminder of the wisdom in arbitrator(s) requiring from parties in disputes regarding claims for delay and disruption, a chronology of events in these initial arbitration stages.[15]

The ICC Rules provide for a similar roster of written submissions in Articles 4 and 5. However, they provide for an extra step. The arbitral tribunal is required to draw up a document defining its terms of reference, to be signed by the tribunal and the parties and submitted to the ICC Court for approval. The approved document then defines the parameters of the arbitration.

B.   DISCLOSURE OF DOCUMENTS

The typical procedure for international arbitrations regarding disclosure is a true blend between the common law and civil law systems. The standard position is to require some level of disclosure of documents, but not anywhere near the scale known to common law. Disclosure tends to be restricted to specifically identified documents or limited categories of documents that are relevant to the issues in the case. This said, disclosure is one of the areas where the origin of parties and their

---

14   See Lucy Reed and Jonathan Sutcliffe, The Americanization of International Arbitration, Mealey's International Arbitration Report Vol. 16 p.37 (April 2001)
15   ICC Construction Arbitration Report, para. 21.

counsel is particularly determinative – although full blown common law-style discovery is extremely rare in international arbitrations. The authors of the ICC Construction Arbitration Report caution of the need to prevent 'arbitration proceedings from being swamped by the mass of documents that are inevitably generated by a construction project'.[16] However, the nature of construction disputes is that they are very fact and document intensive and so, generally, some scope for discovery will be important. The IBA Rules accord with general practice in not excluding internal party documents from potential disclosure.

The process by which disclosure occurs originates in the power usually granted to arbitrators to compel production of documents in a party's possession or control. The order to produce documents may be made of the arbitrator(s)' own volition but is more usually made upon the request of another party. This is the process adopted in Article 3 of the IBA Rules, which is becoming the norm on this aspect of procedure. Article 3 requires each party to submit to the arbitrator(s) and the other parties all documents available to it on which it relies, except those submitted by another party (Article 3(1)). This is intended to avoid trial by ambush, so that the opposing party is aware as early as possible of the documents on which the other party will rely to support its case. In addition, any party may submit a request to produce documents including a description of the requested documents or of a 'narrow and specific requested category of documents that are reasonably believed to exist' and, as a safeguard, a description of the documents' relevance to the case (Articles 3(2)-(3)). The opposing party may either produce the requested documents or object to the request in writing. In the latter case, the arbitrator(s), in consultation with the parties, must consider the issue and may order production of the documents if they are relevant to the case and none of the qualifying reasons for objections applies (Article 3(6)).

Sanctions for non-production of documents is a difficult area in arbitration law. As a practical matter, an arbitrator(s) may choose to draw adverse inferences if a party refuses to provide the document. Article 9(4) of the IBA Rules expressly entitles arbitrator(s) in such circumstances to 'infer that such document would be adverse to the interests of that Party', although in practice it may be difficult to know what inference to draw. Additionally, many jurisdictions have a supportive regime by which their national courts are empowered to support orders of arbitrators. For example, pursuant to section 42 of the English Arbitration Act 1996, a court may make an order requiring a party to comply with an order made by the tribunal; section 26 of the Swedish Arbitration Act 1999 provides that a court application may be made to order a party or person to produce a document as evidence or where a party wishes a witness or an expert to testify under oath; and § 1050 of the German Code of Civil Procedure also provides that the court may assist arbitrators by making such orders.

The extent to which concepts of privilege (see above) affect disclosure obligations varies. This is an area where the main institutional rules (ICC,

---

16   Ibid., para. 52.

UNCITRAL and LCIA) provide no guidance, which contributes to the lack of a uniform approach to privilege in international arbitration. The approach taken by many arbitrators is to analyse the privilege issue on a case-by-case basis, taking into account the circumstances of the case before deciding whether to require disclosure of a document over which privilege has been asserted. The IBA Rules address privilege by requiring arbitrators, upon request of a party or of their own volition, to exclude from evidence or production any document or statement which is the subject of a claim for privilege or other legal impediment under the legal or ethical rules determined by the arbitrators to be applicable (Article 9(2)). The rules of the American Arbitration Association are another set of rules that address privilege, providing that arbitrators are to take into account the applicable rules of legal privilege (for example, confidentiality of communications between a lawyer and client).

C.   EXCHANGE OF WITNESS STATEMENTS

Typical international arbitration procedure regarding witnesses of fact again represents a compromise between the common law and civil law. However, the most commonly adopted procedure in this area is more closely aligned to common law expectations than those of the civil law.

As indicated earlier in this chapter, there exists a tension between the common law and civil law traditions regarding who may be called as witnesses of fact. Although civil law principles on this point are occasionally applied, the general approach in international arbitration is that any person may be a witness, including the parties' own officers and employees. For example, Article 4(2) of the IBA Rules states:

> 'Any person may present evidence as a witness, including a Party or a Party's officer, employee or other representative'.

Article 20.7 of the LCIA Rules is to the same effect. Neither the ICC rules nor the UNCITRAL Arbitration Rules specifically address the issue, but in the absence of any prohibition, in practice there is no restriction on who may be called as a witness. Where parties are permitted to call potentially partisan witnesses, the opposing party will have the right to cross examine witnesses on matters of credibility.

As an alternative, parties may develop and adopt a more hybrid approach to witnesses, such as was used in the Iran US Claims Tribunal. There, interested parties, described as 'party witnesses', were entitled to submit witness statements and give 'information' to the tribunal subject to cross examination, but were not strictly treated as witnesses giving evidence and their evidence was afforded less weight.

It is also now common for counsel to interview potential witnesses about facts relevant to proceedings, and for the parties to exchange witness statements some time before the hearing (see Chapter 12).

D. EXCHANGE OF EXPERT REPORTS

Separately from the evidence given by witnesses of fact, arbitration proceedings frequently involve opinion evidence given by experts. Expert witnesses do not have first hand experience of the facts or issues in the case, but provide their expert opinions on issues in the case based on the facts. Again there is a divergence in practice and expectations regarding experts between common law and civil law. In common law jurisdictions, parties appoint their own experts. In civil law jurisdictions, experts are appointed by the court or other tribunal.

The role of experts in international arbitration varies. Most established arbitration rules provide for experts to be utilized in accordance with either or both traditions. For example, Article 20(3) of the ICC Rules permits the tribunal to hear experts appointed by the parties and Article 20(4) permits the tribunal itself to appoint one or more experts, define their terms of reference and receive their reports. Similarly, Article 5 of the IBA Rules contemplates party-appointed experts and Article 6, tribunal-appointed experts. Article 21 of the LCIA Rules specifically permits arbitrator-appointed experts. Party-appointed experts are not prohibited by the LCIA Rules; they are contemplated by Article 20(7), and are frequently used in practice in LCIA arbitrations.

In some arbitrations, there may be a stage of expert conclaves, where the experts appointed by the parties and/or the tribunal meet together with the aim of agreeing technical issues so as to limit remaining issues in dispute. Article 5(3) of the IBA Rules expressly provides for such a process.

> 'It is desirable that independent experts ... should discuss their views with each other before preparing their reports, as they should eventually agree about most things if they are truly independent. ... The tribunal must ensure that it is clear whether or not agreements between experts bind the parties'.[17]

Where experts are appointed by the parties, their expert reports are normally filed along with the relevant party's witness statements and they are subject to cross examination. Their aim is to lend credibility to the parties' case by the support of independent expert opinions. Article 5(2) of the IBA Rules provides guidance on the contents of an expert report.

See Chapter 12 for further information on the use of experts.

---

17   Ibid., para. 61.

E. EXCHANGE OF SUBMISSIONS

Following the civil law tradition, the principal form of submissions by parties in an international arbitration is written submissions (also called statements, memorials or briefs). If a hearing is held, oral submissions and arguments are kept to a minimum, replaced instead by written submissions. Typically, narrative-style written submissions are delivered to the arbitrator(s) and the opponent party at least a week before the hearing, sometimes longer. There may be more than one round of submissions. Service is normally sequential rather than simultaneous. The practice of exchanging written submissions supports shorter oral openings, which, as a practical matter, reduces the length and cost of the proceedings, and therefore the disruption to the parties' and arbitrators' other commitments.

F. HEARINGS

Arbitral rules allow for disputes to be decided on documents alone. Generally, however, parties are given the right to insist upon a hearing, and international arbitration proceedings invariably involve one or more hearings.[18] The manner of conducting a hearing, like any other aspect of arbitration procedure, is a matter for the parties and/or the arbitrator(s). Hearings may be split, for example, between quantum and liability. This is relatively common in construction disputes, where determining quantum is a large and costly exercise, particularly if quantification on a number of alternate cases is required before liability is known.[19] The governing factor in determining the number, length and content of hearings in an arbitration tends to be efficiency – the aim being to organize the required hearings to promote efficiency for all the people concerned, as well as the process itself. This may mean, for example, organizing hearings on the basis of issues so as most efficiently to utilize the availability and involvement of witnesses and experts concerned with only some of the issues in dispute.

See Chapter 12 for more detail on the conduct of hearings in arbitration proceedings.

G. CLOSING SUBMISSIONS

Closing submissions are typically ordered to be served after the close of the hearing within a timeframe agreed between the parties or set by the arbitrator(s). The

---

18 For example, Art. 20(6) ICC Rules provides that the Arbitral Tribunal may decide the case of documents alone, unless any party requests a hearing. Art. 19 LCIA Rules also contemplates documents-only arbitrations.
19 ICC Construction Arbitration Report, para. 49 provides guidance on whether cases should be split in this way.

timeframe varies from a few days up to, more typically, a few weeks. Closing submissions are not intended to raise new arguments, but to summarize the arguments put in previous submissions and at the hearing.

VI. ADDITIONAL ISSUES

A. Preliminary Issues

Consistent with their endeavours to achieve efficiency in international construction arbitration, parties may wish to consider whether there are any issues in dispute between the parties which may be suitable for treatment as preliminary issues, to be determined ahead of other issues in the case. Normally, an issue will be suitable for preliminary determination only if its resolution has the potential to dispose of some or all of the case. Otherwise, the fracturing of a case in this way is generally inadvisable.

The types of issues which typically lend themselves to preliminary determination include whether the parties are bound by the arbitration agreement, whether the tribunal has jurisdiction to hear the dispute, what is the applicable law of a contract or a dispute, whether an action has been brought within applicable time limitations or whether a condition precedent has been satisfied. However, even these questions are not always appropriate for preliminary determination. For example, sometimes it is more efficient to hear questions of jurisdiction at the same time as substantive matters where a detailed understanding of the substantive matters is required in order to be able to determine whether the arbitrator(s) have the relevant jurisdiction.

B. Security for Costs

An order for security for costs affords a successful respondent protection against the circumstances of being awarded the costs incurred in running its defence, where the claimant does not have the finances to pay such costs. The order requires the claimant to provide security for the respondent's anticipated costs. Where a claimant fails to observe such an order, its claims will usually be stayed.

Security for costs orders are historically viewed as common law (outside of the United States) creatures, particularly due to the acceptance by continental Europe of the Hague Conventions on Civil Procedure 1905 and 1954, which prohibit security for costs being ordered in relation to nationals of signatory states.

The arbitral institutions have differing attitudes towards ordering security for costs, ranging from express provision of such a power to a purposive interpretation of the rules. For example the LCIA Rules specifically provide that the arbitral tribunal shall have the power to order a party to provide security for costs.[20] Pursuant to the ICC Rules, the arbitral tribunal has a broad power to order 'any interim or conservatory measure'.[21] Besides these provisions, there is no further guidance on, for example, the procedure by which a tribunal should enforce an order for security for costs.

Where arbitral tribunals have limited jurisdiction to order security for costs, parties seeking such an order may be required to apply to national courts.

However, there has been a shift in the attitudes taken by the courts with regard to intervening in international arbitrations. In the English case of *Bank Mellat*,[22] Goff LJ held that where parties merely choose to arbitrate in England as a matter of convenience, and there is no more significant connection to England, it would not be appropriate for the English court to order a foreign claimant in arbitration proceedings to provide security for costs, even though it had power to do so.

This position was revised by the case of *Ken-Ren*,[23] in which the House of Lords held that the English court could order a foreign claimant in arbitration proceedings to provide security for costs, even where the arbitration has no connection with England except for the fact that it is the chosen seat. It was held that an English court could support an ICC arbitration in England through its ability to grant an order for security for costs, although such an order should be made only in exceptional circumstances.

Ken-Ren met with widespread disapproval in the arbitration community, largely because of the damage it was perceived to have done to London as a seat for international commercial arbitration. In response, when the English Arbitration Act 1996 was enacted the following year, it expressly established that it is the arbitral tribunal, not the courts, that has power to order security for costs in an international arbitration, unless the parties agree otherwise.[24] This remains the position in England.

C.   INTERIM MEASURES

Interim measures (also known as preservation orders or conservation orders) are made in order to protect parties' rights pending the final settlement of the dispute. They are interim measures to prevent a party from incurring any disadvantage during the period from the start of the arbitral proceedings to the implementation of the award. These orders protect the parties' interests until the award is recognized or enforced. It should, however, be noted that (unlike most court orders of this

---

20   Art. 25.2 LCIA Rules.
21   Art. 23 ICC Rules.
22   *Bank Mellat v Helleniki Techniki SA* [1984] QB 291.
23   *SA Coppée Lavalin NV v. Ken-Ren Chemicals and Fertilizers Ltd* [1995] 1 AC 38.
24   S. 38(3) English Arbitration Act 1996: 'The tribunal may order a claimant to provide security for the costs of the arbitration', as long as the order is not based on the claimant's residence outside the United Kingdom. This is subject to s.38(2), which provides that the tribunal's power can be removed by agreement of the parties.

nature) the jurisdiction of the arbitral tribunal extends only to the parties to the agreement: the tribunal has no power to bind third parties or to compel them, directly or indirectly, to comply with any direction, award or order it may make.

Upon a request for such measures, the tribunal will initially assess whether it has jurisdiction to grant the relief by examining the parties' agreement. The power may be expressly conferred in the parties' agreement to arbitrate or may be contained in institutional rules incorporated into that agreement. For instance, the power to grant interim or conservatory measures is given to tribunals by the ICC Rules,[25] the LCIA Rules[26] and the UNCITRAL Arbitration Rules,[27] unless the parties have agreed otherwise.

However, the arbitrators' power to grant interim relief is always subject to the mandatory laws of the country in which the arbitration is taking place. In addition, a general power contained in institutional rules may be restricted by limitations contained in the parties' agreement. As a result, a tribunal faced with an application for interim relief will have to look at the relevant arbitration rules and the applicable law, as well as to the agreement to arbitrate, in order to determine whether in fact it has the necessary jurisdiction to grant the order requested.

Unfortunately there is a lack of uniform guidance as to the type of relief available in relation to these measures. Under international arbitration rules there is little or no specific direction. Accordingly, arbitrators 'can either use the rules available under the law applicable to the arbitration, or settle the issue without making any reference to the applicable law'.[28] The latter is the more common approach, in which the arbitrators evaluate the facts directly to make their decision on whether or not to grant the relief sought. Urgency is a prerequisite for the ordering of conservatory measures, in that such measures will not be granted where no irreparable damage will be caused to the property or the rights of the parties by waiting for the final resolution of the dispute.

The enforcement of such orders is through the issuing of reminders by the tribunal where a decision is disobeyed. If such disobedience persists, the tribunal is generally 'empowered to take a failure to obey an order for interim measures into account in its final decision, particularly in any assessment of damages'.[29] Despite this, it is generally considered that a tribunal – unlike a court – does not have power to make adverse orders in default of compliance, though in practice failure to comply is unlikely to endear the party in default to the tribunal, or to result in any favours being granted to it.

---

25 Art. 23 ICC Rules.
26 Art. 25 LCIA Rules.
27 Art. 21 UNCITRAL Rules.
28 Julian Lew, *'Commentary on Interim and Conservatory Measures in ICC Arbitration Cases'*, ICC International Court of Arbitration Bulletin 11(1) (2000), 23–30 .
29 See Final award (1998) in case 9593 (available in *'Extracts from ICC Awards referring to Interim and Conservatory Measures'*, ICC International Court of Arbitration Bulletin 11(1) (2000), 107).

In addition to the powers conferred on tribunals, national courts generally retain residual authority, both before and after the tribunal is constituted, and in their supporting role. However, it is more likely that the arbitral tribunal will exercise its power in such circumstances where expressly provided for in the international arbitration rules. Nevertheless, the arbitral process is often considered ill-suited to dealing with conservatory relief because, for example, it may take weeks or months to appoint a tribunal, or where there is a need to make an order binding a third party (e.g. banks in the context of a freezing order) as the tribunal has no jurisdiction to bind anyone other than the parties to the arbitration agreement.

In fact, the whole area of the availability of interim relief in arbitrations has recently been under intensive – and often heated – review by 'Working Group II (Arbitration)', a group established by UNCITRAL to consider the working of the Model Law in this area and to make recommendations for changes to it. The result is a greatly expanded Article of the UNCITRAL Model Law (Article 17) dealing with interim measures, which confirms and sets out in more detail a tribunal's powers to make such orders – possibly even including the power to make 'anti-suit' orders. Unlike the previous version, the new Article 17 now spells out the elements that a party must demonstrate to the tribunal's satisfaction before obtaining an order for interim relief. This includes satisfying the tribunal that the requesting party has a reasonable chance of succeeding on the merits, that damages would not be an adequate remedy and that any harm caused to the subject party by making the order would be substantially outweighed by the harm caused to the requesting party if the order was not made. All of this is very familiar from an English perspective, as too are the requirements that the requesting party notify the tribunal of any change in circumstances on which its application was made, and that the requesting party be responsible for any harm caused to the subject party if the tribunal subsequently finds that the order should not have been made. All of this was (relatively) uncontroversial.

By way of contrast, a most heated issue was the availability of *ex parte* interim relief. In most countries, some form of urgent relief is available from the courts on the application of one party in the absence of the other. In England, for instance, injunctions to prevent specified actions are routinely available from a judge on the application of one party without notice to the other party. The advantage of this process is that notice need be given to the other party only after obtaining the order – by which time it is (in theory) too late for that party to do the act prohibited by the court. At the very least, any attempt to do so will normally attract potentially severe sanctions from the court for breach of its order.

There are a number of issues about extending this procedure into regular commercial arbitrations. Stripped of the rhetoric, these can be reduced into two distinct classes of objections. The first is purely philosophical, the second practical.

The principal philosophical objection is that arbitration is a consensual process and it is fundamentally wrong in principle for one party to have access to the arbitrator on a private basis. Indeed, if this were to take place in other circumstances it would in all probability amount to misconduct, which would form the basis either for a challenge to the validity of the process or award in the courts of the place of the arbitration, or as a basis for resisting enforcement of an award at a subsequent stage.

The practical objections are perhaps more compelling. To start with, the most common purpose for applying to a court for an injunction – particularly an ex parte injunction – is that the respondent cannot be trusted and needs to be restrained from subverting the dispute resolution process. However, while it is helpful if the court making the order has personal jurisdiction over the respondent and can therefore fine or imprison it (or its officers, if a company) if the order is breached, the real benefit of an injunction is that third parties will also recognize the effect of the injunction and comply with its terms. Thus, a court order preventing the disposition of the respondent's funds *should* be respected by the respondent but *will* be respected by any bank or other financial institution within the court's jurisdiction given notice of the order. Similarly, an order preventing the removal of an aircraft or ship from the jurisdiction will be respected by those authorities responsible for providing air traffic control or pilots, and aircraft and port services. In other words, the real benefit of an injunction lies not in its direct effect on the person who is the subject of the order, but in its indirect effect via third parties who control the movement of the item which it is sought to restrain. No arbitration order can possibly have this effect without enforcement by a court – in which case, a party would be better served by starting in court.

Another powerful concern is in reproducing in the context of an arbitration the checks and balances on the process of obtaining ex parte awards which have been built into the procedures for obtaining them. Taking the English process as an example again, an ex parte injunction can be obtained only if supported by sworn testimony in respect of all the relevant facts, including not only the facts that support the application, but all the facts relevant to whether the application should be made. Failure by the applicant to give 'full and frank' disclosure of the relevant circumstances is a ground for immediate lifting of the injunction, and possibly for sanctions on the applicant.

The seriousness of the process is also supported by the requirement that the applicant must in all circumstances give an undertaking to the court to make good any damages the respondent (or others) suffer as a result of the applicant obtaining an injunction in circumstances where it was not entitled to. This undertaking may, in certain cases, have to be backed up by the provision of tangible security. Moreover, not only is an injunction only ever obtained for a relatively short period before further review by a judge (hearing both the applicant and the respondent), but it is *always* open to the respondent to go back to the judge who made the order (or another judge) before that time to explain why the injunction should immediately be lifted.

The practical objections to arbitrators exercising the same powers to grant injunctions as judges can therefore be summarized as follows. First, the powers are in principle ineffective because an award (even if it can be obtained quickly, bearing in mind the need to establish the tribunal) has no coercive effect on the third parties, on whose assistance it is in practice normally so dependent,[30] absent enforcement by a court. Second, granting such powers (even if largely ineffective) to arbitrators, without the comprehensive array of checks and balances built up over years by national courts, is extremely unsatisfactory, even more so when it is considered that any general extension of arbitrators' powers will be capable of being exercised not only by retired judges and other experienced international arbitrators but by any arbitrator, no matter how (in)experienced. Finally, there must be legitimate doubts over arbitrators' ability (or even willingness) to be available at very short notice to hear arguments from the respondent as to why any order made should be lifted.

Despite these powerful objections – and even though the Working Group has not finally reported at the time of writing – the chances are that the UNCITRAL Model Law will be amended to allow arbitrators to issue some form of ex parte orders. Based on the compromise being thrashed out in the Working Group at its 43rd session in October 2005, the key elements – taking account of some of the concerns set out above – are likely to be as follows:

- a party may, unless the parties otherwise agree (i.e. in the agreement to arbitrate) apply to the tribunal in the absence of the other party for an order for interim measures and for a 'preliminary order' directing that the purpose of the order for interim measures not be frustrated;
- a tribunal hearing such an application may make a 'preliminary order' ex parte, but only if it is satisfied that all the elements applicable to the grant of an order for interim measures referred to above have been met *and* that there is a reasonable concern that the purpose of the application for interim measures would have been frustrated by prior disclosure by the applicant to the other party;
- immediately after determining the application, the tribunal will give notice to the party against which the application was made, advising it of the application, the tribunal's decision and any written communications (and indications of any oral communications) passing between the applicant and the tribunal;
- the tribunal will give an opportunity for the party against which any preliminary order is made to present its case at the earliest practicable time;
- a preliminary order made ex parte will lapse after 20 days, but may be adopted in its original or modified form as an order for interim measures by the tribunal after the party against which it has been made has been given an opportunity to present its case;
- the applicant must provide security for damages caused by the making of a preliminary order unless the tribunal considers it inappropriate or unnecessary to do so;

---

30 This is in addition to the general concern that any temporary order of this sort may not be treated as an 'award' and therefore may not be enforceable internationally pursuant to the terms of the New York Convention.

- the applicant shall have a continuing obligation (until the party against which the application has been made has presented its case) to disclose to the tribunal all circumstances relevant to the grant of a preliminary order; and
- a preliminary order shall not be judicially enforceable (unlike an order for interim measures which is, by virtue of a further likely amendment to the UNCITRAL Model Law, to be recognized and enforced by national courts 'irrespective of the country in which it was issued').

Overall the changes proposed – assuming they make it to the final version of the revised UNCITRAL Model Law – are in many ways radical and offer claimants new opportunities to advance their cases. However, the changes are unlikely greatly to affect the construction industry, either immediately or at all. In the first instance, a revised form of UNCITRAL Model Law will not automatically be adopted, even by those countries that have adopted its predecessor. In addition, in those countries where the legislative process is lengthy, it is unlikely that a change of this relatively minor (but at the same time controversial) nature will attract much parliamentary time.

Perhaps as importantly, the reality is that it is unusual for applications for injunctions to be made in construction cases, and those that are made are normally best handled by the courts, as in the Channel Tunnel case,[31] where Eurotunnel sought an injunction preventing its contractor from stopping work on an element of the tunnel works which was critical for the timely completion of the project. This case provides a good example of the difficulties inherent in these sorts of applications. In addition to the basic issue of whether the English courts had jurisdiction under the then current legislation to make any such order (now resolved), there were serious questions about whether it was appropriate to make an order effectively compelling the contractor to continue its work. The judge at first instance indicated he was prepared to make such an order (in response to which the contractor agreed to carry on working), but the House of Lords – the ultimate court of appeal – finally determined that it was inappropriate to make such an order on a temporary basis. This was because, as a result of the time taken to go first to a dispute resolution board and then to arbitration, making such an order would have the effect of finally determining the question, which the court felt it should not be doing in the light of the parties' agreement to arbitrate.

D.   SITE VISITS

Construction arbitrations lend themselves particularly well to site visits, which can be an effective way of engaging the tribunal with the subject matter of a particular dispute. The IBA Rules explicitly authorize this procedure:

---

31   *Channel Tunnel Group Ltd and another v Balfour Beatty Construction Ltd and others* [1993] AC 334.

'Subject to the provisions of Article 9.2, the Arbitral Tribunal may, at the request of a Party or on its own motion, inspect or require the inspection by a Tribunal-Appointed Expert of any site, property, machinery or any other goods or process, or documents, as it deems appropriate. The Arbitral Tribunal shall, in consultation with the Parties, determine the timing and arrangement for the inspection. The Parties and their representatives shall have the right to attend any such inspection'.[32]

The LCIA Rules also explicitly contemplate site visits:

'Unless the parties at any time agree otherwise in writing, the Arbitral Tribunal shall have the power, on the application of any party or of its own motion, but in either case only after giving the parties a reasonable opportunity to state their views: ... to order any party to make any property, site or thing under its control and relating to the subject matter of the arbitration available for inspection by the Arbitral Tribunal, any other party, its expert or any expert to the Arbitral Tribunal'.[33]

Given the broad discretion given the parties and the tribunal in organizing the arbitration process, such a procedure should also be permissible even under arbitration rules that do not explicitly refer to site visits.

VII. ADMINISTRATIVE ISSUES

There are also a number of practical or administrative issues that must be given consideration, including the appropriate venue for the hearing, provision of transcription and translation services. These issues are addressed in more detail in Chapter 12.

---

32 Art. 7 IBA Rules.
33 Art. 22.1(d) LCIA Rules.

Chapter 12
The Conduct of the Hearing

I.   INTRODUCTION

Arbitration is a consensual process, so with the agreement of the parties – and of course the acquiescence of the tribunal – the hearing can take essentially any form, provided that it allows an opportunity for the fair consideration of the parties' respective contentions. Consider the following provisions of major sets of arbitration rules:[1]

> *UNCITRAL Arbitration Rules:*
> 'Subject to these Rules, the arbitral tribunal may conduct the arbitration in such manner as it considers appropriate, provided that the parties are treated with equality and that at any stage of the proceedings each party is given a full opportunity of presenting his case'.[2]
>
> *American Arbitration Association International Arbitration Rules:*
> 'Subject to these rules, the tribunal may conduct the arbitration in whatever manner it considers appropriate, provided that the parties are treated with equality and that each party has the right to be heard and is given a fair opportunity to present its case'.[3]
>
> *ICC Rules:*
>
> 'The proceedings before the Arbitral Tribunal shall be governed by these Rules and, where these Rules are silent, by any rules which the parties or, failing them, the Arbitral Tribunal may settle on … . In all cases, the Arbitral

---

1   These rules will apply by virtue of the parties' agreement, whether in the original arbitration agreement or at the time that a dispute arises and is submitted to arbitration.
2   Art. 15(1).
3   Art. 16(1).

Tribunal shall act fairly and impartially and ensure that each party has a reasonable opportunity to present its case'.[4]

LCIA Rules:
'The parties may agree on the conduct of their arbitral proceedings and they are encouraged to do so, consistent with the Arbitral Tribunal's general duties at all times: (i) to act fairly and impartially as between all parties, giving each a reasonable opportunity of putting its case and dealing with that of its opponent; and (ii) to adopt procedures suitable to the circumstances of the arbitration, avoiding unnecessary delay or expense, so as to provide a fair and efficient means for the final resolution of the parties' dispute.

....

'Unless otherwise agreed by the parties under Article 14.1, the Arbitral Tribunal shall have the widest discretion to discharge its duties allowed under such law(s) or rules of law as the Arbitral Tribunal may determine to be applicable; and at all times the parties shall do everything necessary for the fair, efficient and expeditious conduct of the arbitration'.[5]

At one end of the spectrum, arbitration hearings have proceeded with nearly the degree of formality common in court proceedings whilst, at the other extreme, there are those which are difficult to distinguish at a glance from a conversation around a conference table. In fact, if the parties agree, a dispute submitted to arbitration may even be decided *without* a hearing. Although a 'documents only' proceeding is an option under most sets of arbitration rules, and it undoubtedly offers the advantage of efficiency, it must be said that it is unusual for an international construction dispute to be resolved in this way, and a hearing will be held if either party requests it or the tribunal so determines.[6]

Despite this broad range of options, there is enough commonality among arbitrations in the general run of things that comment can usefully be made about the likely phases of an arbitration hearing and issues that might arise in preparing for and conducting such a hearing. For example, in virtually every arbitration that goes to a hearing, bundles of the core documents on which each party intends to rely are exchanged in advance. In every such case the advocate(s) for each party will make submissions to the tribunal relating both to the relevant law and to the facts of the dispute at hand. In addition, witnesses are commonly called to give and to be examined on their evidence.

Documents, submissions, and presentation of witness evidence are considered in the first three sections of this chapter. The final section deals with hearing practicalities and other issues that might arise.

---

4   Art. 15.
5   Art. 14.
6   See, e.g., Art. 20(2) ICC Rules, Art. 19.1 LCIA Rules, and Art. 15(2) UNCITRAL Arbitration Rules. The AAA International Arbitration Rules, seem to presuppose that there will be a hearing (see Art. 20).

The Conduct of the Hearing

Before turning to each of these topics, it is worth making two general points about the conduct of the hearing. First, in general the procedures and evidentiary rules that are applied in arbitration are more flexible than in court proceedings. This is the case not least for reasons of efficiency: hearing time is precious – both because there is always more that the parties (or their lawyers) want to cover than there is time for, and because hearing days, requiring the attendance of the tribunal, parties, advocates and witnesses and incurring fees for administrative support (hearing venue, stenographers and similar) are very costly. Second, for both of these reasons, the presentations made at the hearing should be carefully tailored in the light of what has gone before in the proceedings.

Typically, and certainly ideally, the hearing will be the culmination of a period of educating the tribunal about the dispute. It will follow the exchange of witness statements, in which each party sets out the facts from their point of view, and pleadings, in which each party argues any legal points in issue and makes submissions on the application of the law to the facts.

II. DOCUMENTS

A. BUNDLES

The backbone of any arbitration hearing is an agreed set, or bundle, of the relevant documents. This results from the disclosure/document exchange process outlined in Chapter 11 above, and should represent a carefully selected subset of those documents exchanged between the parties.

The first point to make about bundles is obvious: the parties need to make sure that all the documents to which they may want to refer at the hearing or in written submissions are included. Although bundles can be, and often are, supplemented up to and during the hearing, they should be provided to the tribunal sufficiently in advance of the hearing to enable the arbitrators to spend some time gaining familiarity with those documents the parties have identified as key. If important documents are not included, obviously the opportunity for the tribunal to begin taking those documents into consideration in their thinking about the case is lost.

The second point is equally basic, if not so obvious: the parties must be careful not to include so many (tangentially relevant) documents in the bundles that they become unwieldy and key documents are buried amid unnecessary volume. Ideally, although in complex cases this may be less feasible, by the end of the process the tribunal will have gained significant familiarity with the key documents – the arbitrators will know their way around the bundles. Being selective and organized in the presentation of documents in the bundles makes this easy for the arbitrators and facilitates the presentation of the parties' cases.

Thought should be given to making the bundles easy to use, with clear consecutive numbering of files and documents, and clear labelling of files. This not only has direct impact on the efficiency of the proceedings, and therefore on cost, but it can also have strategic or substantive significance, if for no other reason than that good points elicited on cross-examination can lose much of their impact if it takes time to locate the relevant documents to support the points. It is often sensible to arrange the core documents that 'tell the story' chronologically, with perhaps other series of bundles containing collected documents on specific themes or of a specific type. That being said, it is possible for bundles to be too user-friendly. In organizing subsets or thematic bundles, especially, take care that you are not thereby rendering transparent strategies that are better revealed during the hearing.

One other matter that bears some thought is the manner in which documents will be referred to in the proceedings. One suggestion from the UNCITRAL Notes on Organizing Arbitral Proceedings is to 'keep a table of contents of the documents, for example, by their short headings and dates, and provide that the parties will refer to documents by those headings and dates'.[7] The simplest possibility may be to number each page of the bundles consecutively, so that each page of each document may be referred to by a single unique number throughout the proceedings. If this approach is adopted, it is good practice to have the bundles agreed and numbered in this fashion as early in the process as possible, and certainly before the parties make any pre-hearing written submissions.

B. AUTHENTICITY OF DOCUMENTS

Generally, arbitral practice avoids the formality of court proceedings, and the hearing proceeds on an unspoken assumption of the authenticity of documents (without the necessity of eliciting witness evidence concerning authenticity), unless authenticity is specifically challenged. In any event, authenticity is one issue that may, and ideally will be resolved by agreement of the parties on bundles. As the UNCITRAL Notes on Organizing Arbitral Proceedings suggest:

> 'The parties may consider submitting jointly a single set of documentary evidence whose authenticity is not disputed. The purpose would be to avoid duplicate submissions and unnecessary discussions concerning the authenticity of documents, without prejudicing the position of the parties concerning the content of the documents'.[8]

A slightly more formal, but still efficient approach to the issue of the authenticity of documents is also suggested in the same Notes:

> 'It may be helpful for the arbitral tribunal to inform the parties that it intends to conduct the proceedings on the basis that, unless a party raises an objection to any of the following conclusions within a specified period of time: (a) a

---

7 UNCITRAL Notes on Organizing Arbitral Proceedings, note 53.
8 See note 53.

document is accepted as having originated from the source indicated in the document; (b) a copy of a dispatched communication (e.g. letter, telex, telefax or other electronic message) is accepted without further proof as having been received by the addressee; and (c) a copy is accepted as correct. A statement by the arbitral tribunal to that effect can simplify the introduction of documentary evidence and discourage unfounded and dilatory objections, at a late stage of the proceedings, to the probative value of documents. It is advisable to provide that the time-limit for objections will not be enforced if the arbitral tribunal considers the delay justified'.[9]

C.   PRESENTATION OF DOCUMENTARY EVIDENCE

The manner in which documentary evidence is to be presented in a given case will largely depend on the particular facts, and the state of the evidence, in that case. As with everything in international arbitration, however, in every case, the presentation of documentary evidence will require a fine balance between drawing key documents to the tribunal's attention and wasting precious hearing time. The old English tradition of extensive oral presentation of evidence – including the reading of documents into evidence – is out of place in international commercial arbitration. It is very costly, and it may not be very effective advocacy. It will far more often be the case that the advocate will bring key documents to the attention of the tribunal, and will then be able to 'let the document speak for itself'.

To the extent, however, that a case turns on a close reading of a particular document, or comparison of the terms of a few documents, it may be worth considering whether visual aids – projecting relevant language on a screen or depicting it on a large-scale exhibit, a simple comparison chart on a sheet of paper, or a more technologically advanced presentation of the materials – would be appropriate for the case. In making this determination, a key consideration will be the background and experiences of the individuals making up the arbitral tribunal.

III.   SUBMISSIONS

A.   WHO MAY APPEAR?

Generally speaking, each party to arbitral proceedings will be represented by a lawyer or, depending on the complexity of the case, a team of lawyers, who will give oral submissions on its behalf. However, whilst this is the most common situation, it is not the case that only lawyers can represent a party at a hearing. Indeed, parties can usually nominate any person they choose as their representative. The English Arbitration Act 1996 provides, on this point, as follows:

---

9   See note 52.

'Unless otherwise agreed by the parties, a party to arbitral proceedings may be represented in the proceedings by a lawyer or any other person chosen by him'.[10]

Similarly, the LCIA Rules state as follows:

'Any party may be represented by legal practitioners or any other representatives'.[11]

The UNCITRAL Arbitration Rules also contain a similar provision.[12]

Accordingly, while engineers or businessmen (for example) are usually called upon at a hearing as expert or fact witnesses, it may, in some circumstances, be more time efficient for them instead to address the arbitral tribunal directly as representatives of the parties.

B. How Long Should Oral Submissions Last?

As a rule, arbitral hearings tend to be relatively short (at least from the perspective of common law practitioners), for a number of reasons. First, a three-member tribunal constituted of senior figures from the arbitration world or elsewhere will have limited availability to meet for extended periods of time. In addition, considerations of cost (the parties will literally be paying for every hour the tribunal sits) play their part in keeping hearing time limited. Finally, whilst common lawyers may be used to and expect long hearings, civil lawyers do not. Thus, even large and complex disputes may have merits hearings lasting only a week or so. If this is the case, then a three-day opening speech is clearly an ineffective use of the time available. Further, it is unlikely that an extended presentation of every single point or document in one party's case will have the persuasive impact of a carefully focused presentation that sets up the case that the party's representatives will be building upon over the course of the hearing.

In addition to the natural restraints placed upon parties' time at the hearing, the arbitral tribunal will usually place time limits (either before or at the beginning of the hearing) on each side's opening submissions, examination-in-chief and re-examination of its own witnesses (in the event that there is any), cross-examination of the other side's witnesses, and closing submissions. Generally, the tribunal will allow each side the same aggregate amount of time for oral submissions, and tribunals will generally make the parties stick to the times allotted to them in order to avoid the risk of one party being left with insufficient time to conclude its presentation and, therefore, run the risk of being prejudiced. The tribunal

---

[10] S. 36 English Arbitration Act 1996.
[11] Art. 18.1 LCIA Rules.
[12] Art. 4 UNCITRAL Arbitration Rules.

will not want to give one party the opportunity at a later date to have the award set aside on the grounds of any alleged unequal treatment accorded to it during the proceedings.

C. WHAT APPROACH SHOULD BE ADOPTED?

Arbitration rules typically allow the tribunal more or less to determine the procedure relating to oral submissions at the hearing, including the order of submissions and whether the parties should present opening and closing submissions. Practices differ depending on the composition and background of the tribunal, but a not uncommon practice is to allow a brief opening statement by each party, followed by the evidence of each side's witnesses. This may involve a short examination-in-chief (even where the witnesses have submitted witness statements which are standing as the evidence-in-chief of those witnesses), with the main focus being on the cross-examination of each witness. Most tribunals will then allow re-examination of the witness. Some tribunals will then allow short closing submissions, whilst others may prefer written post-hearing pleadings to be submitted by the parties, whether on specific issues or on each side's case as a whole.

Alternatively, tribunals may direct the parties towards a symmetrical presentation of their positions, whereby one side presents its claim, the other its defence and counterclaim, the first its defence to counterclaim and rebuttal to defence and the second, finally, any defence to the counterclaim.

Practice differs as to whether the claimant or the respondent has the last word (common law practice, for example, is that it is for the claimant to have the last word).

Clearly the content and length of the oral submissions at the hearing will also depend on the extensiveness of any prior written submissions as well as any directions the tribunal may have given as to specific issues it wishes the parties to address.

D. WRITTEN SUBMISSIONS

Finally, as noted in Chapter 11 above, practice differs between common and civil law practitioners as to whether notes are submitted to the tribunal summarizing the oral pleadings or not. Some counsel are used to giving pleading notes to the tribunal and the other side at the time of the hearing or shortly beforehand. Other (typically common law) counsel submit detailed skeleton arguments in advance of the hearing and go through the key points in their submissions. In the latter case, the tribunal and the parties generally agree upon this course of action in advance of the hearing. It is also common for detailed post-hearing submissions to be provided to the tribunal, pulling together each party's case in the light of the evidence as it develops at the hearing.

E. VISUAL AIDS AND OTHER 'BELLS AND WHISTLES'

The arbitral process has the potential to be used more creatively than court proceedings. It is, however, a matter of carefully judging the tribunal and assessing what will be the most effective way of presenting a case. Visual aids such as PowerPoint presentations and video clips can be very effective with the right audience, but some tribunals may find them overpowering and distracting. In addition, there is a very real danger of a party locking itself into a particular form of presentation when using such aids. Particularly when responding, or with an impatient tribunal, having to follow a pre-ordained path built into a PowerPoint presentation can have damaging results.

IV. PRESENTATION OF WITNESS EVIDENCE

A. WITNESS STATEMENTS

In the ordinary course, the evidence of a witness will first be presented in a written witness statement. These statements may be take the form of sworn affidavits, although more often they are documents signed by the witness based only upon a statement of truth.[13]

The IBA Rules provide the following helpful guidance on the preparation of a witness statement:

> 'The Arbitral Tribunal may order each Party to submit within a specified time to the Arbitral Tribunal and to the other Parties a written statement by each witness on whose testimony it relies ... (the 'Witness Statement'). If Evidentiary Hearings are organized on separate issues (such as liability and damages), the Arbitral Tribunal or the Parties by agreement may schedule the submission of Witness Statements separately for each Evidentiary Hearing.
> Each Witness Statement shall contain:
> (a) the full name and address of the witness, his or her present and past relationship (if any) with any of the Parties, and a description of his or her background, qualifications, training and experience, if such a description may be relevant and material to the dispute or to the contents of the statement;[14]

---

13 See, for example, Art. 20.3 LCIA Rules, (*'Subject to any order otherwise by the Arbitral Tribunal, the testimony of a witness may be presented by a party in written form, either as a signed statement or as a sworn affidavit'*).
14 Art 4.4 IBA Rules.

(b) a full and detailed description of the facts, and the source of the witness's information as to those facts, sufficient to serve as that witness's evidence in the matter in dispute;
(c) an affirmation of the truth of the statement; and
(d) the signature of the witness and its date and place...'.

While the preparation of a witness statement will, in all but the most exceptional cases, be a process in which the party's lawyers will be closely involved, it is important that this involvement be limited to assisting the witness to express his knowledge and recollections clearly and with a view to addressing the issues in dispute in the arbitration (see further below). The most effective witness statements are those which read as though they were written by the person who is giving the statement, and there is nothing more transparent – and less convincing – than a group of witness statements which appear to recite an agreed 'party line' that has been drafted by lawyers, rather than tell their own stories.

B. WITNESS PREPARATION

Preparation of a witness begins with the drafting of the witness statement, a process which should be based on identifying and discussing with the witness his recollections about the events in dispute and all of the relevant documents and other evidence which has come to light in respect of which he is likely to have relevant knowledge. In construction cases, the key document in this process with most engineers is his log or diary – often a manuscript document but almost always written contemporaneously before events have had a chance to become clouded. This process of the lawyers educating themselves about the witness's role in, and knowledge of, the relevant events, and of the witness, as appropriate, re-familiarizing himself with the relevant events, lays a solid foundation for preparing a witness to give oral testimony. If for some reason a written witness statement has not been prepared and submitted, then this process needs to be undertaken in anticipation of the witness's direct testimony.

The witness preparation process will depend to some extent upon the approach that is to be taken to witness examination, discussed below. The amount of time that can and should be devoted to preparation will depend upon a number of factors, including the witness's availability and the relative significance of the witness in the larger context of the case. Nevertheless, a few general points can be made.

First, there is (subject to applicable ethical rules) no impropriety in a party and its lawyers interviewing and helping to prepare a witness to give oral evidence. Although the rules of some national courts or local ethical rules (predominantly in civil law jurisdictions) frown on this practice,[15] it is accepted in many other

---

15 Examples include the Code of Professional Ethics of the Geneva Bar Association (Art. 13); the Austrian Rules on the Exercise of the Legal Profession, Supervision of the Duties of

jurisdictions[16] and commonplace and appropriate in international arbitration.[17] The LCIA Rules, for example, expressly provide:

> 'Subject to the mandatory provisions of any applicable law, it shall not be improper for any party or its legal representatives to interview any witness or potential witness for the purpose of presenting his testimony in written form or producing him as an oral witness'[18]

and the IBA Rules state:

> 'It shall not be improper for a Party, its officers, employees, legal advisors or other representatives to interview its witnesses or potential witnesses'.[19]

Accordingly, one commentator concludes,

> 'it is clear that few, if any arbitral tribunals would consider proper witness preparation to be objectionable. Indeed the procedural requirement for parties to submit written witness statements from all witnesses they intend to call, which has become widespread in international commercial arbitration, effectively necessitates that some form of witness preparation will take place prior to any hearing. In view of this fact, it is undoubtedly preferable that lawyers, who are subject to rules of professional conduct, carry out such witness preparation'.[20]

Second, the purpose of witness preparation is twofold: to ensure the witness's familiarity with the matters in dispute on which he is likely to be examined and to ensure the witness's familiarity and (as much as possible) comfort with the process in which he will participate. As another commentator advises:

Attorneys and Training of Attorney Trainees (Art. 8); and the Belgian Bar Association Rules.

16 See, for example, Rule 4.03 Rules of Professional Conduct, The Law Society of Upper Canada; and Rule 5-310 Rules of Professional Conduct, State Bar of California. Even when contact with witnesses prior to their giving oral evidence is not prohibited, of course, lawyers remain subject to other rules of professional responsibility and ethical standards.

17 Interestingly, Belgian law, which generally prohibits counsel's approach to a witness prior to the witness giving oral evidence, allows this to occur if this is allowed under the procedural laws governing the arbitration.

18 Art. 20.6 LCIA Rules.

19 Art. 4(3) IBA Rules.

20 D. Roney, *'Effective Witness Preparation for International Commercial Arbitration: A Practical Guide for Counsel'*, Journal of International Arbitration 20 (2003) 429, 430. This article sets forth a useful six-step guide on witness preparation. See also Georg von Segesser, *'Witness Preparation in International Commercial Arbitration'*, Bulletin de l'Association Suisse d'Arbitrage 20 (2002) 222–228; and John P. Madden, *'How to Present Witness Evidence in an Arbitration: American Style'*, Bulletin de l'Association Suisse d'Arbitrage 11 (1993) 438–445.

'The role of counsel should be to assist witnesses in developing the confidence and clarity of thought required to testify truthfully and effectively based upon their own knowledge or recollection of the facts. It should be borne in mind that being examined in the 'witness box' is, for most witnesses, an unfamiliar and intimidating experience'.[21]

There is no one right way to fulfil this dual function, but it will be important to provide the witness with all of the pleadings and evidence (including the statements of other witnesses and documentary evidence) relevant to the matters on which he will give evidence. These should be independently reviewed by the witness, and then gone over again in detail in conversations with counsel. The witness should be informed of the strategy each party to the arbitration is likely to be pursuing. This has the benefit not only of making the witness feel (and be) part of the team, but also of equipping the witness to give thoughtful answers to questions raised on examination. It will also be helpful to the witness to experience some 'mock' examination. This is by no means a session for the witness to rehearse his answers to questions, but is rather an opportunity for him to experience cross-examination and to practice the skills – careful listening to the question asked, thinking and only then thoughtfully responding – that will make his oral evidence most effective at the hearing.

It is also important to make sure that the witness knows exactly what he may expect, in terms of the setting in which he will be giving evidence, the process, and whether he will be asked to swear an oath or affirm the truth of his statement. He should be informed of his 'rights' in the process, including the right to ask for clarification of questions and to review documentary evidence that is referred to in his examination, and of his responsibility to answer questions truthfully and to the best of his knowledge. It can be helpful to have a witness sit in the hearing for a brief period prior to giving evidence in order to familiarize himself with the flow of the proceedings. Obviously, if this is permitted under the agreed procedure, it can also be of substantial significance to have witnesses observe the hearings and provide commentary and reactions to the evidence that is being developed as the proceedings unfold.

C. WITNESS EXAMINATION

1. Order of Presentation

The IBA Rules set forth the three standard phases of witness examination: direct examination (or 'examination-in-chief'), cross-examination and re-examination, and the order in which each party's witnesses ordinarily appear:

---

21 Redfern and Hunter, *International Commercial Arbitration*, para. 6–87.

'The Claimant shall ordinarily first present the testimony of its witnesses, followed by the Respondent presenting testimony of its witnesses, and then by the presentation by Claimant of rebuttal witnesses, if any. Following direct testimony, any other Party may question such witnesses, in an order to be determined by the Arbitral Tribunal. The Party who initially presented the witness shall subsequently have the opportunity to ask additional questions on the matters raised in the other Parties' questioning'.[22]

As the IBA Rules note, this normal manner of proceeding may be varied by agreement of the parties or order of the tribunal.[23] Ultimately, failing party agreement, the tribunal will determine the order in which the witnesses are to appear and the manner in which the examination will take place.

2. Oath or Affirmation of Truthful Evidence

The first thing that ordinarily occurs when a witness appears to give evidence is that the chairman of the tribunal (or the sole arbitrator) either asks the witness to swear that he will give truthful evidence or states the requirement that the witness must tell the truth. Practice differs as to which of these approaches will be adopted. The following comment on the practice in ICC arbitrations is valid in respect of international commercial arbitration generally:

'[I]n many cases it does not occur to the ICC arbitrator to put a witness on oath (i.e. to instruct the witness to swear under oath that the testimony about to be given is true). The practice may be more frequent in some common law jurisdictions where witnesses in arbitrations routinely give testimony under oath'.[24]

In fact, national law in many jurisdictions, especially common law jurisdictions, empowers arbitrators to administer an oath.[25] That being said,

'it is much more frequent practice ... for the chairman or sole arbitrator to inform the witness that he has a duty to give truthful testimony and that he may be subject to criminal penalties for false testimony, and to secure the witness' acknowledgement of this duty'.[26]

---

22 Art. 8.2 IBA Rules.
23 Ibid., ('The Arbitral Tribunal, upon request of a Party or on its own motion, may vary this order of proceeding....'.).
24 Paulsson *et al.*, *International Chamber of Commerce Arbitration*, p. 435.
25 See, for example, s. 38(5) English Arbitration Act 1996 and s. 7 US Uniform Arbitration Act (enacted in various forms in many of the States of the United States of America).
26 Paulsson *et al.*, *International Chamber of Commerce Arbitration*, p. 435, n.1.

3. The Process of Examination

a.   *Direct Examination*

Following the procedure outlined in the preceding section, the direct examination of the witness will commence. Usually, a witness statement having been submitted, direct examination will consist only of the witness being asked to identify and affirm the truthful contents of his written statement (which will stand as his evidence in chief). Occasionally, the witness will need to make corrections or clarifications to the evidence contained in his statement, and, less frequently, additional direct evidence will be elicited by questions on a topic that had not come to light at the time the witness statement was prepared.

b.   *Cross-Examination*

In international commercial arbitration, the main focus of the presentation of witness evidence is on the next stage, cross-examination of the witness by the party with which it is not affiliated (or on behalf of which its evidence has not been submitted). Although legal traditions vary among common law and civil law systems, which can give rise to different approaches to cross-examination (with common lawyers being known for more searching examination that may focus as much on issues of credibility as on the factual evidence being elicited in a fashion that would not be usual for those from the civil law tradition), it is easy to overstate this distinction. As distinguished commentators on the ICC arbitration process observe:

> 'Contrary to the rumor that cross-examination is anathema to continental arbitrators, the authors' experience has been that most ICC arbitrators, irrespective of their origin, allow counsel a fair measure of cross-examination on all significant issues brought up in the witness' main statement'.[27]

c.   *The Role of the Tribunal*

Where there may be a more significant difference between common and civil lawyers is in the degree to which the tribunal will actively engage in questioning the witness. No matter what the background of the arbitrators, you should expect that they will have questions for the witness, and prepare the witness for that possibility. However, arbitrators schooled in the continental tradition may be more likely to take quite an active role in directing the examination of the witness. In Austria, Germany and the Netherlands, it is not uncommon for the tribunal to take the lead in questioning along the lines of a more inquisitorial model, with questions by the parties coming only after the tribunal's examination. In the United Kingdom and the United States this would be unusual – more often in these (and other common law) jurisdictions

---

27   Ibid., p. 439.

the tribunal will play a supplemental role in examining witnesses, certainly interjecting questions as necessary for clarification, but otherwise allowing the parties to conduct the examination and saving any extensive questioning on other topics until the parties have concluded their cross-examination.

d. *The Scope of Examination*

Due to the time constraints usually imposed on the arbitral process, and the more relaxed procedural environment, the scope of examination of witnesses in an arbitration is likely to be more focused than it would be in, say, English court proceedings. It would ordinarily be practically unworkable for the hearing to proceed on the basis that the advocate for one party has to put each and every point it wishes to plead to the other party's witnesses. That being said, it us undoubtedly the case that '[a]rbitral tribunals ... give greater weight to the evidence of a witness that has been tested by cross-examination, or by an examination by the arbitral tribunal itself'.[28] And the principal subject-matter for cross-examination should be those key factual matters in dispute in the case.

D. WITNESS CONFERENCING

In some arbitrations, witnesses may be asked to confer with one another in order to narrow the issues in dispute between them. The IBA Rules further contemplate that, by party agreement or upon the order of the tribunal, this conference (referred to in some sources as 'confrontation') should occur as part of the examination of the witnesses before the tribunal:

> 'The Arbitral Tribunal, upon request of a Party or on its own motion, may vary this order of proceeding, including the arrangement of testimony by particular issues or in such a manner that witnesses presented by different Parties be questioned at the same time and in confrontation with each other'.[29]

The authors of the ICC Construction Arbitration Report expressed similar views:

> 'There was much support for the view that factual witnesses should be heard before the experts formally present their reports and are questioned on them, since the questioning of a factual witness may lead an expert to a better understanding and to the modification or withdrawal of an opinion or provisional conclusion. It was also thought that, where the parties are to tender experts or witnesses on the same topic, they should be questioned

---

28 Redfern and Hunter, *International Commercial Arbitration*, para. 6–88.
29 Art. 8.2 IBA Rules.

together so as to clear up misunderstandings that may have arisen between them'.[30]

The advantage of witness conferencing is that it allows on the spot challenge of evidence by those with a factual and technical knowledge (to remove the need for counsel to refer another witness to a statement in a transcript of evidence given earlier in the hearing). The aim is to clarify issues more directly and to shorten the hearing time as the questioning of a series of witnesses on the same issue is avoided. Witness conferencing is a relatively new development, however. There may be the risk of chaos if the procedure is not well controlled, with witnesses speaking out of turn or descending into arguments. It is important that the tribunal should prepare in advance and have a clear understanding of the facts and technical issues so as to be able to orchestrate effective questioning. The tribunal here will act as a 'ring-master' not a 'referee'. As the procedure is relatively novel and there are obviously different ways of approaching it, the tribunal must set clear guidelines to ensure both counsel and the witnesses understand how the process is to work. A pre-hearing meeting with counsel to agree ground rules is sensible. It is also important that the tribunal should ensure that each side's counsel has a fair opportunity of asking questions, and strict control must be exerted to ensure that witnesses do not speak at the same time or become unduly aggressive or domineering.

E.   EXPERT WITNESSES

As noted in Chapter 9, expert witness evidence becomes relevant in cases in which, in order to resolve the dispute presented to them, the tribunal must evaluate not only factual evidence and legal submissions, but matters of opinion on subjects other than the law. This is very often the case in construction disputes, with the result that hearings of construction disputes almost always involve some degree of expert evidence.

The evidence of party-appointed expert witnesses is ordinarily presented and tested by examination in a manner similar to that outlined above. In some cases, however, the tribunal may find it necessary, to prevent the 'battle of the experts' becoming a war of attrition, to limit the number of expert witnesses each party may call. In order to limit the issues to be addressed at a hearing, it is common practice for the experts to meet before the hearing, on a without prejudice basis, to identify those areas in their respective reports where they agree the issues. There may be a direction for a joint report identifying those areas where they have reached agreement and those which remain in dispute, with a brief description of their respective views. Oral evidence at the hearing will then be limited to the outstanding issues.

---

30   ICC Construction Arbitration Report, para. 66.

Alternatively, it may be agreed, or the tribunal may order, that the experts should be heard together. Counsel can then cross examine the opposing experts on the same issues rather than taking the evidence of each expert sequentially. An extension of this approach is for the experts to ask questions of each other. This procedure is sometimes referred to as 'hot tubbing' and can have the advantage of clarifying the issues and shortening the procedure as the questions are not all routed through the parties' counsel. As with witness conferencing with witnesses of fact (addressed above), the tribunal needs to ensure that the process is properly managed and controlled.

F.   VIDEO CONFERENCING

If a witness cannot be present at the hearing, he can give evidence by video conferencing, although this can only be recommended (and possibly not even then) for relatively less important witnesses and if it is absolutely impossible for the witness to attend as, generally speaking, a witness will be more effective in person. Additional problems may include the transmission being disrupted mid-flow, an unseen third party feeding answers to a witness, or concerns regarding the confidentiality of the testimony. All in all, without a broadcast standard video link and supervision of the witness at the remote studio, the process of examining and cross-examining a witness by video link is far from ideal.

Nevertheless, video conferencing is fully compatible with a number of institutional rules.[31] It will be important, however, to check the mandatory rules at the seat of arbitration allow for video conferencing. If they do not allow for video conferencing but such prohibition is overlooked, this may jeopardize the enforcement of the award.

Practical considerations surrounding witness video conferencing include agreeing a date, time and place with the witness; ensuring that the equipment works (possibly by conducting a test run); assessing the approximate duration of the conference; and agreeing whether it is necessary for there to be supervisors present with the witness at the time of his testimony.

V.   PRACTICALITIES AND OTHER ISSUES

A.   ATTENDANCE

Arbitration hearings take place in private. Access to the hearings is generally restricted to the parties, their representatives and any witnesses and experts during their testimony, as well as the tribunal and any secretary to the tribunal. Hearings are not open to the public, and outsiders may be present only if both parties and the

---

31 See, for example, Art. 20(1) ICC Rules.

tribunal agree. This position is reflected in, for example, the ICSID arbitration rules, which state as follows:

> 'The Tribunal shall decide, with the consent of the parties, which other persons besides the parties, their agents counsel and advocates, witnesses and experts during their testimony, and officers of the Tribunal may attend the hearings'.[32]

A party cannot be excluded from the hearing unless it disrupts the hearing so much that the tribunal deems it impossible to continue in an orderly fashion.

It will be a matter for agreement between the parties (failing which an order of the tribunal) whether those who are called as witnesses may be present in the hearing prior to giving their evidence.

B.   SCHEDULING OF THE HEARING

1. Duration

The length of the hearing has already been discussed above. Another point to consider, however, is whether to opt for a single hearing or several separate hearings. Single hearings are generally preferable if all the issues can be heard in a relatively short period of time, as they involve less travel costs and the opportunity for the tribunal to fully immerse itself in the case. However, for hearings that will take longer, it may be difficult to schedule a single hearing, and having a number of hearings may be preferable where the dispute raises a number of issues and/or involves multiple parties, which can easily be divided between separate hearings.

2. Cancellation Penalties

At the outset of the arbitration, in the course of agreeing the arbitrators' fees, the tribunal will usually notify the parties of their intention to charge for time reserved for hearings that do not, for whatever reason, take place. These cancellation penalties are a recognition of the fact that, the arbitrators having blocked time to sit on one case, cannot often at short notice arrange for other work to fill that time if the hearing is cancelled or postponed.

The amount of cancellation penalties tends to differ depending on the amount of notice the parties give to the arbitrators (e.g., 50 per cent of the tribunal's fees to be paid upon cancellation of the hearing with a month's notice, 80 per cent with a couple of week's notice, etc.). The issue of cancellation fees is, however, a sensitive one. It is by no means universal for cancellation fees to be requested or, if requested,

---

32   The UNCITRAL Arbitration Rules also provide that *'Hearings shall be heard in camera unless the parties agree otherwise'* (Art. 25.4).

agreed. Arbitrators who are professionals in their principal field of practice will not, of course, suffer any loss of earnings through having a case collapse or settle if they are in full time salaried (or equivalent) employment. Not all arbitral institutions recognize the practice. It is, however, difficult for only one of two parties to take a stand on this issue, particularly where the arbitrator is a full time, sole practitioner and the relevant institutional rules permit (or do not prohibit) it.

C.   LOGISTICS OF THE HEARING

1.   Who Organizes the Hearing?

One or both of the parties may be responsible for the organization of the hearing (normally the claimant, with the agreement of the other party). Alternatively, this task may fall to the sole or presiding arbitrator or to an individual designated by the tribunal as secretary to the tribunal. Sometimes, but less frequently, in the case of an arbitration administered by an institution such as the LCIA or ICC, the institution makes the arrangements for the hearing.

The cost and task of organizing a hearing should not be underestimated. For example, a hearing will typically involve organizing a number of rooms (breakout rooms for both sides, the tribunal and the transcribers, as well as the hearing room itself) in addition to accommodation for each side, their witnesses and experts, as well as the tribunal, the tribunal's secretary, the transcribers and any translators.

2.   Venue for Hearing

When considering the venue for the hearing there are numerous factors to be taken into account. Most importantly, are the conference facilities suitable and adequate? Factors to bear in mind include the size of the room, the availability of breakout rooms for the parties, arbitrators and possibly witnesses, facilities for meals and refreshments, availability of equipment such as overhead projectors, telephones, faxes, photocopiers and PCs with internet access, as well as technicians to deal with any problems that may arise. Access to law libraries may also be available.

It is worth considering how long the venue is open, including evening and weekend access, as many locations have very specific hours with no room to manoeuvre, whereas others are happy to accommodate circumstances in which hearings may overrun or where the parties specifically wish to have use of the venue beyond normal office hours.

Issues in relation to the payment of a deposit need to be borne in mind as most venues will require such a payment. The cancellation policy needs to be reviewed as some venues require substantial notice to be given if payment is to be returned.

At the most basic of levels, the location of the venue itself needs to be chosen with care so as to ensure it is convenient for all parties, their witnesses and the arbitrators. Travel time and costs need to be considered in relation to the location of the accommodation.

Security is also an issue, in order to avoid having to remove document bundles at the end of each day.

3. Arrangement of Hearing Room

The hearing room itself should be spacious with separate long tables for each of the counsel teams, party representatives and the tribunal. The transcribers and translators may also need separate tables and, in our experience, the transcribers often also take up a separate room as their equipment can be disruptive to the hearing.

The room will need to accommodate several sets of the hearing bundles (the arbitrators', the parties', one for the use of the witnesses). Each side's counsel will need enough room behind its table to be able to keep its documents readily to hand.

4. Stenographers or Transcribers

It is not essential to have a transcript of the hearing and in some cases the tribunal may instead prepare a summary (which tends not to include the testimony of the witnesses). However, in our experience, 'in disputes of any complexity' it is generally a good idea to have professional transcribers prepare a verbatim record of the hearing. Occasionally this facility is dispensed with, if the venue is ill-suited to accommodating the necessary equipment.

In the event that transcribers are used, it is a good idea to provide them with a list of parties and key names and terms before the hearing as well as any handouts as the hearing progresses.

Another consideration is whether the parties require the transcribers to produce a daily transcript of the hearing or, instead, whether a full transcript after the hearing has ended is sufficient. Although generally more expensive, the first option enables counsel to review how the day has gone, as well as preparing for submissions going forward.

Generally, the parties will review the transcript and provide the transcribers with amendments, which the transcribers will use to produce a correct, agreed, version. This process, whilst seemingly straightforward, can become difficult if parties attempt to use it as an excuse to literally re-write the record.

The costs of the transcribers are added to the costs of the arbitration.

5. Translators

It is usually the responsibility of the party whose witness is in need of a translator to make the necessary arrangements. In some cases, the parties may agree on a translator who can provide services for witnesses presented by both sides.

Testimony of witnesses will tend to be more reliable if given in their mother tongue and, therefore, if all of the members of the tribunal do not understand that language, it may be preferable to have a translator. Some witnesses, on the other hand, may not speak the language of the arbitration perfectly, but may speak it well enough to give evidence directly. It is a matter for the parties and their advocates to judge.

The choice of interpreter is important, as it will be from that person that the tribunal will be hearing the witness's evidence. Therefore, it is important to choose someone who has experience of international arbitration, who is competent and also independent (it will do your party no good if the tribunal is given the impression that the translator is partial).

You will also need to consider whether translation should be simultaneous or consecutive and, if consecutive, budget for the additional time this evidence will talk.

D. *Ex Parte* Hearings

1. What is an *Ex Parte* Hearing?

If one party refuses or fails to appear, the hearing can proceed *ex parte*.[33] Refusal by a party can be express, e.g., where a party refuses to respond to correspondence or states that it will not participate in the arbitration, or implied, e.g., where the party does not say outright that it will not attend the hearing but, instead, creates an unreasonable delay that the tribunal decides to treat as a refusal or an abandonment of the right to participate. What constitutes an unreasonable delay for these purposes is a matter for the tribunal.

The tribunal will, in any event, set out in the award the circumstances surrounding the non-appearance of the party in question, in order to demonstrate that the party was given ample opportunity to appear and that the award is still enforceable notwithstanding the *ex parte* nature of the hearing.

2. What is the Procedure in an *Ex Parte* Hearing?

Arbitral tribunals cannot render awards similar to default judgments. A tribunal has to consider the merits of the dispute in front of them and come to a considered and reasoned decision on the same. Thus, even if one party does not appear, the other party still has to prove its case to the satisfaction of the tribunal. This means that a hearing may be necessary, albeit a shorter one than would be the case if both parties were appearing.

---

33 See Art. 21(2) ICC Rules, Art. 15.8 LCIA Rules, and Art. 28(2) UNCITRAL Arbitration Rules.

Where an *ex parte* hearing does take place, there is no need for the tribunal to represent the non-appearing party. However, it is the case that, in such circumstances, the tribunal will often test the arguments of the appearing party more rigorously than it would otherwise have done.

Chapter 13
Effect of the Award

I. INTRODUCTION

In the absence of a settlement between the parties, the culmination of the arbitral process is the tribunal's decision. Broadly, for this decision to be enforceable – particularly internationally – it needs to be in the form of an 'award'. However, there is no internationally accepted definition of when a decision of the tribunal constitutes an award. Obviously, if the decision is not enforceable the whole arbitration has almost certainly been an expensive waste of time. The question of when a tribunal's decision constitutes an award is, therefore, worthy of consideration.

To start with, it is worth observing that a tribunal may make numerous decisions in the course of an arbitration, not all of which will be awards. In addition, there may be several different types of awards. For instance, during the course of an arbitration and before the final decision is reached, a tribunal may make interim or partial awards, covering a wide range of issues. These issues may include the jurisdiction of the tribunal to determine the dispute, interim measures (such as the protection or preservation of the subject matter of the dispute pending its determination) or orders for the provision by one party of security for the other's costs of the arbitration. The tribunal may also make procedural orders, which can be distinguished from awards.

While, as indicated above, this distinction is of most concern when it comes to determining whether the decision of the tribunal is enforceable, in ICC cases, where all awards are scrutinized by the ICC Court in accordance with Article 27 of the ICC Rules, there is an additional need to determine whether a tribunal's decision is an award. See, for instance the French case of *Société Cubic Defense Systems, Inc. v Chambre de Commerce Internationale*.[1] In this case, the French

---

1   Tribunal de Grande Instance de Paris, May 21, 1997, confirmed by the Cour d'Appel, September 15, 1998 and by the Cour de Cassation, February 20, 2001. For a German

court rejected a plea that a procedural decision of an arbitral panel was in reality a disguised award, which should have been scrutinized and corrected by the ICC. Had the court found that the procedural decision was in fact an award, at minimum there would have been further delay while the tribunal reconsidered its decision, reformulated it as an award and then referred the award to the ICC for review and issue. Arguably, had the irregularity not been corrected, this could have formed the ground for a challenge to the validity of any award on the merits of the case ultimately issued by the tribunal.

Notwithstanding this area of uncertainty in the classification of decisions, there is a generally accepted proposition that awards are decisions of the tribunal which 'finally determine the substantive issues with which they deal'.[2] An award need not deal with all of the issues in dispute (in which case it might properly be described as a 'partial award') but the essence of an award is that it must be dispositive of one or more of the substantive issues between the parties. So, typically, an early decision on whether the tribunal has jurisdiction to determine the dispute would be given as a (partial) award on jurisdiction, the intention being that on this point the tribunal has (from its own perspective) finally ruled on its own ability to determine the disputes referred to it. This degree of finality within the arbitration allows the parties to proceed with the rest of the proceedings, secure in the knowledge that at some later stage the tribunal will not turn around and effectively bring the proceedings to an end by deciding that, after all, it had no jurisdiction to embark on the process.

A.   TYPES OF AWARD

Although monetary awards are the most common remedy granted by arbitral tribunals, they are not the only remedy available.[3] Arbitral awards may also grant extensions of time; correct interim contractual decisions by engineers, architects and the like; order restitution, specific performance, declaratory relief, permanent injunctions, punitive damages (where the jurisdiction permits); adapt contracts, fill gaps in contracts; order rectification of contracts and make decisions on interest and costs.

---

perspective, see Rolf Trittmann, *'When should Arbitrators Issue Interim or Partial Awards and/or Procedural Orders'*, Journal of International Arbitration 20 (2003), 255–265.

2   Redfern and Hunter, *International Commercial Arbitration*, para. 8–06.
3   For a fuller discussion on remedies, see John Collier and Vaughan Lowe, *The Settlement of Disputes in International Law: Institutions and Procedures*, (Oxford University Press, 1999), pp. 248–253.

B.   *RES JUDICATA* EFFECT OF THE AWARD

A valid award will bind the parties to the arbitration. Usually, an award has no effect whatever on those who were not parties to the arbitration, and neither confers rights nor imposes obligations upon third parties. However, there are exceptions. Arbitration awards can bind third parties, but only in circumstances where those parties' rights depend on the rights of the parties to the arbitration. For instance, an assignee of a contract is generally regarded as being bound by the terms of any arbitration clause contained in the assigned agreement even though (as a matter of English and some other laws) an assignee can take only the benefit of a contract and is not bound to perform its terms. Similarly, an insurer exercising its rights of subrogation would be bound to resolve disputes in accordance with any arbitration agreement the insured had entered into, and would be bound by the results of any such arbitration.

As between the parties, the award has the effect of *res judicata*.[4] This is the principle that a matter, finally adjudicated by a competent court or arbitral tribunal, may not subsequently be reopened or challenged by the original parties or their successors in interest. This doctrine, found in many common and civil law systems,[5] prevents any of the parties to a dispute trying to re-open litigation by bringing another action against the same party in respect of the same issues as those determined in the original proceedings. The doctrine of *res judicata* is based on public policy considerations, including the desire to ensure that there is a conclusion to the dispute and that a party to the dispute is not troubled more than once in relation to the same issue, and thus generally to promote efficient and final dispute resolution. Indeed, in the English legal system it is not only matters which were actually resolved in proceedings which are treated as having this degree of finality but also matters which could and should have been resolved in the proceedings.[6] The full rigour of this principle probably applies only to proceedings in the English courts, but its potential application to arbitration proceedings should not be ignored in English law/English seat arbitrations.

The corollary to the principle of *res judicata* is that there is no concept of precedent (or *stare decisis*)[7] in arbitrations, domestic or international. Therefore, a prior decision by one arbitral tribunal has no legal authority beyond its effect on

---

4   Literally *'the thing has been judged'*.
5   For example, Art. 1703 of the Belgium Code Judiciaire of 19 May 1998, reads as follows: *'Unless the award is contrary to ordre public or the dispute was not capable of settlement by arbitration, an arbitral award has the authority of res judicata…and may no longer be contested before the arbitrators'*.
6   *Henderson v Henderson* (1843) 3 Hare 100, 115.
7   Literally *'to stand by that which is decided'*. This is the principle that in deciding the case before him, a judge, in certain circumstances, is bound to stand by previously decided cases and to accept and follow the principles of particular precedents. It differs from *res judicata* in that it relates to a principle or rule of law being treated as settled, whereas *res judicata* relates to the finality of a decision of a particular dispute between particular parties.

the parties to the arbitration (and those claiming through them) as *res judicata*. It is binding only on the parties to that particular arbitration and does not form a part of some greater whole to be relied upon by another party in a similar situation. That said, it is often the case that prior arbitration awards are cited in subsequent arbitrations between different parties, albeit with varying degrees of success. The need for this can be found in a number of different areas. For instance, for many years the FIDIC suite of contracts has required disputes under those standard forms to be resolved by arbitration in accordance with the ICC Rules. As in many cases these international standard forms are not amended to allow disputes to be determined – in public – by courts, the only source of materials on their interpretation is in arbitration awards. Another need for the introduction of the decisions of earlier tribunals arises out of the (fortunately relatively rare) forms of arbitration clauses where the substantive law of the contract is defined as (or as including) the decisions of international tribunals.[8]

There are at least two obvious problems with the practice of referring to previous arbitration awards. The first is that arbitrations are generally regarded as being private matters with, the consequence that there is no systematic recording and publication of awards. Therefore, the number of accessible decisions is always limited, with an additional concern being that the cases that are published and available for future reference may well have entered the public domain for reasons unconnected with their significance as potential legal authorities (for example, because the successful party publishes the award to advertise its success).

The second problem is that there is no quality control over the vast majority of arbitral awards.[9] Since they have no need to be concerned about the establishment of a coherent, internally consistent set of principles, arbitrators (rightly) are free to determine a case entirely on its merits, the peculiarities of that specific case, the requirements of the arbitration clause and the substantive and the procedural law. This combination of factors is specific to individual decisions and is clearly not conducive to the development of a reliable system of precedent, as is found within common law systems, by the appeal process, the level at which the decision was made (first instance, first appeal, second appeal etc.) and also the reputation of the specific judge or judges in the area of law or practice giving rise to the dispute.

Only, perhaps, in the emerging body of law relating to the interpretation of bilateral investment treaties (BITs) are these problems addressed. This arises out of the increasing availability of a published set of awards on a series of similar BITs governed not exclusively by national laws but by the principles of international

---

8 For instance, the contract for the construction of the cross-Channel tunnel link between England and France contained a governing law clause which (in material part) said *'The construction, validity and performance of the Contract shall in all respects be governed by and interpreted in accordance with the principles common to both English law and French law, and in the absence of such common principles by such general principles of international trade law as have been applied by national and international tribunals'*.

9 The ICC Court scrutinizes awards before they are delivered to the parties, but it is the only one of the major commercial arbitral institutions to do so.

law. A further element is the fact that where such BIT disputes are resolved by arbitration in accordance with the ICSID Arbitration Rules, there is an internal review process which, while expressly intended not to be an appeal on the merits of the case, does provide a degree of quality control and consistency across a number of cases.[10]

C. NOTIFICATION OF THE AWARD

Once the award has been made, most international and institutional rules of arbitration provide for the delivery of the award to the parties.[11] Normally this is only after payment of all sums due to the arbitrators and the arbitral institution. Where this is not done, the parties, or at least the party who expects to win, will no doubt make it his business to obtain a copy and notify the other side, if necessary paying whatever is necessary to settle outstanding fees if advances paid to date have not proved to be sufficient.

Of course, this may not be the end of the matter. In some countries, steps must be taken to register or deposit the award to make it effective and advice should always be taken on this and other local requirements of the place where the arbitration is taking place. If the arbitration award is not complied with voluntarily (and while many awards are honoured, the trend is, perhaps, for more to be resisted), enforcement against the assets or the officers of the losing party will become necessary. This is considered in further detail below.

D. OPTIONS FOR THE LOSING PARTY

A party who receives an unfavourable award has four options. These are:

(i) to accept the award and comply with its terms;
(ii) to attempt to use the award as a basis for settlement;
(iii) to challenge the award with a view to having it set aside; or
(iv) to do nothing, wait until the winning party brings enforcement proceedings and then resist enforcement of the award.

This chapter considers options (iii) and (iv) in some depth. Little need be said about the first two options. The first option is self-explanatory, while the second turns on

---

10   Reed, Lucy, Jan Paulsson and N Blackaby, *Guide to ICSID Arbitration*, (Kluwer Law International, 2004).
11   For example, Art. 28(1) ICC Rules provides that, *'[o]nce an Award has been made, the Secretariat shall notify to the parties the text signed by the Arbitral Tribunal provided always that the costs of the arbitration have been fully paid to the ICC by the parties or by one of them'.* See also Art. 32.6 UNCITRAL Arbitration Rules and Rule 48(1) ICSID Arbitration Rules.

the commercial realities of the situation the parties find themselves in at the end of the arbitration, such as whether they might have common commercial interests in the future, the possibility of challenging the award, the chances of the losing party resisting enforcement and, of course, the ability of the losing party to meet the award or (if not a monetary award) to perform its terms.

II. CHALLENGING THE AWARD

A. THE MEANING OF 'CHALLENGE'

The term 'challenge' encompasses both the common law notion of an 'appeal' against an award, and the civil law notion of 'recourse' to a court of law against an award.[12]

B. THE PURPOSE OF A CHALLENGE

Challenging an award affords the losing party a means of attempting to have the award modified or even set aside. If an award is set aside by the court of the place where the arbitration takes place then it is usually treated as invalid and should be unenforceable. This is certainly true of enforcement in the country where the award has been set aside and may also be true where the winning party tries to enforce the award in another jurisdiction. For instance, Article V(1)(e) of the New York Convention provides that recognition and enforcement of an award *'may'* be refused if the award has been 'set aside or suspended by a competent authority of the country in which, or under the law of which, that award was made'. However, courts in Austria, Belgium, the United States, and France have all enforced awards which have been set aside by the courts of the place where they were made.[13]

Of course, one reason for the adoption of arbitration as a means of dispute resolution is to obtain a final and binding award without the risk of being mired down in additional proceedings, whether in the courts of one or other of the parties, the courts of the place of arbitration or the courts of the place of enforcement.

---

12 Other terms that the reader may come across are, for example, 'review' or 'annul'.
13 For a more detailed discussion of the public policy and other issues in this area, see Hamid G. Gharavi, *The International Effectiveness of the Annulment of an Arbitral Award*, International Arbitration Law Library, 15 (Kluwer Law International, 2002), pp. 77–117. See also Petrochilos, *Procedural Law in International Arbitration*, chapter 7; and Georgios Petrochilos, *'Enforcing awards annulled in their State of Origin under the New York Convention'*, International and Comparative Law Quarterly 48 (1999), 856. The case of *Chromalloy Aeroservices v The Arab Republic of Egypt* 939 F. Supp. 907 (DDC 1996) is at present the sole case permitting the enforcement of an annulled award in the United States. See also *International Bechtel Co Ltd v Department of Civil Aviation of the Government of Dubai* 300 F. Supp. 2d 112 (DDC. 2004). For a French perspective, see *Unichips v Gesnouin*, Paris, 12 February 1993, [1993] Rev Arb 255.

Accordingly, the bases for challenge to the substance of the award are becoming increasingly limited and narrowly delineated, as a result of converging internationally accepted standards. After all, every submission to arbitration contains an implied (or, in the case of some institutional arbitrations, express) promise by each party to abide by the award of the arbitrator, and to perform his award. It is on this promise that the claimant proceeds when he takes action to enforce the award.[14] A challenge against an award should not, therefore, consist of a review of the relative merits of that award, although where the arbitration process has failed, perhaps as a result of bias or other lack of due process in the proceedings, correction of the injustice is obviously appropriate and normally available.

C.   PRIOR EXHAUSTION OF OTHER AVAILABLE OPTIONS

Before making an application for challenge of the award to the relevant national court, the challenging party must first consider whether there are any other available remedies which it would be sensible to consider. Indeed, in some cases it may even be a pre-requisite to an application for challenge that all other remedies have been exhausted. Examples of these other remedies can be found in the rules governing the arbitration or the national laws of the place of the arbitration, which may provide a mechanism for the tribunal that made the award (or in some cases the institution or another tribunal) to correct it in some way or to provide an additional award.

However, in most cases, the provisions under which a tribunal can *correct* parts of the award are necessarily limited in scope. For example, under Article 29 of the ICC Rules, a tribunal may correct, 'clerical, computational or typographical errors or any errors of a similar nature contained in an [a]ward'.[15] The UNCITRAL Arbitration Rules contain similar provisions for correction of minor errors and interpretation.[16] So too does the UNCITRAL Model Law (Article 33), which forms the basis for an increasing number of national laws on arbitration.[17]

In addition to the correction of minor errors of the type discussed above, it may also be possible that the institutional rules applicable to the arbitration in question permit a review of the merits of the award. Although uncommon in the arbitration rules of the major commercial arbitration institutions and specialist rules for the resolution of construction disputes, such provisions do exist. For instance, the current arbitration rules of the Arbitration Foundation of South Africa provide (Article 22) for the possibility of an appeal to another arbitral tribunal, at least where

---

14   See generally *Bremer Oeltransport GmbH v Drewry* [1933] 1 KB 753.
15   Art. 29.2 ICC Rules also allows a tribunal to issue an interpretation of the award at the request of one of the parties.
16   Arts 35–37 UNCITRAL Arbitration Rules.
17   See also, for instance, Arts 17.1 and 27 LCIA Rules. In England this power to correct trivial typographical errors is conferred expressly by s. 57 English Arbitration Act 1996. Likewise, the United States Federal Arbitration Act provides that the relevant district court may make an order modifying or correcting errors (s. 11 (as amended 1970)).

the parties have indicated that an interim or final award may be the subject of such an appeal.

It should also be noted that a party dissatisfied by an ICSID tribunal award may apply to have the award interpreted, revised or annulled.[18] Indeed, ICSID arbitral awards are particularly noteworthy as this internal review mechanism is the only method of challenging the award: a party cannot bring a separate challenge in court.[19] However, despite this extensive array of powers to adjust awards and the provision of an internal 'appeals' procedure, even this system falls short of a full review of the merits. The power to interpret an award, whether by the same tribunal or, where that it is not possible (perhaps because of the death or incapacity of a tribunal member), by a separate tribunal,[20] is not dissimilar to the powers to correct minor errors found in the rules referred to above. The power to revise is an unusual one, depending on the subsequent discovery by the applicant of a fact which would decisively affect the award and which was not known to the applicant at the time of the award. In addition, the applicant must not have been negligent in not knowing about that fact. Such situations are, understandably, rare.

This leaves only the power of annulment. The ICSID Arbitration Rules make it clear that this power is about procedural failings by the tribunal. The grounds for annulment are: the tribunal was not properly constituted, the tribunal has 'manifestly exceeded its powers', there was corruption on the part of a member of the tribunal, there was a serious departure from a fundamental rule of procedure or the award failed to state the reasons on which it was based. While successful annulment applications were once relatively common and their availability remains a useful check on the conduct of tribunals and the development of principles of international law, the trend appears to be for successful applications to be on the decline. This is so notwithstanding the ingenuity of the arguments of losing parties to bring substantive review of the merits of the case into one of the permitted grounds for annulment, most particularly that a tribunal in failing to apply the law properly or taking proper account of the evidence put before it in reaching the correct decision on the facts 'manifestly exceeded its powers'.

D. CHALLENGING AN AWARD IN COURT

If, having considered the provisions for interpretation or correction of the award or the internal review procedures (or indeed, if there are no internal review procedures), a party is still not satisfied, that party must then consider upon what grounds it can bring a challenge in court.

---

18 Arts 50–52 ICSID Arbitration Rules.
19 Art. 26 Washington Convention provides that 'Consent of the parties to arbitration under this Convention shall, unless otherwise stated, be deemed consent to such arbitration to the exclusion of any other remedy'.
20 Interpretation of an ICSID award by a separate tribunal was carried out in the case of *Wena Hotels Limited v Arab Republic of Egypt* (Case No. ARB/98/4).

Effect of the Award

1. In Which Court Should an Application be Made?

It is conventional wisdom that a challenge to the award must usually be made in the courts of the place where the award was made, which is to say, the seat of the arbitration.[21] From time to time, however, courts of countries other than the place of the arbitration have determined that they too have jurisdiction to decide upon the validity of the award. Thus, prior to the passing of the Indian Arbitration and Conciliation Act 1996, the Indian courts had determined that the previous arbitration law gave them jurisdiction to review the conduct of arbitrations taking place outside India, at least where the agreement to arbitrate was governed by Indian law.[22] On essentially the same wording in their own arbitration laws, the courts in Pakistan adopted a similar position.[23] With the adoption of the UNCITRAL Model law in both these countries the position has at least in theory been brought into line with international norms, though no doubt from time to time similar cases will arise in other jurisdictions as international arbitration is introduced but before it becomes widely accepted.

2. When Should an Application be Made?

If a party is considering challenging an award in a national court, it must act promptly or it could lose the right to challenge. It is important to take local legal advice on this point as the time limits vary greatly from country to country. For example, an application to have the award set aside in a country which has adopted the UNCITRAL Model Law without amendment has to be made within three months of the date upon which the party making the application received the award (Article 34). Yet if an application for correction or interpretation has been made under Article 33, time does not start to run for any challenge until the application for correction or interpretation has been disposed of by the tribunal, giving a considerable time for a challenge to be mounted. By way of comparison, English law provides that a challenge must be made within 28 days of the award.[24] For an international perspective, in domestic arbitrations in Hong Kong, any challenge must be made within 21 days of the award;[25] and in municipal French arbitrations, there is a one-month time limit to lodge the appeal.[26]

---

21 This is also the position adopted by Art. 34 UNCITRAL Model Law. The award may only be set aside by the 'competent authority' specified by the state (see Art. 6).
22 See the Indian Supreme Court decision in *National Thermal Power Corporation v The Singer Co* 3 Supreme Court Cases (1992) 551–573.
23 *Hitachi Limited et al. v Mitsui & Company Deutschland and Rupali Polyester et al.* [1998] Supreme Court Monthly Review 1618–1687.
24 s. 70(3) English Arbitration Act 1996.
25 Rules of the Supreme Court of Hong Kong, Order 73, Rule 5.
26 Art. 1489 New Code of Civil Procedure.

E. THE GROUNDS FOR CHALLENGE UNDER THE UNCITRAL MODEL LAW

A party challenging an award in court must prove one of the grounds listed in the relevant arbitration law. There are, however, several generic grounds upon which a challenge may usually be based. With the increasing acceptance of the UNCITRAL Model Law either in its entirety or as the checklist against which other national laws are drafted it is worth looking at the specific grounds set out in the UNCITRAL Model Law as being representative of the types of challenge which can generally be made.[27] It is also no coincidence that there is a high degree of commonality between the grounds for challenge to an award contained in the UNCITRAL Model Law and the New York Convention grounds on which recognition and enforcement of an award may be refused (see further below).

1. Application for Setting Aside Under the UNCITRAL Model Law

Articles 34(2)(a) and (b) UNCITRAL Model Law provide an exclusive list of grounds under which a party may set aside an arbitral award in court. The UNCITRAL Model Law intended only the courts of the place of the arbitration to have jurisdiction over the validity of the arbitration award. It also intended that the normal remedy was for the award to be set aside, although Article 34(4) does provide that the award may be remitted to the tribunal for further consideration in appropriate cases.

Under the UNCITRAL Model Law, there are four grounds pursuant to which a party can apply to a court to have an award set aside.[28] The grounds are based unequivocally on the existence (or otherwise) of an agreement to arbitrate, natural justice and procedural fairness (or, as it is sometimes called, 'due process' or 'legality'). Looking at each in turn.

Article 34(2)(a)(i) allows for an application to set aside an award where:

'a party to the arbitration agreement referred to in Article 7 was under some incapacity; or the said agreement is not valid under the law to which the parties have subjected it or, failing any indication thereon, under the law of this State'.

This is obviously a reflection of the basis of arbitration which, depends on there having been an agreement between the parties to resolve their disputes in this way.

---

27 For a more detailed discussion or other possible grounds of challenge, see generally Lew, Julian D.M, Loukas A. Mistelis, and Stefan M. Kroll, *Comparative International Commercial Arbitration*, (Kluwer Law International, 2003), para. 303–316; and Garnett *et al.*, *A Practical Guide*, pp. 113–121.
28 Arts 34(2)(a)(i)-(iv).

However, it is worth noting that an arbitration agreement's validity is to be judged according to the law of the place of arbitration if the parties have not subjected (expressly or by implication) that agreement to a particular law. This can lead to rather arbitrary results if, for instance, the parties not only failed to identify the law governing the agreement to arbitrate but also the place of arbitration. In such a case the ICC, for instance, would select the place of arbitration and thus not only the court to which challenges would be made but also the standards by which the validity of the agreement to arbitrate would fall to be determined by that court.

Article 34(2)(a)(ii) allots an award to be set aside if:

'the party making the application was not given proper notice of the appointment of an arbitrator or of the arbitral proceedings or was otherwise unable to present his case'.[29]

Lack of due process is possibly one of the most important grounds for recourse against an arbitral award and amounts to violation of the principles of fairness. It is vital that minimum procedural standards are observed to ensure that the parties are treated with equality and are given a fair hearing, with a full and proper opportunity to present their respective cases. Difficulties normally arise not because basic procedural principles are contested but as a result of disagreement over how these principles are implemented. For instance, tribunals often wish to restrict the length of hearings by giving each party the same (limited) time to present its case. Yet is it fair, if one party has three times as many witnesses as the other and its witness statements have been admitted as evidence, for its opponent to be given only the same amount of time to present its case, despite the fact that this will require the cross-examination of three times as many witnesses? As is so often the case, the principles are easy to state but often difficult to apply in practice.

For this reason, national courts are left to determine, in an *ad hoc* fashion, exactly what is required to constitute a 'fair hearing'. Standards may vary from country to country, further emphasizing the need for cultural awareness of the norms of both the place where the arbitration takes place (for the purposes of maintaining or resisting a challenge) and the possible place of enforcement.[30] For example, in the United States, lack of an oral hearing may be regarded as a breach of due process and a sufficient ground to set aside the award. Yet, instructively, American courts have held that a United States corporation's due process rights were not infringed by a tribunal's decision not to reschedule a hearing for the convenience of the American corporation's witness[31] and that a respondent's lack of participation in an arbitration did not infringe its due process rights under United States law where

---

29   Art. 18 UNCITRAL Model Law regulates the parties' equal treatment and affords them full opportunity to present their respective cases.
30   See further below for an analysis of the parallel grounds in the New York Convention.
31   *Parsons & Whittemore Overseas Co. v Societe Generale de L'Industrie du Papier*, RAKTA 508 F 2d 969 (2d Cir 1974).

it received notice of the proceedings but offered no explanation of its failure to participate.[32]

Article 34(2)(a)(iii) allows an award to be set aside where:

'the award deals with a dispute not contemplated by or not falling within the terms of the submission to arbitration, or contains decisions on matters beyond the scope of the submission to arbitration, provided that, if the decisions on matters submitted to arbitration can be separated from those not so submitted, only that part of the award which contains decisions on matters not submitted to arbitration may be set aside'.

The arbitral tribunal's duty to restrict its ruling to the ambit of the dispute at hand can be found in Article 16 of the UNCITRAL Model Law , which is concerned with the tribunal's jurisdiction or competence. It is noteworthy, however, that Article 16(2) also requires that a party which discovers that the arbitral tribunal exceeded its jurisdiction must immediately raise an objection (and no later than the submission of the statement of defence). A party failing to comply with Article 16(2) risks being precluded from relying on this ground for annulment.[33]

Article 34(2)(a)(iii), therefore covers situations where an award was made by a tribunal with jurisdiction but which exceeded its powers under that jurisdiction. This argument most often arises where the subject matter of a case develops, either at the instance of the claimant or the respondent, after the matter has first been referred to arbitration. Of course, the ICC and other arbitration rules prescribe how this 'case creep' is to be dealt with. The ICC Rules, for instance, broadly lock the parties into the cases they have presented prior to the agreement of the Terms of Reference, with the addition of subsequent claims only being permitted with the consent of the tribunal, taking all the circumstances – including the stage at which the new claims are raised – into account.[34]

Finally, Article 34(2)(a)(iv) provides for set aside where:

'the composition of the arbitral tribunal or the arbitral procedure was not in accordance with the agreement of the parties, unless such agreement was in conflict with a provision of this Law from which the parties cannot derogate, or, failing such agreement, was not in accordance with this Law'.

Again reflecting the contractual, consensual nature of the arbitration process, the tribunal must be composed in accordance with the agreement of the parties or the governing law of the arbitration. However, if the agreement of the parties regarding

---

32 *Biotronik Mess-und Therapiegeraete GmbH & Co. v Medford Medical Instrument Co.* 415 F. Sup. 133, 140 (DNJ 1976).
33 See *Report of the United Nations Commission on International Trade Law on the work of its eighteenth session*, 3–21 June 1985, (A/40/17), para. 288, available at: http://www.uncitral.org/uncitral/en/commission/sessions/18th.html (accessed 19 August 2005).
34 Art. 19 ICC Rules.

the composition of the tribunal or the procedure conflicts with the governing law of the arbitration, it will be disregarded.

In addition to the above four grounds, there are a further two that may be raised either by a party or by a national court seised of the matter on its own initiative.[35] The first is that:

> 'the subject matter of the dispute is not capable of settlement by arbitration under the law of the place where the arbitration took place'.

If the court of the place of arbitration finds that 'the subject-matter of the dispute is not capable of settlement by arbitration' under its domestic laws, it can set the award aside. It should be noted that it is the law of the place of the arbitration that determines this issue and not the law chosen to govern the arbitration agreement or the law governing the agreement giving rise to the dispute. What does and does not fall within the class of matters not capable of being resolved by arbitration varies from place to place. However, some matters which typically may not be resolved by arbitration are these which involve the status of things as against third parties, e.g. patents. Also generally regarded as being outside the scope of arbitration are matters which involve the resolution of matters regarded as being in contravention of the criminal law of the country in which the matter complained of took place and/or against the criminal law of the country in which the arbitration is being conducted. This ground has all too often been seen to be a fertile way for respondents to escape their obligations to resolve their disputes by international arbitration, with many claims, not all spurious, of illegal conduct being raised by way of 'defence' and then relied on as a means of avoiding the agreement to arbitrate altogether.

The last ground Article 34(2)(b)(ii) allows an award to be set aside where 'the award is in conflict with the public policy' of the seat of the arbitration. This last ground has something of a chequered recent history, not least as a result of its potential for exploitation by respondents and sympathetic judiciary in countries newly adopting the UNCITRAL Model Law. The root cause of this is that the term 'public policy' has no accepted international definition.[36] Indeed, when adopting and interpreting the UNCITRAL Model Law every state can, quite legitimately, have its own concept of what is required by its 'public policy'.[37] Although inspired by the French concept of *'ordre public'*, the term 'public policy' is wider than the French doctrine because it covers more than merely the principles of procedural justice or fairness (or legality). Despite this considerable uncertainty, however, it

---

35 Arts 34(2)(b)(i) – (ii) UNCITRAL Model Law.
36 See, for example, Duncan Miller, *'Public Policy in International Commercial Arbitration in Australia'*, Arbitration International 9 (1993), 167–196.
37 See below for the New York Convention's public policy provision. For a very detailed discussion on public policy, see Vesselina Shaleva, *'The 'Public Policy' Exception to the Recognition and Enforcement of Arbitral Awards in the Theory and Jurisprudence of the Central and East European States and Russia'*, Arbitration International 19 (2003), 67–93.

is generally accepted that that the term should cover only those principles of law and justice (both substantive and procedural) that are fundamental to the principles of the state concerned. That said, the boundaries of this ground for challenge are not well defined. Along with the 'not arbitrable' ground discussed above, public policy has been used to protect the public interest in matters which parties might, perhaps, wish to keep private. Thus a United States court has recorded that public policy has been used as a reason for limiting the domain of arbitration in antitrust cases.[38] Antitrust issues are not solely a private matter but have the potential to affect millions of people; so, while a court may allow an antitrust arbitration to continue, it will not countenance the breaching of the legitimate interest in the enforcement of those antitrust laws.

The records of the discussions leading up to the adoption of the UNCITRAL Model Law [39] also make it clear that the public policy provision is intended, among other things, to cover the possibility of setting aside an award if the arbitral tribunal has been corrupted in some way, or if it has been misled by corrupt evidence. This was, apparently, considered necessary because doubts were raised as to whether the other provisions adequately covered all the circumstances in which awards might be set aside. It must be borne in mind, however, that what one national court considers public policy may not be the same as the public policy of another national court.

2. Other Grounds for Challenge

The potential grounds for challenge will not always be limited to those in the UNCITRAL Model Law. Every losing party will therefore need to consider to what extent the national laws of the country in which the challenge is to be made provide for a measure of control over the arbitral process.

The extent of this control will vary from country to country, taking it beyond the scope of this chapter to discuss all the different ways in which challenges may be made around the world. However, in order to give a flavour of what may or may not be possible, in civil law countries it is uncommon for a court to have the power to review the merits of a case.[40] In contrast, in some common law countries such as England, there is the possibility of a challenge on a point of law, though parties to both domestic and international arbitrations taking place in England may – and frequently do – contract out of this provision, for instance by the adoption of any

---

38 *Mitsubishi Motors Corp. v Soler Chrysler-Plymouth Inc.* 473 US 614, 628 (1985).
39 See the report of the UNCITRAL Secretary General, *'Analytical compilation of comments by governments and international organizations on the draft text of a model law on international commercial arbitration'*, (A/CN.9/263), para. 29–35, available at: http://www.uncitral.org/uncitral/en/commission/sessions/18th.html (accessed 19 August 2005).
40 See the NCCP in France (Art. 1483) and Portuguese Law No 31/86 (Art. 29). Interestingly the current law in South Africa, perhaps as a result of common law influences, does permit review of the merits of arbitration awards.

of the major institutional arbitration rules.[41] It is also a statutory requirement that before leave to appeal is given by an English court (unless the application is made by consent of all the parties) the court will have to be satisfied that the tribunal had in all probability reached the wrong conclusion or that the question of law is one of general public importance and the conclusion reached by the tribunal is open to serious doubt. These threshold requirements seriously limit the possible impact of this section on the finality of most arbitration awards in England. As a result of these varying criteria, it is important that a party wishing to bring a challenge to the award takes local law advice as soon as possible after the award is rendered.

F. REMEDIES AVAILABLE AS A RESULT OF A SUCCESSFUL CHALLENGE

As noted above, the basic remedy on a challenge to an award is to have the award rendered a nullity, leaving the parties to resolve the underlying dispute before another tribunal, with the attendant time and cost disadvantages that this will entail. Although Article 34(4) of the UNCITRAL Model Law (which permits the court to allow the tribunal to take such steps as would eliminate the grounds the court might otherwise have for setting aside the award) mitigates to some extent the 'all or nothing' approach of the UNCITRAL Model Law, under which an award is either set aside or left to stand, the potential remedies depend on the powers and provisions of the national laws which the court applies.

Possible options available as a result of a successful challenge[42] adopted in various jurisdictions around the world include the court:

(i) rejecting the challenge and confirming the award. Once the award is confirmed, the winning party can proceed to enforcement;
(ii) varying the award;
(iii) accepting the challenge and remitting or referring it back to the arbitral tribunal for reconsideration, amendment or clarification (this is in addition to the power contained in Article 34 of the UNCITRAL Model Law, referred to above, to allow time – before the court rules on the challenge – for the tribunal to put its house in order); or
(iv) setting aside the award (either completely or in part).

The practice of remitting the award to the tribunal for further review is more widespread in countries where the common law has had some influence including, for instance, Bermuda, Hong Kong, Eire, Malaysia, and Sri Lanka. In England, for example, setting aside an award is considered a drastic measure, in view of the consequences of such an action, and would be justified only if remission – the

---

41 S. 69 English Arbitration Act 1996.
42 Although there is no statistical data, it is generally accepted that most challenges are unsuccessful.

primary method of judicial intervention under the English Arbitration Act 1996[43] – did not achieve a just result.[44] In civil law jurisdictions remission is less common.

Setting aside an award in its entirety is the most common remedy following a successful attack on an award. If the award is set aside, it becomes unenforceable in the country in which it was made. Usually it will also be unenforceable elsewhere, although there are cases where the award has been set aside in one country (the seat of the arbitration) yet was still enforceable in another country.[45]

If the award has been set aside, the parties need to consider whether they are still bound by the arbitration agreement. This may depend on the grounds on which the award has been set aside. If the court has held that the arbitration agreement is invalid then the parties will no longer be bound. However, if the court has only set aside the award for procedural, jurisdictional or permitted substantive grounds, the arbitration agreement may still be effective.

G.   CONCLUSION

As can be seen, challenging an award is not a straightforward process and the grounds for challenge, the timing and the procedure varies considerably from jurisdiction to jurisdiction. There is, however, a definite drift towards harmonization of the rules governing the challenge of awards. There are three possible explanations for this development. First is the hope that a better understood set of grounds for challenge will reduce the scope for errors in the arbitral process, which in turn will reduce the need and opportunity for judicial review. Second (though not unconnected), is the pervasiveness of international conventions such as the UNCITRAL Model Law and the New York Convention, which champion 'arbitrator autonomy' and also provide an increasing degree of conformity across jurisdictions. Thirdly, countries' interest in attracting arbitration business to their territories drives a less interventionist approach to meet actual or perceived desires for finality in the arbitration process. The corollary of this is, of course, that scope for challenge based on peculiarities of national laws is likely to decline.

---

43   Ss 68(3) and 69(7) English Arbitration Act 1996.
44   Jurisdiction to remit an award also exists under the English Arbitration Act 1996 where a challenge to the award has been made for serious irregularities (s. 68(3)(a)) or error of law (s. 69(7)(c)).
45   For example, in France (see *Omnium de Traitement et de Valorisation v Hilmarton*, Cour de Cassation, 10 June 1997, YBCA XXII (1997), 696–698 and the United States (see *Chromalloy Gas Turbine Corp v Arab Republic of Egypt supra*). See also Redfern and Hunter, *International Commercial Arbitration*, Chapter 10.

III. ENFORCEMENT OF THE AWARD

It is thought that the vast majority of arbitral awards are (still) performed voluntarily – perhaps saying as much about the perceived effectiveness of the various national and international procedures for compelling enforcement as about the nature of the parties engaged in resolving their disputes through arbitration. While it is difficult to provide precise statistics because arbitration is a private process, it is generally reckoned by leading arbitration institutions that 80–90 per cent of their awards are performed voluntarily. [46]

When, however, the losing party refuses to carry out the award, or engages in delaying tactics, the successful party will have to take steps to enforce performance. As has already been observed, arbitration awards are not generally self-executing. A tribunal usually has no power to enforce the award; only in exceptional circumstances does the arbitral process also involve some form of actual or perceived coercion. For instance, pressure may be applied by the threat of adverse publicity. For example, it is not uncommon in disputes in the commodity sector for notices to be given to all members of the relevant association informing them of any member's refusal to abide by an award made under the association's rules. Similarly, a losing party to an ICSID arbitration may be concerned that by failing to perform the award, its status with the World Bank could be adversely affected and its access to funds made more difficult as a result. In addition, in relation to Internet domain name arbitrations, the award is effectively self-executing since the institutions empowered to allocate domain names will enforce awards from one of a number of recognized arbitration procedures without the need for any judicial intervention. The construction industry however is not made up of a sufficiently coherent body of institutions as to allow effective non-judicial process to take place.

Despite these exceptions, the basic position is that arbitral tribunals cannot themselves compel compliance with their awards; once the final award has been validly handed down, the tribunal's role is at an end. An arbitrator's award, unlike an order or judgment of a court, does not at once entitle the successful party to levy execution against, say, the assets of the unsuccessful party. Judicial intervention of some sort is almost always required.

A. METHODS OF ENFORCING AN AWARD

As might be imagined, the steps required to enforce an award vary from country to country. The party seeking enforcement should therefore take legal advice from lawyers in the relevant jurisdiction(s), i.e. the jurisdiction(s) in which the losing

---

46 In addition enforcement has been refused in, perhaps, fewer than 5 per cent of the cases under the New York Convention A. J. See Pierre-Yves Gunter, *'Enforcement of arbitral award, injunctions and orders'*, Arbitration and Dispute Resolution Law Journal (1999), 265–279. See below for a detailed examination of the New York Convention.

party (or its officers, if a company) or its assets are located. What follows, therefore, is a review of the general provisions applicable to the recognition and enforcement of awards.

B. RECOGNITION AND ENFORCEMENT

Although the terms 'recognition' and 'enforcement' are often used in the same breath, they are in fact distinct concepts and, frequently, distinct procedures. Indeed, there are cases in which an award has been recognized but not enforced.[47] There may also be circumstances in which a successful respondent may wish to have the award in its favour recognized (for example, as determining the dispute or for use as a shield to other proceedings on the same matter) even though there can be no question of enforcement of the award in the particular circumstances.

Recognition refers to the process by which a national court acknowledges the existence of the arbitration, and recognizes the decision of the tribunal in the award rendered. In many cases these will be summary proceedings confirming the award (often called *exequatur*). By having a court recognize an award, a successful claimant will establish that it is in a position to enforce it. A successful respondent will do the same, as well as preventing the claimant from successfully bringing court proceedings in respect of matters that have already been decided by the award.

Enforcement usually occurs at the same time as recognition or follows shortly thereafter.[48] When a court is asked to enforce an award, it is not merely recognizing its legal effect but also ensuring its terms are carried out. Enforcement therefore goes one step further than recognition, introducing an element of compulsion: the losing party must perform its obligations. At this point, the processes of court litigation and arbitration merge, since an arbitration award is almost invariably enforced in the same way as a judgment of the court concerned. Thus, a money award will be both recognized and enforced as a money judgment and enforcement may take any of the usual forms including the attachment and sale of assets, the transfer of debts owed to the losing party, the appointment of court officers over the assets of the losing party to gather in and realize sufficient sums to meet the award or, in extreme cases, the insolvent liquidation of the losing party.

C. DOMESTIC AND FOREIGN AWARDS

A distinction must also be drawn between domestic and foreign awards, a distinction that is not always easy to make.

---

47 See, e.g., *Dallal v Bank Mellat* [1986] 1 All ER 239; *Bank Mellat v Helleniki Techniki SA* [1983] 3 All ER 428; *Fidelitas Shipping Co Ltd v V/O Exportchleb* [1965] 2 All ER 4.
48 For example, Art. 28(6) ICC Rules states that 'the parties undertake to carry out any Award *without delay*' (emphasis added).

1. Domestic Awards

Enforcement of an award in a domestic arbitration (i.e. one where all of the parties are resident in the place of the arbitration) is governed solely by the national law of the country in which the arbitration is taking place. However, domestic enforcement provisions may still be relevant to international arbitration, inasmuch as awards arising from international arbitration may be enforced under the procedures used for domestic awards.[49] Such a situation will arise where the winning party seeks to have the arbitral award recognized or enforced in the country that was the seat of the arbitration.

Recognition and enforcement procedures under national laws are usually fairly straightforward, with the requirements for enforcement of domestic awards often simpler and less onerous than for foreign awards.[50] Again, it is important to seek local legal advice as to the exact requirements as these will vary from country to country. In England, for example section 66 of the English Arbitration Act 1996 provides for automatic recognition of an arbitral award so the only proceedings necessary are for enforcement. In other countries, however, this is not the case. As an example, under French law an application for recognition has to be made before enforcement[51] and in Switzerland the award must first be deposited and then registered with the relevant court or state authority before it becomes enforceable. If the recognition or enforcement of a domestic award is refused, this may lead to the setting aside of the award.[52]

2. Foreign Awards

Foreign awards are awards that were made outside the country in which recognition or enforcement is sought. Unlike foreign court decisions (which are generally not easily enforceable outside certain well defined blocks of countries, such as the European Union and the British Commonwealth), arbitration awards can – at least in theory – be readily enforced across national boundaries. This is as a result of the large number of states which have acceded to the New York Convention. This treaty, agreed in 1958 and now acceded to by over 130 countries,[53] now largely

---

49   Although it must be noted that it is up to the country in question to determine whether it considers an award to be domestic or international in nature.
50   For example, German law requires only submission of a copy of the domestic award (§1064(1) ZPO). See also the Netherlands Code of Civil Procedure (Art. 1063), which limits the grounds for refusal to enforce an award.
51   See Art. 1498 Code of Civil Procedure 1981
52   For example, under s. 66 English Arbitration Act 1996. Nevertheless, there is still a chance, in limited cases, that an award set aside in one country may still be enforced in another. See note 47. See *'Setting aside awards – challenging incomplete awards'*, Arbitration Law Monthly 2 (December/January 2002), 1–2.
53   There are currently 137 counties who have ratified the convention. For a complete list of ratifications, see: http://www.uncitral.org/uncitral/en/uncitral_texts/arbitration/NYConvention_status.html (accessed 16 November 2005).

governs the recognition and enforcement of foreign awards, to the virtual exclusion of earlier treaties[54] and, although to a slightly lesser extent, the limited number of bilateral and 'local' multinational treaties dealing with the same matters.

IV.   THE NEW YORK CONVENTION

The New York Convention is a significant improvement on earlier arbitral protocols and conventions. The enforcement mechanism of the New York Convention is dealt with in Articles III to VI.

The New York Convention simplified the enforcement of foreign arbitral awards by:[55]

(i) shifting the burden of proof from the party seeking enforcement to the party opposing it;[56]
(ii) limiting the defences available to a party resisting enforcement to those listed in the New York Convention itself; and
(iii) eliminating the old 'double-*exequatur*' requirement of earlier treaters, which required that an award had to be recognized in the court of the country in which the award was made and also by the country that was enforcing the award.

The New York Convention has often been highly praised[57] and now constitutes the backbone of the international regime for the recognition and enforcement of foreign arbitration awards. Various aspects of the New York Convention are considered in more depth below.

A.   SCOPE OF APPLICATION OF THE NEW YORK CONVENTION

The scope of application of the New York Convention is set out in Article I(1). Broadly, the New York Convention applies to all foreign awards, that is, all those awards 'which are not considered as domestic awards in the State where their recognition and enforcement are sought'.[58] Thus, the New York Convention

---

54 For instance, the Geneva Protocol on Arbitration Clauses of 1923 and the 1927 Geneva Convention on the Execution of Foreign Arbitral Awards. For a fuller historical perspective, see Redfern and Hunter, *International Commercial Arbitration*, para. 10–20*ff*.

55 See generally Albert Jan Van Den Berg, *'Non-domestic arbitral awards under the 1958 New York Convention'*, Arbitration International 2 (1986), 191–219; Michael Pryles, *'Foreign awards and the New York Convention'*, Arbitration International 9 (1993), 259–273.

56 Reversing the burden of proof from that in the Convention on the Execution of Foreign Arbitral Awards signed at Geneva on 26 September 1927 (the Geneva Convention of 1927).

57 For example, it has been described as the 'most effective instance of international legislation in the history of commercial law', see Michael John Mustill, *'Arbitration: History and Background'*, Journal of International Arbitration 6 (1989), 43–56.

58 Art. I(1) New York Convention.

potentially applies regardless of whether the award was rendered inside or outside another Contracting State's territory.

However, despite this potentially wide reaching application, Article I(3) permits Contracting States to make two 'reservations', as they are called, so that in that state the New York Convention only applies:

(i) on the basis of reciprocity (i.e. only to arbitration awards made in another Contracting State); and
(ii) to awards rendered in *commercial matter*s, 'considered as commercial under the national law of the State making such declaration'.[59]

Sixty-seven countries acceding to the New York Convention have made the reciprocity reservation. Despite this, the large number of countries which have now signed up to the New York Convention means that the potential adverse affects of this reservation have become very much reduced, with few economically significant countries not within its ambit.[60] Even so, problems do occasionally arise as a result of the relevant country not having acceded to the New York Convention[61] or (more difficult to identify) not properly incorporating the provisions of the New York Convention into its domestic law, even where it has acceded.[62]

B. RECOGNITION AND ENFORCEMENT

The New York Convention requires both recognition and enforcement of awards to which it relates. Contracting States bound by the New York Convention are obliged to respect the binding effect of the awards and so to enforce awards to which the New York Convention applies in accordance with their national procedural rules. In this connection, Article III of the New York Convention imposes an obligation on Contracting States not to impose greater inconvenience or larger fees for such enforcement than are imposed in connection with the recognition and enforcement of domestic awards of the relevant country.

---

59 The Convention does not define a 'commercial matter' as can be seen; this is determined by the laws of the enforcing court. See the Tunisian case, *Taieb Haddad and Hans Barett v Société d'Invesstissement Kal*, Tunisian Cour de Cassation, 10 November 1993, excepts published in Yearbook of Commercial Arbitration XXIII (1998), 770–773.
60 However, there are some exceptions. For instance, at the time of writing, the United Arab Emirates has not acceded to the New York Convention, though it is expected to in the near future.
61 See, for instance, *Texaco Pananma Inc. v Duke Petroleum Transport Corp.*, 3 September 1996 95 Civ. 3761 (LMM); International Arbitration Report 11(9) (1996) pp. D1 – D2, 'The New York Convention of 1958 relied on by the respondent does not apply, since [sic] respondent is a Liberian corporation and Liberia is not a signatory party to that Convention'.
62 For example, Pakistan was an early signatory to the New York Convention but did not amend its domestic law to reflect its terms until 14 July 2005.

C.   FORMALITIES FOR ENFORCEMENT

The formalities specified in the New York Convention for recognition and enforcement of a foreign award are straightforward.[63] The party seeking recognition and enforcement must provide to the relevant court:

(i) the duly authenticated[64] original award or a duly certified copy thereof;[65] and
(ii) the original agreement to arbitrate or a duly certified copy thereof.

Where the award and the arbitration agreement are not in the official language of the country in which enforcement and recognition are being sought, Article IV(2) of the New York Convention also provides that certified translations must be provided. Once these documents have been supplied, the New York Convention requires that award should be enforced – *unless* the party against which enforcement is sought can bring itself within the scope of one of the exclusive grounds for non-enforcement listed in the New York Convention.

D.   RESISTING ENFORCEMENT UNDER THE NEW YORK CONVENTION

Before listing the exclusive grounds for the refusal to recognize and enforce an award under the New York Convention, the following three points should be noted:

(i) the New York Convention does not permit any review of the merits of an award to which it applies;[66]
(ii) even if one or more of the of the grounds for non-enforcement exists, the enforcing court is (arguably) not bound to refuse enforcement – it *'may'* refuse enforcement;[67] and
(iii) the grounds were intended to be applied restrictively because the general rule is that foreign awards must be recognized and enforced.

The word *'may'* in (ii) above has been placed in italics because although in the English and Spanish text of the New York Convention the language used implies that the judges may use their discretion, the wording of the French version implies that the judges have no discretion.[68] This is unfortunate considering the thrust of

---

63   See Art. II(2) New York Convention.
64   Which is to say that the signature must be attested to be genuine.
65   Which is to say that the copy must be attested to be a true copy of the original.
66   Art. V New York Convention. Arts 35 and 36 UNCITRAL Model Law governing recognition and enforcement of awards are almost identical to the New York Convention. The UNCITRAL Model Law does not permit a review on the merits either (see above).
67   Arts V(1) and (2) New York Convention.
68   See generally Jan Paulsson, *'May or Must Under the New York Convention: An Exercise in Syntax and Linguistics'*, Arbitration International 14 (1998), 227–230.

the New York Convention is to achieve uniformity, and no doubt arises out of the difficulty in achieving exactly equivalent meanings in each one of the five equally authentic texts – Chinese, English, Spanish, French and Russian – in which the New York Convention has been drafted.

Overall, the New York Convention is clearly recognized as having a pro-enforcement bias, with the effect that the grounds for non-enforcement are to be construed narrowly and only accepted in serious cases.[69] Consequently, the preferred view is that the court in which an application for enforcement is made is not under an obligation to refuse recognition or enforcement if it finds that any of the grounds under the New York Convention are proved to exist. It simply has the discretion to do so. That said, where a ground for non-enforcement is proved to the satisfaction of the enforcing court, enforcement will in fact normally be refused.

Lastly, a note of caution must be sounded. It should be observed that it is somewhat inevitable that the grounds in the New York Convention are liable to be interpreted divergently by various disparate courts seized in any particular matter. This has been brought to the attention of UNCITRAL, which has set up a committee to review compliance with the New York Convention by those States which have acceded to it. Despite this, current differences in interpretation will continue to exist and new ones will no doubt arise. As with other areas, local law advice needs to be obtained whenever an application is made or resisted.

E. THE NEW YORK CONVENTION GROUNDS FOR REFUSAL OF RECOGNITION AND ENFORCEMENT

There are seven grounds listed in the New York Convention for the refusal of recognition and enforcement of an arbitral award. As mentioned earlier in this chapter, there is a pleasing symmetry (with only a slight variation in the wording used) between Article 34 of the UNCITRAL Model Law (which deals with challenges to an award in the place of the arbitration) and Article V of the New York Convention. As the New York Convention makes these seven grounds the only permitted ones for non-enforcement, the courts of a Contracting State may not base their refusal to enforce an award on any other grounds. Those courts should also interpret the specified grounds restrictively, to give effect to the pro-enforcement bias of the New York Convention discussed above.

The New York Convention provides that recognition and enforcement may be refused if the resisting party can prove the existence of one or more of the grounds set out below: First, Article V(1)(a) provides for refusal where:

---

[69] See further below. See also *Parsons & Whittemore Overseas Co. v Société Generale de L'Industrie du Papier supra* (where it was argued that an award may not be enforceable because it was made in manifest disregard of the law).

'the parties to the [arbitral agreement] were, under the law applicable to them, under some incapacity, or the said agreement is not valid under the law to which the parties have subjected it or, failing any indication thereon, under the law of the country where the award was made'.

This first ground relates to the material or formal (in)validity of the arbitration agreement and concerns both capacity and formality. The New York Convention cannot and does not set out to establish whether parties to the agreement to arbitrate have the requisite capacity to enter such agreement. This issue is to be decided by the laws applicable to them, in practice the laws of their place of domicile or incorporation or the laws of the place where the arbitration is taking place. It is, however, for the court conducting the enforcement procedure to determine (adopting its own conflicts of laws rules) what it regards as the law applicable to the parties for these purposes.

The formal validity of the arbitration agreement may also be invoked as a ground for refusal to enforce. The parties' contractual autonomy is preserved by their freedom to choose an applicable law to govern their agreement to arbitrate being respected. However, as with the equivalent provision of the UNCITRAL Model Law, [70] this provides a default position in the event of their failure to do so. The difference in the default position between the UNCITRAL Model Law and the New York Convention could, in unusual circumstances (i.e. where the award is made in a different country from the country where the arbitration took place) give rise to differing results to a challenge based on this ground and an application to refuse enforcement. Interesting though these differences are, in practice, this ground has rarely been a ground for non-enforcement of an award.

The second New York Convention ground (Article V(1)(b)) applies where:

'the party against whom the award is invoked was not given proper notice of the appointment of the arbitrator or of the arbitration proceedings or was otherwise unable to present his case'.

Like the UNCITRAL Model Law (discussed above), this is often considered the most important ground for refusal of enforcement under the New York Convention. The main thrust of this provision is to ensure that the requirements of 'due process' (albeit a rather unclear international standard) are observed and that the parties are, put simply, given a fair hearing. Due process can be described as the existence of those elements necessary to ensure that the arbitration itself is properly conducted, with proper notice given as to the appointment of an arbitrator and procedural fairness afforded to the parties. This latter point incorporates the right to a fair hearing conducted in a competent manner (namely, notice of the hearing; an opportunity to attend the hearing; the right to be present throughout the hearing; an

---

70 Art. 34(2)(a)(i).

opportunity to present one's argument and evidence; and the opportunity to controvert an opponent's case).

Only serious breaches of fairness will provide grounds for non-enforcement under this head.[71] A serious breach would include, for example, where a party is unable to present its case[72] or where an arbitral tribunal refuses to allow a party to cross-examine experts appointed by the tribunal.[73]

Article V(1)(c) of the New York Convention allows refusal of recognition and enforcement where:

> 'the award deals with a difference not contemplated by or not falling within the terms of the submission to arbitration, or it contains decisions on matters beyond the scope of the submission to arbitration, provided that, if the decisions on matters submitted to arbitration can be separated from those not so submitted, that part of the award which contains decisions on matters submitted to arbitration may be recognized and enforced'.

Although a ground that has rarely been successful in the past,[74] invoking questions of jurisdiction is becoming ever more frequent as the first port of call in relation to the grounds for a refusal of enforcement of an arbitral award, provided the relevant part of the award cannot be separated from the parts that do not comply with the terms of the arbitration agreement.

Jurisdictional issues as a ground for challenging an award are discussed above in respect to the UNCITRAL Model Law;[75] and it has been noted that the right to raise an issue of jurisdiction may be lost if the matter became apparent in the course of the arbitration and no objections were taken at the time.

Under Article V(1)(d), recognition and enforcement may be refused if:

> 'the composition of the arbitral authority or the arbitral procedure was not in accordance with the agreement of the parties, or, failing such agreement, was not in accordance with the law of the country where the arbitration took place'.

---

71   See Albert Jan Van den Berg, *The New York Arbitration Convention of 1958*, (Kluwer, 1981), 298.
72   See *Iran Aircraft Industries v Avco Corp.*, 980 F 2d 141 (2nd Cir. 1992) – this was a rare example in which an American court refused to enforce an arbitral award.
73   See *Paklito Investment Pty Ltd. v Klockner East Asia Ltd.* [1993] 2 HKLR 39 (Hong Kong).
74   See *Virgilio De Agostini and Loris and Enrico Germani v Milliol SpA, Pia and Gabriella Germani and Andrea De Agostini* Corte di Appello di Milano, 24 March 1998; Yearbook Commercial Arbitration XXV (2000), pp. 641–1164, 739–50; Dimitri Santoro, *'Forum Non Conveniens: a Valid Defense under the New York Convention?'*, ASA Bulletin 21(4) (2003), 713–735.
75   See note [33] above. In relation to the New York Convention, see generally Redfern, *'Jurisdiction Denied: The Pyramid Collapses'*, Journal of Business Law 1 (1986), 15–22.

An agreement by the parties regarding the composition of the arbitral tribunal or the arbitral proceedings supersedes the national rules of the country where the arbitration took place, except, perhaps, where it conflicts with that country's fundamental legal principles. The law of the country where the arbitration took place usually comes into play only in the absence of an agreement.[76]

Finally, Article V(1)(e) allows a court to refuse to recognize or enforce an award where:

> 'the award has not yet become binding on the parties, or has been set aside or suspended by a competent authority of the country in which, or under the law of which, that award was made'.

This ground for refusal of recognition and enforcement of an arbitral award (which, like the others, also appears in the UNCITRAL Model Law) has given rise to more controversy than any of the previous grounds. This ground has led to the situation in which an award that has been set aside, and so is unenforceable in its country of origin, may be refused enforcement under the New York Convention in one country, but granted enforcement in another.

This problem arises because – unlike Article 34 of the UNCITRAL Model Law examined above – the New York Convention does not in any way restrict the grounds on which an award may be set aside or suspended by the courts of the country in which, or under the law of which, that award was made. The burden is, however, on the party resisting enforcement to:

> '… prove that the suspension of the award has been effectively ordered by a court in the country of origin. The automatic suspension of the award by operation of law in the country of origin…is not sufficient'.[77]

Even then, given the discretion allowed to the enforcing court not to refuse enforcement even where a ground is made out, there is scope for the enforcing court to exercise its own value judgments when faced with resistance on this ground. It is by no means unknown for such a court to form its own view on the merits of the grounds relied on by the courts of the country which set the award aside.

In addition to the grounds discussed above, the New York Convention provides that recognition and enforcement of an arbitral award may also be refused if the *competent authority* in the country in which enforcement is sought finds that:

> 'the subject-matter of the difference is not capable of settlement by arbitration under the law of that country'.[78]

---

76 See *Chen Hang Chu v China Treasure Enterprise Ltd* [2000] 2 HKC 814.
77 Van den Berg, *The New York Arbitration Convention of 1958*, p. 352.
78 Art. V(2)(a) New York Convention.

The preliminary question to be addressed is what is the law governing arbitrability?[79] Again, unlike the position in relation to a challenge in a country which has adopted the UNCITRAL Model Law (where arbitrability is governed by the law of the place of the arbitration and challenge) the non-arbitrability of an award as a ground for non-enforcement of an award under the New York Convention is determined according to the law of the country where recognition and enforcement is sought. Thus, to survive a challenge to a successful arbitration, a party must be able to demonstrate that the subject matter of the dispute is arbitrable not only in the place of the arbitration but also in every place that enforcement may take place. Fortunately, while increasingly raised as a ground for challenge or as a means of resisting enforcement, the number of occasions where a construction dispute is genuinely not arbitrable in accordance with accepted norms will be few and far between.

Lastly, the New York Convention allows recognition and enforcement to be refused where the competent authority in the country where recognition and enforcement are sought finds that:

> 'the recognition or enforcement of the award would be contrary to public policy of that country'.

Public policy has been discussed as one of the grounds permitting a challenge to an arbitration award in the courts of the place of the arbitration. Here again, an unsuccessful party has 'another bite of the cherry' if it can manage to persuade the relevant court that the award would be contrary to the public policy of the country in which enforcement is sought. Often, though not invariably, this will be the place where the unsuccessful party is domiciled or incorporated.

Putting to one side the occasional, less justifiable instances when this ground has been successfully employed, and accepting that the term 'public policy'[80] is difficult to define,[81] it can be adequately, though not comprehensively, summed up as the:

> 'fundamental rules of natural law, the principles of universal justice, [the central rules of] public international law and the general principles of morality accepted by what is referred to as civilized nations'.[82]

---

79 See Homayoon Arfazadeh, *'Arbitrability under the New York Convention: the Lex Fori Revisted'*, Arbitration International 17(1) (2001), pp. 73–87.
80 See Hans Smit, *'Comments on Public Policy in International Arbitration'*, American Review of International Arbitration 13 (2002), 65–67.
81 See above.
82 Van den Berg, *The New York Arbitration Convention of 1958*, p. 361.

Decisions from courts in some parts of the world have shown a readiness to limit – occasionally severely – the public policy defence to enforcement.[83] Nonetheless, the boundaries of *national* public policy are mutable and this is perhaps why the public policy exception is the most frequently invoked basis for refusal of an arbitral award because of its potentially broad, amorphous scope. Courts, however, have generally taken a narrow view of the public policy exclusion.[84] For an English perspective, see the *Soleimany* case,[85] concerning the illegality of a contract under English law. This is believed to be the first case in which the English courts – in this case, the Court of Appeal – have refused to enforce an arbitral award on the grounds of public policy in arbitration matters. The case has been criticized for failing to recognize the distinction between foreign and domestic public policy.[86]

Put plainly, the twin grounds of arbitrability and public policy effectively 'free' the forum from being obliged to enforce an award it views with distaste, and that is so even if the arbitral award in question was untainted by procedural shortcomings.

V. OTHER BILATERAL AND MULTILATERAL CONVENTIONS

While the New York Convention is regarded as the primary vehicle for the recognition and enforcement of foreign arbitration awards, there do exist other international agreements that are potentially applicable. This section is intended to serve only as an introduction to some of these conventions. This review is not intended to be exhaustive and parties are advised to take local law advice to consider whether there are any other regional conventions that may be helpful if there is any doubt about the applicability of the New York Convention to their particular circumstances.

A. BILATERAL CONVENTIONS

In addition to the multilateral conventions referred to above and below, certain States have executed bilateral conventions on the recognition and enforcement of arbitral awards. Switzerland, for instance, has executed such bilateral conventions with Spain, Czechoslovakia (as was), Germany, Italy, Sweden, Belgium, Austria

---

83 See *Corporacion Transnacional de Inversiones SA de CV v STET International SpA* [2000] 49 O.R. (3d) 414, O.J. No. 3408 (to be contrary to public policy, the award must *'fundamentally offend the most basic and explicit principles of justice and fairness......or evidence intolerable ignorance or corruption on the part of the tribunal'*). See also *'Enforcement of foreign awards'*, Arbitration Law Monthly 1(4) (May 2001), 6–7.
84 See *Hakeem Seriki, 'Enforcement of foreign arbitral awards and public policy: a note of caution'*, Arbitration and Dispute Resolution Law Journal 3 (2000), 192–207, 195–98.
85 *Soleimany v Soleimany* [1999] 3 All ER 847.
86 See *Seriki, 'Enforcement of foreign arbitral awards'*, 203.

Effect of the Award 307

and Liechtenstein.[87] The result is that bilateral conventions may be useful when trying to enforce an award in another contracting country.[88] These conventions usually operate on the basis of reciprocity, and form an extensive network of rights and obligations.

The use of bilateral conventions generally arises in two circumstances. The first is in the (increasingly rare) circumstances where the New York Convention does not apply. The second, perhaps less common, is where the criteria for enforcement under the bilateral convention are more favourable than under the New York Convention. Given the pro-enforcement bias of the New York Convention the number of instances where this might occur are likely to be limited.

B.   MULTILATERAL CONVENTIONS

1.   The 1965 Washington Convention

In 1965, the Washington Convention established ICSID. Its purpose was to provide a mechanism for the resolution of international investment disputes. The Washington Convention provides its own enforcement procedures.[89] These procedures provide that each contracting state must recognize and enforce an ICSID award as if it were a final judgment of its own national courts.[90] The Washington Convention is of considerable value when within its strict jurisdictional criteria but provides no assistance to parties to arbitrations not conducted in accordance with its terms.

2.   The 1975 Panama Convention[91]

This Convention relates to South and Central American countries and the United States.[92] The Panama Convention appears to have been heavily influenced by the New York Convention and many of the provisions are identical.[93]

---

87 See Pierre-Yves Gunter, *'Enforcing Arbitral Awards, Injunctions and Orders'*, available at: http://www.psplaw.ch/Publications/AMBAR-1.html (accessed 16 November 2005). In the United States, similar (as opposed to identical) conventions are known as Friendship, Commerce and Navigation treaties.
88 See generally Rudolf Dolzer and Margrete Stevens, *Bilateral Investment Treaties*, (Brill Academic Publishers, 1995), chapter 5.
89 Arts 53 and 54 Washington Convention.
90 Art. 54(1) Washington Convention.
91 *Inter-American Convention on International Commercial Arbitration 1975*, done at Panama on 30 January 1975, available at: http://www.jurisint.org/en/ins/154.html (accessed 23 August 2005).
92 It has been signed by 19 and ratified by 18 States (Yearbook Commercial Arbitration XXVIII (2003), pp. 1345–1346). An annually updated list of the Contracting States is reproduced in Part V-D of each Yearbook.
93 Note that the Panama Convention refers to 'execution' of an award rather than enforcement.

3. The Amman Convention

The Amman Convention[94] is modelled closely on the Washington Convention of 1965. The Amman Convention, which created the Arab Centre of Commercial Arbitration, was adopted with the active involvement of all the Arab countries and on the basis of each of their respective recommendations. The seat of the Centre is Rabat, Morocco. As all pleadings and submissions *must* be in Arabic, it is largely inaccessible on an international sphere where English is the prevailing language.

Broadly speaking, the Amman Convention provides that the High Court of each contracting state has jurisdiction to grant enforcement of arbitral awards. The Amman Convention provides that awards may only be refused enforcement by the courts of a contracting state where the award is contrary to the public policy of that contracting state. The Amman Convention further provides that an award made by the Rabat Centre is not subject to appeal before the judicial authorities of the country where enforcement is sought.

The Amman Convention provides three reasons for setting aside awards. The first is, that the arbitral tribunal clearly acted *ultra vires* its powers, the second, that a court judgment establishes that there is an extant fact which would substantially alter the arbitral award (this will be construed *contra proferentem*) and the third, that there was undue influence on one (or more) of the arbitrators, which could have affected his decision.[95]

VI. CONCLUSION

In view of the above discussion, it is difficult to deny that the recognition and enforcement of an arbitral award is greatly assisted by a plethora of arbitral conventions – notably the New York Convention. Parties are advised to consider the question of the recognition and enforcement of an award at the outset of the arbitral process to avoid the unenviable situation of receiving a favourable award that subsequently cannot be enforced.

---

94 *Arab Convention on Commercial Arbitration 1987*, done at Amman on 14 April 1987, available at: http://www.jurisint.org/en/ins/155.html (accessed 23 August 2005).
95 See generally Abdul Hamid El-Ahdab, 'Enforcement of Arbitral Awards in the Arab Countries', Arbitration International 11(2) (1995), 169–181.

Chapter 14
Closing Thoughts

I. INTRODUCTION

The previous chapters describe the manner in which arbitrations arising out of construction disputes are conducted today, with (in some cases) indications of where innovation is occurring. Clearly there have been changes over the last 15 or 20 years, with the greatest change being in the reduction, though not, perhaps, the complete demise, of the construction cases where oral hearings ran for months rather than weeks. The days when (in one of the authors' experience) a construction case settled after 70 sitting days are now largely behind us. What, then, does the future hold in store? That there will be further change is not in doubt. Change, likely as not, will come about in three areas.

II. THE RESPONSE TO ADR

The first area of change will be the response to the continued pressure from end users – clients – to improve efficiencies, with alternative dispute resolution procedures making further inroads into the number of cases resolved in arbitration if efficiencies are not delivered. It is, perhaps, illuminating that in a text addressing international construction arbitration as it is practised today a significant number of chapters of this book have focused on stages prior to an arbitration. In the authors' experience this reflects very much an existing trend towards finding alternative ways of resolving disputes and, during the progress of the project, looking for ways to avoid disputes and manage claims resolution effectively. The reasons are obvious: participants in the industry wish to avoid the costs and wasted time involved in protracted dispute resolution and to find ways to facilitate the settlement of disputes at the earliest opportunity wherever possible.

As a result, the challenge is for lawyers and arbitrators to respond to the market's pressures to devise more efficient procedures. As examined in this book, in the authors' view such issues should be addressed at the outset when drafting the construction contract. Attention should be given to techniques for claims administration and appropriate dispute resolution procedures. Alternatives to arbitration should be addressed and discussed with clients at this stage. And great care should be taken in drafting dispute resolution provisions to avoid pitfalls that may slow down the process or which lead to jurisdictional wrangles. There is little worse or more unproductive than a dispute about how a dispute is to be resolved, a process which, in addition, invariably favours the party without the merits on its side. If, however, alternative dispute procedures fail, lawyers and the tribunal should be creative in devizing efficient and cost effective techniques for managing the arbitration so as to achieve a fair and speedy resolution of the disputes.

In the light of this it is unsurprising that arbitral institutions have responded to market pressures. We have seen the development of rules and guides for the use of Dispute Review Boards (discussed in Chapter 5). Arbitral institutions now also offer standard wording and supporting services for mediation, conciliation, expertise or other forms of dispute resolution as described in Chapters 3 and 6. Arbitration institutions are also recognizing that the 'traditional' period of 18 months to two years for a standard construction dispute is too long and are looking at ways in which their administration – and the administration of the arbitrators they appoint – can be improved so as to achieve a swifter result.[1] As noted in Chapter 3, the International Centre for Dispute Resolution, the international division of the American Arbitration Association prides itself on is choice of arbitrators and the speed with which they are appointed and aims for a cycle time from start to finish of its arbitrations of under a year.[2]

Similarly, new techniques have been developed by tribunals and the parties' advisors for reducing the cost and time spent in hearings. Written submissions now play a significant role with oral opening and closing submissions being confined to summaries of the key points. Some arbitrators favour the use of witness conferencing to bring together all of the witnesses relevant to a particular issue at the same time rather than hearing witnesses on a sequential basis (discussed in Chapters 8 and 12). So called 'hot tub' procedures are another example, again discussed in Chapter 12 whereby the experts are questioned by the tribunal and opposing counsel together with the possibility for the experts to ask each other questions on matters in disputes.

As a result, reports of the sudden death of construction arbitration are exaggerated. Indeed statistics demonstrate international arbitrations are increasing. One key reason is the advantage that arbitration has over litigation, as discussed in Chapter 3. The key points are the ease of enforcement under international

---

[1] See, for instance, the Institution of Civil Engineers *Arbitration Procedure 2006*, published in Febraury 2006, which provides, in addition to a construction forced conventional procedure, 'short' (44 day) and 'expidited' (100 day) arbitration procedures for small and medium sized cases.

[2] Note that the ICDR caters for a wide range of commercial disputes including those arising out of construction and project agreements.

conventions and the opportunity to select a neutral tribunal in a neutral arbitration-friendly venue. Another reason is an increasing trend towards parties' use of the arbitration provisions of Bilateral Investment Treaties (see Chapters 3 and 13), some of which will, in substance, be construction claims. The future of international arbitration as a tool for resolving disputes in the construction and project sector therefore looks assured. The challenge is, as indicated above, to deliver in terms of efficiency and economics in the period prior to the issue of the award what international arbitration largely delivers once an award has been made.

III. INTRODUCTION OF NEW TECHNOLOGY

The second area of change, perhaps complementing the first, is the impact that improved communications and new technology will – at least if accepted by the parties, witnesses and tribunals – have on the process. Properly deployed, new technology can reduce the inherent difficulties in assembling everybody together in one place for a hearing. This is not to say that the technology is in place already. Anyone who has attempted to interview a witness over a standard commercial video link – or, worse, a 'web-cam' from an out-of-the-way site – will appreciate the difficulties. Little short of a full scale broadcast quality link provides the necessary intangible information to the audience to enable a proper assessment of the witness's demeanour and the weight to be put on his responses. For this reason, video and teleconferencing facilities are best restricted to procedural matters where what is relevant is the strength of the arguments and not an assessment of the likely veracity of the speaker. Even then, a party physically present in the same room as the tribunal has a considerable advantage over his remote opponent as a result of feedback gleaned from the body language and instinctive responses of the tribunal. These are almost invariably missed over even the best telephone or video connections.

But this is not to say that the technology will not (soon) be available to allow remote management of full hearings. Only that it is not here yet. What is here is the electronic or 'virtual' case file, where all the papers are filed electronically on a secure server and accessed by the parties and the tribunal over an internet link from wherever they happen to be. Even if the human factor means that (for some) the key documents will be downloaded, printed out, highlighted with a yellow marker and annotated in multi-coloured inks, this is still a great advantage for all concerned who are prepared to work on at least this basis. No longer is it necessary for the tribunal, the parties, their counsel or any of the witnesses to be co-located with the hard copies of voluminous files – still less for them to be all in the same location at the same time for the case to proceed.

One example of the current state of the technology can be found in the ICC's NetCase, which is offered at no additional charge by the ICC for any ICC arbitration provided the parties – and all the arbitrators – agree. The ICC is at pains to make it clear that this is not an 'online arbitration' system but a technology aid to assist the parties to a conventional arbitration. Thus it is not intended (as some online arbitration systems are) fully to replace the existing procedures, with all aspects of the case being heard as if an electronic 'paper arbitration'. Instead its objectives are to provide a secure (encrypted) medium for the transfer and filing of and access to the arbitration documents. In addition it provides instant up-to-date information on the status and progress of the case, with matters such as the procedural timetable and directions and state of the payments made to the ICC being immediately available. Though of potential considerable value in their current form, systems such as NetCase are also ripe for development into international case management and presentation tools, with full documentation being made available – and searchable – in a format common to all participants.

Technology will also impact on the preparation of cases in another way when the electronic file is preceded by the fully electronic project. Only 15 or so years ago it was relatively uncommon for project documents to be available (except to the party which created them) in electronic format. The result, unless relatively unambitious systems were put in place from the project's inception, was that any attempt to produce an electronic database for the purposes of dispute resolution was, if not doomed to failure, inordinately expensive. Indeed, it was not unknown for attempts to introduce electronic document management systems after disputes became apparent to be severely scaled back or abandoned as the task of dealing with the backlog of paper based documents became overwhelming. Today that has all changed, with (as discussed in Chapter 10) the vast majority of project documentation in major projects being both created and transmitted in electronic, fully machine-readable formats. This paradigm shift in the way projects are documented means that whilst overall volumes of documents have probably increased (the curse of both e-mail and the ease with which drafts can be created but not destroyed), the tools to assess and deploy these documents with reduced human intervention are rapidly becoming more sophisticated.

And in one key area – that of programme or schedule analysis – the potential exists for shared on-line, up-to-date programming information. This will provide parties keen to embrace the possibilities this offers with a sound foundation on which delay analysis can be quickly, reliably and economically undertaken.

IV. CULTURAL CONTRIBUTIONS

The third and final area of change will come about as a result of cultural contributions from parts of the world where construction – and inevitably construction disputes – will feature largely in the next 15 or 20 years. The emergence of China as the world's third largest economy with a vibrant domestic and international construction industry is one such area. The other, arising out of the need to create large amounts of energy related infrastructure, is formed by those countries influenced by the Sharia.

There will, however, be no resulting step change in the nature or use of arbitration since the practice of arbitration in both areas is deep-rooted and of great antiquity. Most likely, changes will come about as a result of the fusion of traditional means of resolving disputes with arbitration as it has recently been practiced in an international context.

Of the two areas, Chinese practices are the more likely to have an immediate impact. Not least this is because there is an established arbitration centre in China which has, by numbers if by no other metric, a larger case load than any other international arbitration institution. In particular, the ability of the arbitrator to mediate disputes at the instance of either party in the course of an arbitration without prejudice to his continued ability to function as the arbitrator may well strike a chord with some parties. Indeed, quite probably with the parties to the case which settled after 70 sitting days referred to above. That case settled at that moment (and only at that moment) because the arbitrator of his own volition applied some fairly heavy-handed reality checking to the parties by indicating what he thought he would rule on one key point in the case. It seems unlikely that that case would have been allowed to continue for so long nowadays and certainly not if run as a CIETAC arbitration. And so much the better, might be a proper conclusion.

Annex 1
Advantages and Disadvantages of Arbitration and Litigation

Set out below is a summary of the primary advantages and disadvantages of arbitration and litigation. These points are discussed in greater detail in Chapter 3.

Arbitration	Litigation
Advantages:	**Advantages:**
– Relatively easy enforcement if the jurisdiction in question is a signatory to the New York Convention – 'Neutral' forum – Choose own arbitrator(s) – Procedural flexibility and party autonomy – Finality of award ('one shot')	– Powers of state courts (e.g. to order interim relief; subpoena witnesses) – Power to join/consolidate proceedings
Disadvantages:	**Disadvantages:**
– Limited powers of arbitrators (e.g. to order interim relief or joinder/consolidation) – Multi-party disputes may be unwieldy	– Harder to enforce internationally – 'Home court' advantage? – No choice of judge – Set procedure – Possible multiple appeals

Annex 2
Sample Model Arbitration Clauses

The following model arbitration clauses are currently recommended for use in contracts by the institutions noted below. The rules of a particular institution should be consulted prior to including a model clause in any agreement to ensure: (i) that the most current recommended model clause language is utilized; and (ii) that the rules are appropriate for the particular contract/project in question. It may also be advisable to seek specialist advice.

LCIA RULES

Any dispute arising out of or in connection with this contract, including any question regarding its existence, validity or termination, shall be referred to and finally resolved by arbitration under the LCIA Rules, which Rules are deemed to be incorporated by reference into this clause.

The number of arbitrators shall be [one/three].

The seat, or legal place, of arbitration shall be [City and/or country].

The language to be used in the arbitral proceedings shall be [...].

The governing law of the contract shall be the substantive law of [...].

RULES OF ARBITRATION OF THE INTERNATIONAL CHAMBER OF COMMERCE

All disputes arising out of or in connection with the present contract shall be finally settled under the Rules of Arbitration of the International Chamber of Commerce by one or more arbitrators appointed in accordance with the said Rules.

AMERICAN ARBITRATION ASSOCIATION'S INTERNATIONAL DISPUTE RESOLUTION PROCEDURES

Any controversy or claim arising out of or relating to this contract, or the breach thereof, shall be determined by arbitration administered by the International Centre for Dispute Resolution in accordance with its International Arbitration Rules.

or

Any controversy or claim arising out of or relating to this contract, or the breach thereof, shall be determined by arbitration administered by the American Arbitration Association in accordance with its International Arbitration Rules.

The parties may wish to consider adding:
 (a) *The number of arbitrators shall be (one or three).*
 (b) *The place of arbitration shall be (city and/or country); or*
 (c) *The language(s) of the arbitration shall be _____.*

AMERICAN ARBITRATION ASSOCIATION'S CONSTRUCTION INDUSTRY ARBITRATION RULES

Any controversy or claim arising out of or relating to this contract, or the breach thereof, shall be settled by arbitration administered by the American Arbitration Association under its Construction Industry Arbitration Rules, and judgment on the award rendered by the arbitrator(s) may be entered in any court having jurisdiction thereof.

Annex 3
Important Drafting Considerations for Dispute Resolution Clauses

The following is a non-exhaustive list of some of the more important considerations and issues that may need to be addressed in drafting dispute resolution clauses in contracts relating to international construction projects. A more detailed discussion of these and related issues is contained in Chapter 3.

PRELIMINARY ISSUES

– How many tiers of dispute resolution should there be? What tiers should be included?

NEGOTIATION

– Should there be a provision for mandatory face-to-face negotiations? If so, should the contract identify the negotiators for each party, or should they be decided upon at the time of the dispute? Should there be more than one level of negotiation? What will happen if one party refuses to negotiate within the time allowed for such negotiations?

ADR

– Should there be ADR (Alternative Dispute Resolution) procedures included the contract? If so, when is the ADR procedure deemed to commence? If it is after court or arbitral proceedings have already commenced, what is the effect of the ADR procedure on any ongoing proceedings? Will they be stayed? If so, how and when can they be restarted in the event that the ADR procedure breaks down?
– Should there be a bona fide attempt to make ADR a condition precedent to proceeding to the next stage in the dispute resolution process? (Note: if the

provisions of the UK Construction Act 1996 (discussed above) apply to the project, it is impossible to prevent a party 'leapfrogging' to adjudication at any time.)
- Should the dispute resolution clause itself designate (and define) a specific ADR procedure such as mediation, conciliation, mini-trial or expert opinion, or leave the exact nature of the process for subsequent agreement and simply state that the parties shall agree the exact form of ADR once a dispute arises? (Note: in some jurisdictions, including England, such an 'agreement to agree' may not be legally enforceable, though the appointment of a third body (such as CEDR) to make the decision on the parties' behalf generally will be recognized as binding).
- What mechanism should be used for agreeing/appointing the neutral? Should that person's qualifications be identified/agreed in advance? Should the clause provide for an outside agency to be involved in appointing the person in question?
- Should the clause set out bespoke detailed procedural rules for the ADR process, adopt those of an outside agency/service provider, or leave procedural matters for subsequent agreement by the parties or person they appoint?
- Should there be joint or separate meetings between the parties and the neutral person they appoint?
- Will the neutral person appointed to assist in the ADR be allowed, or required, to make recommendations or evaluations of the parties' respective cases and the likely outcome of any subsequent litigation or arbitration?
- Should the ADR clause contain express confidentiality provisions and/or specifically reserve the parties' strict legal rights in case the ADR procedure breaks down?
- Depending upon the nature of the dispute, should there be express provision for parties to continue to perform their substantive obligations during the ADR procedure (e.g. requiring the contractor to continue working, pending resolution of the dispute)? Will such clauses be enforceable under the relevant laws? Will such a clause have the (possibly unintentional) effect of preventing the exercise of termination rights during the currency of the process?
- Is an express provision restricting the use of information disclosed in the ADR process in any subsequent court or arbitration proceedings required (e.g. restricting its use to the enforcement of any resulting settlement agreement)?
- Should there be an express provision limiting the role of the neutral in any subsequent court or arbitration proceedings?
- Should the clause deal expressly with the authority of the parties' representatives to negotiate, make admissions, etc during the ADR process?
- What is the timetable for the procedure, and what are the specified procedural steps?
- Should the clause exclude particular types of disputes and/or remedies from the ADR procedure?
- How are the costs and fees of the chosen form of ADR to be dealt with?

ADJUDICATION

- Should there be adjudication? If so, should it be an individual or panel of qualified people?
- Should the adjudicator or panel be identified in the contract (i.e. a standing panel or board) or should there be a mechanism to appoint an adjudicator/panel at the time each dispute arises (ad hoc)? Will the adjudicator/panel make interim binding decision or only non-binding recommendations?
- Will the parties create their own bespoke adjudication provisions or adopt those of an institution (e.g. the ICC Dispute Board Rules)?
- How will the costs and expenses of the adjudicator/panel be allocated? To what extent will the adjudicator/panel have the power to take the initiative in ascertaining and investigating facts and law related to the dispute?
- How much time will the adjudicator/panel have to provide its decision?
- Will the adjudicator/panel be required to make regular visits to site during the duration of the project?
- Will the adjudicator/panel be entitled to appoint its own advisors to assist on matters of legal interpretation or areas outside its expertise? If so, will prior consent of the parties be required?
- Will the adjudicator/panel have the power to award interest and/or costs?
- What, if any, powers will the adjudicator/panel have in the event that one party refuses to participate in a reference?
- What, if any, provisions will be included in respect of confidentiality and the ability of the adjudicator/panel members to act as witnesses in subsequent proceedings?

ARBITRATION

- Should there be arbitration or litigation as the final tier? (Relevant considerations include the enforceability of an arbitral award, which in turn involves considering the location of assets against which you may want to enforce.)
- If arbitration is selected, how many arbitrators should there be?
- Should the qualifications of the arbitrator(s) be specified?
- Who should appoint the tribunal in default of agreement by the parties?
- Should the procedural rules of the arbitration be ad hoc or institutional?
- Where should the seat of the arbitration be located? How do the courts in that jurisdiction approach intervention in arbitral proceedings?
- Should there be an express provision addressing confidentiality?
- How should possible rights of appeal be addressed?
- Should consolidation and joinder provisions be included?

LITIGATION

- Should disputes be resolved in the court(s)?
- If so, which court(s)?
- How will a judgment be enforced?
- You will also need to address submission to exclusive/non-exclusive jurisdiction of the court(s), and make provision for an agent for service of process (where necessary).

OTHER CONSIDERATIONS

Issues on which it may be necessary to seek local law advice include:

- Is the jurisdiction that is to be the seat of arbitration a signatory to the New York Convention? What about the jurisdiction(s) in which you are likely to want to enforce an arbitral award?
- Does the law of the place of performance, or the governing law, if it is different, provide rights to suspend performance, despite contractual provisions o the contrary (for example, in a situation of non-payment)?
- If a state or state agency/emanation is the counterparty, are there any relevant bilateral or multilateral treaties? If so, how best might you exploit treaty protection? For example, can the project vehicle be incorporated in a state that enjoys good investor protection arrangements under BITs or MITs with the state in which the project is to take place?

Annex 4
World Bank Standard Bidding Documents for the Procurement of Works

CLAUSE 20 – CLAIMS, DISPUTES AND ARBITRATION

20.1   CONTRACTOR'S CLAIMS

If the Contractor considers himself to be entitled to any extension of the Time for Completion and/or any additional payment, under any Clause of these Conditions or otherwise in connection with the Contract, the Contractor shall give notice to the Engineer, describing the event or circumstance giving rise to the claim. The notice shall be given as soon as practicable, and not later than 28 days after the Contractor became aware, or should have become aware, of the event or circumstance.

If the Contractor fails to give notice of a claim within such period of 28 days, the Time for Completion shall not be extended, the Contractor shall not be entitled to additional payment, and the Employer shall be discharged from all liability in connection with the claim. Otherwise, the following provisions of this Sub-Clause shall apply.

The Contractor shall also submit any other notices which are required by the Contract, and supporting particulars for the claim, all as relevant to such event or circumstance.

The Contractor shall keep such contemporary records as may be necessary to substantiate any claim, either on the Site or at another location acceptable to the Engineer. Without admitting the Employer's liability, the Engineer may, after receiving any notice under this Sub-Clause, monitor the record-keeping and/or instruct the Contractor to keep further contemporary records. The Contractor shall permit the Engineer to inspect all these records, and shall (if instructed) submit copies to the Engineer.

Within 42 days after the Contractor became aware (or should have become aware) of the event or circumstance giving rise to the claim, or within such other period as may be proposed by the Contractor and approved by the Engineer, the

Contractor shall send to the Engineer a fully detailed claim which includes full supporting particulars of the basis of the claim and of the extension of time and/or additional payment claimed. If the event or circumstance giving rise to the claim has a continuing effect:

(a) this fully detailed claim shall be considered as interim;
(b) the Contractor shall send further interim claims at monthly intervals, giving the accumulated delay and/or amount claimed, and such further particulars as the Engineer may reasonably require; and
(c) the Contractor shall send a final claim within 28 days after the end of the effects resulting from the event or circumstance, or within such other period as may be proposed by the Contractor and approved by the Engineer.

Within 42 days after receiving a claim or any further particulars supporting a previous claim, or within such other period as may be proposed by the Engineer and approved by the Contractor, the Engineer shall respond with approval, or with disapproval and detailed comments. He may also request any necessary further particulars, but shall nevertheless give his response on the principles of the claim within such time.

Each Payment Certificate shall include such amounts for any claim as have been reasonably substantiated as due under the relevant provision of the Contract. Unless and until the particulars supplied are sufficient to substantiate the whole of the claim, the Contractor shall only be entitled to payment for such part of the claim as he has been able to substantiate.

The Engineer shall proceed in accordance with Sub-Clause 3.5 [Determinations] to agree or determine (i) the extension (if any) of the Time for Completion (before or after its expiry) in accordance with Sub-Clause 8.4 [Extension of Time for Completion], and/or (ii) the additional payment (if any) to which the Contractor is entitled under the Contract.

The requirements of this Sub-Clause are in addition to those of any other Sub-Clause which may apply to a claim. If the Contractor fails to comply with this or another Sub-Clause in relation to any claim, any extension of time and/or additional payment shall take account of the extent (if any) to which the failure has prevented or prejudiced proper investigation of the claim, unless the claim is excluded under the second paragraph of this Sub-Clause.

20.2 APPOINTMENT OF THE DISPUTE BOARD

Disputes shall be referred to a DB for decision in accordance with Sub-Clause 20.4 [Obtaining Dispute Board's Decision]. The Parties shall appoint a DB by the date stated in the Contract Data.

The DB shall comprise, as stated in the Contract Data, either one or three suitably qualified persons ('the members'), each of whom shall be fluent in the language for communication defined in the Contract and shall be a professional experienced in the type of construction involved in the Works and with the interpretation of contractual documents. If the number is not so stated and the Parties do not agree otherwise, the DB shall comprise three persons, one of whom shall serve as chairman.

If the Parties have not jointly appointed the DB 21 days before the date stated in the Contract Data and the DB is to comprise three persons, each Party shall nominate one member for the approval of the other Party. The first two members shall recommend and the Parties shall agree upon the third member, who shall act as chairman.

The agreement between the Parties and either the sole member or each of the three members shall incorporate by reference the General Conditions of Dispute Board Agreement contained in the Appendix to these General Conditions, with such amendments as are agreed between them.

The terms of the remuneration of either the sole member or each of the three members, including the remuneration of any expert whom the DB consults, shall be mutually agreed upon by the Parties when agreeing the terms of appointment of the member or such expert (as the case may be). Each Party shall be responsible for paying one-half of this remuneration.

If a member declines to act or is unable to act as a result of death, disability, resignation or termination of appointment, a replacement shall be appointed in the same manner as the replaced person was required to have been nominated or agreed upon, as described in this Sub-Clause.

The appointment of any member may be terminated by mutual agreement of both Parties, but not by the Employer or the Contractor acting alone. Unless otherwise agreed by both Parties, the appointment of the DB (including each member) shall expire when the discharge referred to in Sub-Clause 14.12 [Discharge] shall have become effective.

20.3   FAILURE TO AGREE ON THE COMPOSITION OF THE DISPUTE BOARD

If any of the following conditions apply, namely:

(a) the Parties fail to agree upon the appointment of the sole member of the DB by the date stated in the first paragraph of Sub-Clause 20.2, [Appointment of the Dispute Board],
(b) either Party fails to nominate a member (for approval by the other Party) or fails to approve a member nominated by the other Party, of a DB of three persons by such date,
(c) the Parties fail to agree upon the appointment of the third member (to act as chairman) of the DB by such date, or
(d) the Parties fail to agree upon the appointment of a replacement person within 42 days after the date on which the sole member or one of the three members

declines to act or is unable to act as a result of death, disability, resignation or termination of appointment,

then the appointing entity or official named in the Contract Data shall, upon the request of either or both of the Parties and after due consultation with both Parties, appoint this member of the DB. This appointment shall be final and conclusive. Each Party shall be responsible for paying one-half of the remuneration of the appointing entity or official.

20.4   OBTAINING DISPUTE BOARD'S DECISION

If a dispute (of any kind whatsoever) arises between the Parties in connection with, or arising out of, the Contract or the execution of the Works, including any dispute as to any certificate, determination, instruction, opinion or valuation of the Engineer, either Party may refer the dispute in writing to the DB for its decision, with copies to the other Party and the Engineer. Such reference shall state that it is given under this Sub-Clause.

For a DB of three persons, the DB shall be deemed to have received such reference on the date when it is received by the chairman of the DB.

Both Parties shall promptly make available to the DB all such additional information, further access to the Site, and appropriate facilities, as the DB may require for the purposes of making a decision on such dispute. The DB shall be deemed to be not acting as arbitrator(s).

Within 84 days after receiving such reference, or within such other period as may be proposed by the DB and approved by both Parties, the DB shall give its decision, which shall be reasoned and shall state that it is given under this Sub-Clause. The decision shall be binding on both Parties, who shall promptly give effect to it unless and until it shall be revised in an amicable settlement or an arbitral award as described below. Unless the Contract has already been abandoned, repudiated or terminated, the Contractor shall continue to proceed with the Works in accordance with the Contract.

If either Party is dissatisfied with the DB's decision, then either Party may, within 28 days after receiving the decision, give notice to the other Party of its dissatisfaction and intention to commence arbitration. If the DB fails to give its decision within the period of 84 days (or as otherwise approved) after receiving such reference, then either Party may, within 28 days after this period has expired, give notice to the other Party of its dissatisfaction and intention to commence arbitration.

In either event, this notice of dissatisfaction shall state that it is given under this Sub-Clause, and shall set out the matter in dispute and the reason(s) for dissatisfaction. Except as stated in Sub-Clause 20.7 [Failure to Comply with Dispute Board's Decision] and Sub-Clause 20.8 [Expiry of Dispute Board's Appointment], neither Party shall be entitled to commence arbitration of a dispute unless a notice of dissatisfaction has been given in accordance with this Sub-Clause.

If the DB has given its decision as to a matter in dispute to both Parties, and no notice of dissatisfaction has been given by either Party within 28 days after it received the DB's decision, then the decision shall become final and binding upon both Parties.

20.5   AMICABLE SETTLEMENT

Where notice of dissatisfaction has been given under Sub-Clause 20.4 above, both Parties shall attempt to settle the dispute amicably before the commencement of arbitration. However, unless both Parties agree otherwise, arbitration may be commenced on or after the 56th day after the day on which a notice of dissatisfaction and intention to commence arbitration was given, even if no attempt at amicable settlement has been made.

20.6   ARBITRATION

Unless settled amicably, any dispute in respect of which the DB's decision (if any) has not become final and binding shall be finally settled by international arbitration. Unless otherwise agreed by both Parties:

(a) arbitration proceedings shall be conducted in accordance with the rules of arbitration provided for in the Particular Conditions,
(b) if no rules of arbitration are provided therein, the dispute shall be finally settled under the Rules of Arbitration of the International Chamber of Commerce,
(c) the dispute shall be settled by three arbitrators, and
(d) the arbitration shall be conducted in the language for communications defined in Sub-Clause 1.4 [Law and Language].

The arbitrator(s) shall have full power to open up, review and revise any certificate, determination, instruction, opinion or valuation of the Engineer, and any decision of the DB, relevant to the dispute. Nothing shall disqualify the Engineer from being called as a witness and giving evidence before the arbitrator(s) on any matter whatsoever relevant to the dispute.

Neither Party shall be limited in the proceedings before the arbitrator(s) to the evidence or arguments previously put before the DB to obtain its decision, or to the reasons for dissatisfaction given in its notice of dissatisfaction. Any decision of the DB shall be admissible in evidence in the arbitration.

Arbitration may be commenced prior to or after completion of the Works. The obligations of the Parties, the Engineer and the DB shall not be altered by reason of any arbitration being conducted during the progress of the Works.

20.7   FAILURE TO COMPLY WITH DISPUTE BOARD'S DECISION

In the event that a Party fails to comply with a DB decision which has become final and binding, then the other Party may, without prejudice to any other rights it may have, refer the failure itself to arbitration under Sub-Clause 20.6 [Arbitration]. Sub-Clause 20.4 [Obtaining Dispute Board's Decision] and Sub-Clause 20.5 [Amicable Settlement] shall not apply to this reference.

20.8   EXPIRY OF DISPUTE BOARD'S APPOINTMENT

If a dispute arises between the Parties in connection with, or arising out of, the Contract or the execution of the Works and there is no DB in place, whether by reason of the expiry of the DB's appointment or otherwise:

(a) Sub-Clause 20.4 [Obtaining Dispute Board's Decision] and Sub-Clause 20.5 [Amicable Settlement] shall not apply, and
(b) the dispute may be referred directly to arbitration under Sub-Clause 20.6 [Arbitration].

APPENDIX A – GENERAL CONDITIONS OF DISPUTE BOARD AGREEMENT

1.   DEFINITIONS

Each 'Dispute Board Agreement' is a tripartite agreement by and between:

(a) the 'Employer';
(b) the 'Contractor'; and
(c) the 'Member' who is defined in the Dispute Board Agreement as being:
   (i) the sole member of the 'DB' and, where this is the case, all references to the 'Other Members' do not apply, or
   (ii) one of the three persons who are jointly called the 'DB' (or 'Dispute Board') and, where this is the case, the other two persons are called the 'Other Members'.

The Employer and the Contractor have entered (or intend to enter) into a contract, which is called the 'Contract' and is defined in the Dispute Board Agreement, which incorporates this Appendix. In the Dispute Board Agreement, words and expressions which are not otherwise defined shall have the meanings assigned to them in the Contract.

2.   GENERAL PROVISIONS

Unless otherwise stated in the Dispute Board Agreement, it shall take effect on the latest of the following dates:

(a) the Commencement Date defined in the Contract,
(b) when the Employer, the Contractor and the Member have each signed the Dispute Board Agreement, or
(c) when the Employer, the Contractor and each of the Other Members (if any) have respectively each signed a dispute board agreement.

This employment of the Member is a personal appointment. At any time, the Member may give not less than 70 days' notice of resignation to the Employer and to the Contractor, and the Dispute Agreement shall terminate upon the expiry of this period.

3.   WARRANTIES

The Member warrants and agrees that he/she is and shall be impartial and independent of the Employer, the Contractor and the Engineer. The Member shall promptly disclose, to each of them and to the Other Members (if any), any fact or circumstance which might appear inconsistent with his/her warranty and agreement of impartiality and independence.

When appointing the Member, the Employer and the Contractor relied upon the Member's representations that he/she is:

(a) experienced in the work which the Contractor is to carry out under the Contract,
(b) experienced in the interpretation of contract documentation, and
(c) fluent in the language for communications defined in the Contract.

4.   GENERAL OBLIGATIONS OF THE MEMBER

The Member shall:

(a) have no interest financial or otherwise in the Employer, the Contractor or Engineer, nor any financial interest in the Contract except for payment under the Dispute Board Agreement;
(b) not previously have been employed as a consultant or otherwise by the Employer, the Contractor or the Engineer, except in such circumstances as were disclosed in writing to the Employer and the Contractor before they signed the Dispute Board Agreement;

(c) have disclosed in writing to the Employer, the Contractor and the Other Members (if any), before entering into the Dispute Board Agreement and to his/her best knowledge and recollection, any professional or personal relationships with any director, officer or employee of the Employer, the Contractor or the Engineer, and any previous involvement in the overall project of which the Contract forms part;
(d) not, for the duration of the Dispute Board Agreement, be employed as a consultant or otherwise by the Employer, the Contractor or the Engineer, except as may be agreed in writing by the Employer, the Contractor and the Other Members (if any);
(e) comply with the annexed procedural rules and with Sub-Clause 20.4 of the Conditions of Contract;
(f) not give advice to the Employer, the Contractor, the Employer's Personnel or the Contractor's Personnel concerning the conduct of the Contract, other than in accordance with the annexed procedural rules;
(g) not while a Member enter into discussions or make any agreement with the Employer, the Contractor or the Engineer regarding employment by any of them, whether as a consultant or otherwise, after ceasing to act under the Dispute Board Agreement;
(h) ensure his/her availability for all site visits and hearings as are necessary;
(i) become conversant with the Contract and with the progress of the Works (and of any other parts of the project of which the Contract forms part) by studying all documents received which shall be maintained in a current working file;
(j) treat the details of the Contract and all the DB's activities and hearings as private and confidential, and not publish or disclose them without the prior written consent of the Employer, the Contractor and the Other Members (if any); and
(k) be available to give advice and opinions, on any matter relevant to the Contract when requested by both the Employer and the Contractor, subject to the agreement of the Other Members (if any).

5. GENERAL OBLIGATIONS OF THE EMPLOYER AND THE CONTRACTOR

The Employer, the Contractor, the Employer's Personnel and the Contractor's Personnel shall not request advice from or consultation with the Member regarding the Contract, otherwise than in the normal course of the DB's activities under the Contract and the Dispute Board Agreement. The Employer and the Contractor shall be responsible for compliance with this provision, by the Employer's Personnel and the Contractor's Personnel respectively.

The Employer and the Contractor undertake to each other and to the Member that the Member shall not, except as otherwise agreed in writing by the Employer, the Contractor, the Member and the Other Members (if any):

(a) be appointed as an arbitrator in any arbitration under the Contract;
(b) be called as a witness to give evidence concerning any dispute before arbitrator(s) appointed for any arbitration under the Contract; or
(c) be liable for any claims for anything done or omitted in the discharge or purported discharge of the Member's functions, unless the act or omission is shown to have been in bad faith.

The Employer and the Contractor hereby jointly and severally indemnify and hold the Member harmless against and from claims from which he is relieved from liability under the preceding paragraph.

Whenever the Employer or the Contractor refers a dispute to the DB under Sub-Clause 20.4 of the Conditions of Contract, which will require the Member to make a site visit and attend a hearing, the Employer or the Contractor shall provide appropriate security for a sum equivalent to the reasonable expenses to be incurred by the Member. No account shall be taken of any other payments due or paid to the Member.

6. PAYMENT

The Member shall be paid as follows, in the currency named in the Dispute Board Agreement:

(a) a retainer fee per calendar month, which shall be considered as payment in full for:
   (i) being available on 28 days' notice for all site visits and hearings;
   (ii) becoming and remaining conversant with all project developments and maintaining relevant files;
   (iii) all office and overhead expenses including secretarial services, photocopying and office supplies incurred in connection with his duties; and
   (iv) all services performed hereunder except those referred to in sub-paragraphs (b) and (c) of this Clause.

The retainer fee shall be paid with effect from the last day of the calendar month in which the Dispute Board Agreement becomes effective; until the last day of the calendar month in which the Taking-Over Certificate is issued for the whole of the Works.

With effect from the first day of the calendar month following the month in which the Taking-Over Certificate is issued for the whole of the Works, the retainer fee shall be reduced by one third. This reduced fee shall be paid until the first day of the calendar month in which the Member resigns or the Dispute Board Agreement is otherwise terminated.

(b) a daily fee which shall be considered as payment in full for:
   (i) each day or part of a day up to a maximum of two days' travel time in each direction for the journey between the Member's home and the site, or another location of a meeting with the Other Members (if any);
   (ii) each working day on Site visits, hearings or preparing decisions; and
   (iii) each day spent reading submissions in preparation for a hearing.
(c) all reasonable expenses including necessary travel expenses (air fare in less than first class, hotel and subsistence and other direct travel expenses) incurred in connection with the Member's duties, as well as the cost of telephone calls, courier charges, faxes and telexes: a receipt shall be required for each item in excess of 5 per cent of the daily fee referred to in sub-paragraph (b) of this Clause;
(d) any taxes properly levied in the Country on payments made to the Member (unless a national or permanent resident of the Country) under this Clause 6.

The retainer and daily fees shall be as specified in the Dispute Board Agreement. Unless it specifies otherwise, these fees shall remain fixed for the first 24 calendar months, and shall thereafter be adjusted by agreement between the Employer, the Contractor and the Member, at each anniversary of the date on which the Dispute Board Agreement became effective.

If the parties fail to agree on the retainer fee or the daily fee, the appointing entity or official named in the Contract Data shall determine the amount of the fees to be used.

The Member shall submit invoices for payment of the monthly retainer and air fares quarterly in advance. Invoices for other expenses and for daily fees shall be submitted following the conclusion of a site visit or hearing. All invoices shall be accompanied by a brief description of activities performed during the relevant period and shall be addressed to the Contractor.

The Contractor shall pay each of the Member's invoices in full within 56 calendar days after receiving each invoice and shall apply to the Employer (in the Statements under the Contract) for reimbursement of one-half of the amounts of these invoices. The Employer shall then pay the Contractor in accordance with the Contract.

If the Contractor fails to pay to the Member the amount to which he/she is entitled under the Dispute Board Agreement, the Employer shall pay the amount due to the Member and any other amount which may be required to maintain the operation of the DB; and without prejudice to the Employer's rights or remedies. In addition to all other rights arising from this default, the Employer shall be entitled to reimbursement of all sums paid in excess of one-half of these payments, plus all costs of recovering these sums and financing charges calculated at the rate specified in Sub-Clause 14.8 of the Conditions of Contract.

If the Member does not receive payment of the amount due within 70 days after submitting a valid invoice, the Member may (i) suspend his/her services (without notice) until the payment is received, and/or (ii) resign his/her appointment by giving notice under Clause 7.

7. TERMINATION

At any time: (i) the Employer and the Contractor may jointly terminate the Dispute Board Agreement by giving 42 days' notice to the Member; or (ii) the Member may resign as provided for in Clause 2.
If the Member fails to comply with the Dispute Board Agreement, the Employer and the Contractor may, without prejudice to their other rights, terminate it by notice to the Member. The notice shall take effect when received by the Member.
If the Employer or the Contractor fails to comply with the Dispute Board Agreement, the Member may, without prejudice to his other rights, terminate it by notice to the Employer and the Contractor. The notice shall take effect when received by them both.
Any such notice, resignation and termination shall be final and binding on the Employer, the Contractor and the Member. However, a notice by the Employer or the Contractor, but not by both, shall be of no effect.

8. DEFAULT OF THE MEMBER

If the Member fails to comply with any of his obligations under Clause 4 (a) – (d) above, he shall not be entitled to any fees or expenses hereunder and shall, without prejudice to their other rights, reimburse each of the Employer and the Contractor for any fees and expenses received by the Member and the Other Members (if any), for proceedings or decisions (if any) of the DB which are rendered void or ineffective by the said failure to comply.
If the Member fails to comply with any of his obligations under Clause 4 (e) – (k) above, he shall not be entitled to any fees or expenses hereunder from the date and to the extent of the non-compliance and shall, without prejudice to their other rights, reimburse each of the Employer and the Contractor for any fees and expenses already received by the Member, for proceedings or decisions (if any) of the DB which are rendered void or ineffective by the said failure to comply.

9. DISPUTES

Any dispute or claim arising out of or in connection with this Dispute Board Agreement, or the breach, termination or invalidity thereof, shall be finally settled by institutional arbitration. If no other arbitration institute is agreed, the arbitration shall be conducted under the Rules of Arbitration of the International Chamber of Commerce by one arbitrator appointed in accordance with these Rules of Arbitration.

PROCEDURAL RULES

Unless otherwise agreed by the Employer and the Contractor, the DB shall visit the site at intervals of not more than 140 days, including times of critical construction events, at the request of either the Employer or the Contractor. Unless otherwise agreed by the Employer, the Contractor and the DB, the period between consecutive visits shall not be less than 70 days, except as required to convene a hearing as described below.

The timing of and agenda for each site visit shall be as agreed jointly by the DB, the Employer and the Contractor, or in the absence of agreement, shall be decided by the DB. The purpose of site visits is to enable the DB to become and remain acquainted with the progress of the Works and of any actual or potential problems or claims, and, as far as reasonable, to endeavour to prevent potential problems or claims from becoming disputes.

Site visits shall be attended by the Employer, the Contractor and the Engineer and shall be co-ordinated by the Employer in co-operation with the Contractor. The Employer shall ensure the provision of appropriate conference facilities and secretarial and copying services. At the conclusion of each site visit and before leaving the site, the DB shall prepare a report on its activities during the visit and shall send copies to the Employer and the Contractor.

The Employer and the Contractor shall furnish to the DB one copy of all documents which the DB may request, including Contract documents, progress reports, variation instructions, certificates and other documents pertinent to the performance of the Contract. All communications between the DB and the Employer or the Contractor shall be copied to the other Party. If the DB comprises three persons, the Employer and the Contractor shall send copies of these requested documents and these communications to each of these persons.

If any dispute is referred to the DB in accordance with Sub-Clause 20.4 of the Conditions of Contract, the DB shall proceed in accordance with Sub-Clause 20.4 and these Rules. Subject to the time allowed to give notice of a decision and other relevant factors, the DB shall:

(a) act fairly and impartially as between the Employer and the Contractor, giving each of them a reasonable opportunity of putting his case and responding to the other's case, and

(b) adopt procedures suitable to the dispute, avoiding unnecessary delay or expense.

The DB may conduct a hearing on the dispute, in which event it will decide on the date and place for the hearing and may request that written documentation and arguments from the Employer and the Contractor be presented to it prior to or at the hearing.

Except as otherwise agreed in writing by the Employer and the Contractor, the DB shall have power to adopt an inquisitorial procedure, to refuse admission to hearings or audience at hearings to any persons other than representatives of the Employer, the Contractor and the Engineer, and to proceed in the absence of any party who the DB is satisfied received notice of the hearing; but shall have discretion to decide whether and to what extent this power may be exercised.

The Employer and the Contractor empower the DB, among other things, to:

(a) establish the procedure to be applied in deciding a dispute,
(b) decide upon the DB's own jurisdiction, and as to the scope of any dispute referred to it,
(c) conduct any hearing as it thinks fit, not being bound by any rules or procedures other than those contained in the Contract and these Rules,
(d) take the initiative in ascertaining the facts and matters required for a decision,
(e) make use of its own specialist knowledge, if any,
(f) decide upon the payment of financing charges in accordance with the Contract,
(g) decide upon any provisional relief such as interim or conservatory measures, and
(h) open up, review and revise any certificate, decision, determination, instruction, opinion or valuation of the Engineer, relevant to the dispute.

The DB shall not express any opinions during any hearing concerning the merits of any arguments advanced by the Parties. Thereafter, the DB shall make and give its decision in accordance with Sub-Clause 20.4, or as otherwise agreed by the Employer and the Contractor in writing. If the DB comprises three persons:

(a) it shall convene in private after a hearing, in order to have discussions and prepare its decision;
(b) it shall endeavour to reach a unanimous decision: if this proves impossible the applicable decision shall be made by a majority of the Members, who may require the minority Member to prepare a written report for submission to the Employer and the Contractor; and
(c) if a Member fails to attend a meeting or hearing, or to fulfil any required function, the other two Members may nevertheless proceed to make a decision, unless:
    (i) either the Employer or the Contractor does not agree that they do so, or
    (ii) the absent Member is the chairman and he/she instructs the other Members not to make a decision.

Annex 5
American Arbitration Association Dispute Resolution Board Guide Specifications

1.01   GENERAL

A.   DEFINITIONS

1. American Arbitration Association – Neutral not-for-profit provider of Dispute Resolution Board (DRB) services, internationally.
2. Board – See Dispute Resolution Board (DRB).
3. Contract – The construction Contract of which this Specification section is part.
4. Dispute – A claim, change order request, or other issue that remains unresolved following negotiation between authorized representatives of the Owner and Contractor.
5. Dispute Resolution Board (DRB) – Three neutral individuals mutually selected by the Owner and Contractor to consider and recommend resolution of Disputes referred to it.

B.   SUMMARY

1. A Dispute Resolution Board (DRB) will be established to assist in the resolution of Disputes in connection with, or arising out of, performance of the work of this Contract.
2. Either the Owner or Contractor may refer a Dispute to the Board. Such referral should be initiated prior to the initiation of other dispute resolution procedures or filing of litigation by either party.
3. Promptly thereafter, the Board will impartially consider the Dispute(s) referred to it. The Board will provide a non-binding written recommendation for resolution of the Dispute to the Owner and the Contractor.

C.   SCOPE

1. This Specification describes the purpose, procedure, function and features of the DRB. A Three-Party Agreement among the Owner, Contractor and the three Board members using the form and content of Attachment A will formalize creation of the Board and establish the scope of its services and the rights and responsibilities of the parties. In the event of a conflict between this Specification and the Three-Party Agreement, the latter governs.

D.   PURPOSE

1. The Board, as an independent third party, will assist in and facilitate the timely resolution of disputes between the Owner and the Contractor.
2. Creation of the Board is not intended to promote Owner or Contractor default on the responsibility of making a good-faith effort to settle amicably and fairly their differences by indiscriminate referral to the Board.

E.   THREE-PARTY AGREEMENT

1. All three DRB members and the authorized representatives of the Owner and Contractor shall execute the DRB Three-Party Agreement within 14 days after the selection of the third member.

F.   CONTINUANCE OF WORK

1. Both parties shall proceed diligently with the work and comply with all applicable Contract provisions while the DRB considers a Dispute.

G.   TENURE OF BOARD

1. The Board will be deemed established on the date of establishment stated in the Three-Party Agreement.
2. The Board will be dissolved as of the date of final payment to the Contractor or, should any disputes be pending as of that date, the date on which the Board issues its recommendations regarding those disputes, unless earlier terminated or dissolved by mutual agreement of the Owner and Contractor. The Board's jurisdiction will continue for a period of 30 days beyond the date of its recommendations for the limited purpose of responding to a request for clarification or in the event that a party introduces new evidence.

1.02   MEMBERSHIP

A.   GENERAL

1. The DRB will consist of one member nominated by the Owner and approved by the Contractor, one member nominated by the Contractor and approved by the Owner, and a third member nominated by the first two members and approved by both the Owner and the Contractor. Unless otherwise agreed by the Owner and Contractor, all members shall be selected from a list provided by the American Arbitration Association, compiled from its International Roster of DRB Members. The third member will serve as Chair unless the Owner and Contractor otherwise agree.

B.   CRITERIA

1.   Experience

a. It is desirable that all DRB members be experienced with the type of construction involved in the project, interpretation of Contract documents and resolution of construction disputes.
b. The goal in selecting the third member is to complement the experience of the first two and to provide leadership of the Board's activities.

2.   Neutrality

a. It is imperative that the Board members be neutral, act impartially and be free of any conflict of interest.
b. For purposes of this subparagraph (1.02.B.2), the term 'member' also includes the member's current primary or full-time employer, and 'involved' means having a Contractual relationship with either party to the Contract, such as by being a subcontractor, architect, engineer, construction manager or consultant.
c. The following are disqualifying relationships for prospective members:
   1. An ownership interest in any entity involved (with) the Contract, or a financial interest in the Contract, except for payment for services as a member of the DRB;
   2. Previous employment by, or financial ties to, any party involved in the Contract, including fee-based consulting services, within a period of ten years prior to award of the Contract, except with the express written approval of both parties;

3. A close business or personal relationship with any key members of any entity involved in the Contract which, in the judgment of either party, could suggest partiality; or
4. Prior involvement in the project of a nature that could compromise that member's ability to participate impartially in the Board's activities.

C. SELECTION OF THE BOARD

1. Request for Assistance

a. Within 14 days of the effective date of the Contract, the Owner and Contractor shall file a Request for Dispute Resolution Board (DRB) Assistance with the American Arbitration Association. The Request for DRB Assistance shall include a description of the construction project including name, location and approximate Contract price and Contract time; guidelines regarding DRB member compensation and expenses, if any, the names, mail and email addresses, telephone and facsimile numbers of the Owner and the Contractor and their representatives, the names and addresses of all design professionals, consultants and first-tier subcontractors then known, together with the AAA filing fee.

2. AAA Inquiry

a. Upon receipt of a properly filed Request for DRB Assistance, the AAA shall promptly schedule a telephone conference call with the Owner and Contractor to discuss desired qualifications of DRB members.

3. List of Proposed Board Members

a. Within 14 days after the information is provided by the Owner and Contractor, the AAA shall send the Owner and Contractor an identical list of persons selected from its International Roster of DRB members, including detailed biographical information and disclosures regarding each listed person.

4. Pre-Appointment Disclosure

a. Prior to their being listed for review by the Owner and the Contractor, proposed Board members shall disclose to the AAA any circumstance likely to affect impartiality, including any bias or any financial or personal interest in the project or any past or present relationship with the parties to the Contract, including subcontractors, design professionals and consultants.

5. Nomination and Acceptance of First Two Members

a. Unless agreed otherwise, the Owner and the Contractor shall each nominate a proposed Board member from the list and convey the nominee's name to the AAA and the other party within 14 days after receipt of the list from the AAA.
b. The Owner and the Contractor shall have 14 days within which to accept, in writing to AAA and the other party, the other party's nominee.
c. No reasons for non-acceptance need be stated. In the event of non-acceptance, the nominating party shall submit another nomination within 14 days of receipt until two mutually acceptable members are named.

6. Nomination and Acceptance of Third Member

a. Upon acceptance of both of the first two members, the AAA will notify them of their appointment, request that they begin selection of the third member and furnish them with the list of persons, biographical statements and disclosures originally sent to the parties. The first two members will endeavor to nominate a third member who meets all the criteria listed above. The third member shall be nominated within 14 days after the first two members are notified to proceed with his/her selection. The nominee's name will be conveyed to the AAA, who will notify the Owner and Contractor. The Owner and the Contractor shall have 14 days within which to accept, in writing to AAA and the other party, the third nominee. No reasons for non-acceptance need be stated. In the event of non-acceptance, the first two members will be requested to submit another nomination within 14 days of receipt of notice of non-acceptance from the AAA.
b. In the event of an impasse in selection of the third member from nominees of the first two members, the third member shall be selected by mutual agreement of the Owner and the Contractor within 14 days of the last non-acceptance notice. In so doing, they may, but are not required to, consider nominees offered by the first two members.

D. ALTERNATIVE PROCEDURE FOR SELECTION OF SINGLE-MEMBER BOARD

1. General

a. If the Contract specifies, or the Owner and the Contractor agree, a single-member Board will be established as provided in this Section 102.D

2. **Procedure**

a. Upon receipt of a properly filed Request for DRB Assistance detailing the agreement of the Owner and the Contractor to a single-member Board, the AAA shall promptly schedule a telephone conference call with the Owner and the Contractor to discuss desired qualifications of the Board member.
b. Within 14 days after the information is provided by the Owner and Contractor, the AAA shall send the Owner and Contractor an identical list of persons selected from its International Roster of DRB members, including detailed biographical information and disclosures regarding each listed person.
c. Proposed Board members shall disclose to the AAA any circumstance likely to affect impartiality, including any bias or any financial or personal interest in the project or any past or present relationship with the parties to the Contract, including subcontractors, design professionals and consultants.
d. The Owner and the Contractor shall each have 14 days in which to strike names not preferred, number the remaining names in order of preference, and return the list to the AAA. The Owner and the Contractor may strike up to three (3) names each.
e. From among the persons who have been approved on both lists, and in accordance with the designated order of mutual preference, the AAA shall invite the acceptance of the Board member.
f. If, for any reason, an appointment cannot be made from the original list, the AAA shall have the authority to send an additional list. If no names are available from that list, the AAA shall have the authority to make the appointment from among other members of its International Roster of DRB members, without the submission of additional lists.

E. POST-APPOINTMENT DISCLOSURE

Board members have a continuing duty to disclose to the AAA any circumstance likely to affect impartiality, including any bias or any financial or personal interest in the project or any past or present relationship with the parties to the Contract, including subcontractors, design professionals and consultants. Upon receipt of such information, the AAA shall communicate the information to the parties and, if it deems it appropriate to do so, to the Board members and others.

F. BOARD MEMBER CHALLENGE PROCEDURE

Any objection for cause of the Owner or Contractor to the continued service of a Board member shall be made to the AAA. The AAA shall determine whether the Board member should be disqualified and shall inform the Owner and Contractor of its decision, which shall be conclusive.

G.   VACANCIES

If for any reason a Board member is unable to perform the duties of the office, the AAA may, on proof satisfactory to it, declare the office vacant. The new Board member(s) shall be selected in the same manner as the original member. In the event of a vacancy after a dispute has been submitted and hearings commenced, the remaining Board members may continue with the hearing and determination of that dispute, unless the parties agree otherwise.

1.03   OPERATION

A.   GENERAL

1. The DRB shall adopt the operating procedures detailed in the attached Schedule A or formulate new or revised operating procedures consistent with this Specification. Notice of adoption of Schedule A or the Board's proposal for new/revised DRB Operating Procedures shall be provided by the Dispute Resolution Board to the Owner and the Contractor within 28 days after the effective date of the Three-Party Agreement.
2. Any DRB proposal for new/revised procedures shall be discussed and concurred in by all parties at the first Board Meeting.

B.   REPORTS AND INFORMATION

The Board will be kept informed of construction activity and other developments by means of timely transmittal of relevant information prepared by the Owner and the Contractor in the normal course of construction, including but not limited to periodic reports and minutes of project progress meetings.

C.   PERIODIC MEETINGS AND VISITS

1. The Board will visit the project site and meet with representatives of the Owner and the Contractor at regular intervals. The frequency and scheduling of these visits will be every three months or as agreed upon among the Owner, the Contractor and the Board, depending on the progress of the work.
2. Each meeting shall consist of an informal roundtable discussion and field observation of the work. The roundtable discussion will be attended by authorized representatives of the Owner and Contractor. During the discussion,

the Board may facilitate conversation among and between the parties in order to resolve any pending claims which may become disputes.
3. The field observations shall cover all active segments of the work. The Board shall be accompanied by authorized representatives of both the Owner and Contractor.

1.04   REVIEW OF DISPUTES

A.   GENERAL

1. The Owner and the Contractor will cooperate to ensure that the Board considers disputes promptly, taking into consideration the particular circumstances and the time required to prepare appropriate documentation.

B.   PREREQUISITES TO REVIEW

A dispute is subject to referral to the Board when:

1. Either party believes that bilateral negotiations are not likely to succeed or have reached an impasse, and,
2. If the Contract provides for a prior decision(s), such a decision(s) has been issued. The parties shall cooperate to timely comply with any pre-review requirements and may waive such requirements by written agreement.

C.   REQUESTING REVIEW

1. Either party may refer a Dispute to the Board. Requests for Board Review shall be submitted in writing to the Chair of the Dispute Resolution Board within 14 days of the final decision required prior to Board review. The Request for Board Review shall set forth in writing the nature of the dispute, the factual and contractual basis of the dispute and all remedies sought, together with all documents that support each element of the claim.
2. A copy of the Request for Board Review shall be simultaneously provided to the other party by the referring party.
3. Within 28 days after the Request for Board Review has been filed, the opposing party shall submit in writing to the Chair of the DRB a Response to Request for Board review, including the factual and contractual basis of any defense, together with all documents that support each element of the defense. If the responding party wishes to counterclaim, the responding party shall, within 28 days after the Request for Board review has been filed, submit, in writing to the Chair of the DRB, a Counterclaim setting forth in writing the factual

and contractual basis of the counterclaim and all remedies sought, together with all documents that support each element of the Counterclaim. A copy of the Response and/or Counterclaim shall be simultaneously provided to the other party by the responding party. Within 28 days after a Counterclaim is filed, the party opposing the Counterclaim shall submit, in writing to the Chair of the Dispute Resolution Board, a Response to the Counterclaim setting forth the factual and contractual basis of any defense, together with all documents which support each element of the Response to the Counterclaim. A copy of the Response to the Counterclaim shall be simultaneously provided by the filing party to the other party.

D. SCHEDULING REVIEW

1. Within seven days receipt of the Response to Request for Board Review or Response to Counterclaim, whichever comes later, the Chair will, in consultation with the Owner and the Contractor, establish dates for any additional pre-hearing submissions and schedule a hearing date. The hearing will generally be conducted at the time of the next regularly scheduled Site visit.
2. In addition, the DRB may convene a preliminary hearing by conference call for the purpose of addressing information exchange, the order of proceedings at the hearing, bifurcation of merit and quantum issues and such other matters that the DRB believes will expedite the hearing process.

E. HEARING LOCATION

1. Normally, the hearing will be held at the job Site. Any location that would be convenient and have the necessary access to facilities and documentation would also be acceptable.

F. HEARING PROCEDURES

1. The Dispute Resolution Board shall adopt the Hearing Procedures detailed in the attached Schedule B or develop new or revised Hearing Procedures consistent with this Specification. Hearing Procedures shall be provided by the Dispute Resolution Board to the Owner and the Contractor within 28 days of the effective date of the Three-Party Agreement.

G.   HEARING ATTENDANCE

1. The Owner and the Contractor shall have authorized representatives at all hearings. The Dispute Resolution Board may establish rules for the participation of legal counsel and experts at hearings. Unless the DRB permits, counsel may not (a) examine directly or cross-examine any participants; (b) object to questions or factual statements during the hearing or (c) make motions or offer arguments.

H.   DELIBERATIONS

1. After the hearing is concluded, the Board will confer to formulate its recommendations. All Board deliberations shall be conducted in private, with all individual views kept strictly confidential from disclosure to others.

I.   RECOMMENDATION

1. The Board's recommendation for resolution of the dispute will be provided in writing to both the Owner and the Contractor within 14 days of the completion of the hearings. In difficult or complex cases, and in consideration of the Board's schedule, this time may be extended by mutual agreement of all parties.

J.   ACCEPTANCE OR REJECTION

1. Within 14 days of receiving the Board's recommendation, or such other time specified by the Board, both the Owner and the Contractor shall provide written notice to the other and to the Board of acceptance or rejection of the Board's recommendation. The failure of either party to respond within the specified period shall be deemed an acceptance of the Board's recommendation. If, with the aid of the Board's recommendation, the Owner and the Contractor are able to resolve their dispute, the Owner will promptly process any required Contract modifications.

K. CLARIFICATION AND RECONSIDERATION

1. Should the dispute remain unresolved because of a bona fide lack of clear understanding of the Board's recommendation, either party may request that the Board clarify specified portions of its recommendation.
2. If new information has become available, either party may request that the Board reconsider its recommendation in light of the new information.

L. ADMISSIBILITY

1. If the Board's recommendation does not resolve the dispute, the written recommendation, including any minority report, will [not] be admissible as evidence [to the extent permitted by law] in any subsequent dispute resolution proceeding or forum [.] [to establish (a) that the Dispute Resolution Board considered the dispute, and (b) the Board's recommendation that resulted from the process.]

1.05   ALTERNATIVE DISPUTE RESOLUTION

A. The Owner and Contractor may, by agreement at any time during review of a Dispute by the Board, refer the dispute to the American Arbitration Association for mediation or any other form of alternative dispute resolution. In such an agreement, the Owner and Contractor shall specify the Dispute that is being referred and, in the event of settlement, shall advise the Board regarding such settlement, after which the Board shall have no further authority to proceed with that matter.

1.06   BOARD MEMBER FEES AND EXPENSES

A. The fees and expenses of the three members of the Board shall be shared equally by the Owner and the Contractor. Unless otherwise agreed by the parties and the Board, the Contractor shall pay the invoices of all Board members after approval by both parties. The Contractor will then bill the Owner for 50 per cent of such invoices.
B. The Owner will, at its expense, prepare and mail progress reports and provide conference facilities and copying services as reasonably required for Board operations.
C. If the Board desires special services such as legal or other consultation, accounting, data research and the like, both parties must agree, and the costs will be shared by them as mutually agreed.

1.07   ADMINISTRATIVE ASSISTANCE OF AAA

A.   AAA ADMINISTRATION

AAA will prepare and provide notices of meetings, transmit meeting minutes and Board recommendations and collect and disburse Board member fees and expenses in accordance with attached Schedule C.

©2004 American Arbitration Association. All Rights Reserved.

Annex 6
ICC Dispute Board Rules, Dispute Board Clauses and Model Dispute Board Member Agreement

ICC DISPUTE BOARD RULES

[In force as from 1 September 2004 with Standard ICC Dispute Board Clauses and Model Dispute Board Member Agreement.]

STANDARD ICC DISPUTE BOARD CLAUSES

ICC offers parties three different kinds of Dispute Board under its Dispute Board Rules. Parties should select the clause that corresponds to the type of Dispute Board they wish to use. ICC does not favour any one of these three types of Dispute Board over the others.
 While ICC recommends the use of the standard clauses, the parties should verify their enforceability under applicable law.

ICC DISPUTE REVIEW BOARD FOLLOWED BY ICC ARBITRATION IF REQUIRED

The Parties hereby agree to establish a Dispute Review Board ('DRB') in accordance with the Dispute Board Rules of the International Chamber of Commerce (the 'Rules'), which are incorporated herein by reference. The DRB shall have [one/three] member[s] appointed in this Contract or appointed pursuant to the Rules.
 All disputes arising out of or in connection with the present Contract shall be submitted, in the first instance, to the DRB in accordance with the Rules. For any given dispute, the DRB shall issue a Recommendation in accordance with the Rules.
 If any Party fails to comply with a Recommendation when required to do so pursuant to the Rules, the other Party may refer the failure itself to arbitration under the Rules of Arbitration of the International Chamber of Commerce by one or more arbitrators appointed in accordance with the said Rules of Arbitration.

If any Party sends a written notice to the other Party and the DRB expressing its dissatisfaction with a Recommendation, as provided in the Rules, or if the DRB does not issue the Recommendation within the time limit provided in the Rules, or if the DRB is disbanded pursuant to the Rules, the dispute shall be finally settled under the Rules of Arbitration of the International Chamber of Commerce by one or more arbitrators appointed in accordance with the said Rules of Arbitration.

ICC DISPUTE ADJUDICATION BOARD FOLLOWED BY ICC ARBITRATION IF REQUIRED

The Parties hereby agree to establish a Dispute Adjudication Board ('DAB') in accordance with the Dispute Board Rules of the International Chamber of Commerce (the 'Rules'), which are incorporated herein by reference. The DAB shall have [one/three] member [s] appointed in this Contract or appointed pursuant to the Rules.

All disputes arising out of or in connection with the present Contract shall be submitted, in the first instance, to the DAB in accordance with the Rules. For any given dispute, the DAB shall issue a Decision in accordance with the Rules.*

If any Party fails to comply with a Decision when required to do so pursuant to the Rules, the other Party may refer the failure itself to arbitration under the Rules of Arbitration of the International Chamber of Commerce by one or more arbitrators appointed in accordance with the said Rules of Arbitration.

If any Party sends a written notice to the other Party and the DAB expressing its dissatisfaction with a Decision, as provided in the Rules, or if the DAB does not issue the Decision within the time limit provided for in the Rules, or if the DAB is disbanded pursuant to the Rules, the dispute shall be finally settled under the Rules of Arbitration of the International Chamber of Commerce by one or more arbitrators appointed in accordance with the said Rules of Arbitration.

[Parties may, if they wish, provide for review by ICC of a DAB's Decisions by inserting the following text in place of the asterisk above: The DAB shall submit each Decision to ICC for review in accordance with Article 21 of the Rules.]*

ICC COMBINED DISPUTE BOARD FOLLOWED BY ICC ARBITRATION IF REQUIRED

The Parties hereby agree to establish a Combined Dispute Board ('CDB') in accordance with the Dispute Board Rules of the International Chamber of Commerce (the 'Rules'), which are incorporated herein by reference. The CDB shall have [one/three] member[s] appointed in this Contract or appointed pursuant to the Rules.

All disputes arising out of or in connection with the present Contract shall be submitted, in the first instance, to the CDB in accordance with the Rules. For any given dispute, the CDB shall issue a Recommendation unless the Parties agree that it shall render a Decision or it decides to do so upon the request of a Party and in accordance with the Rules.*

If any Party fails to comply with a Recommendation or a Decision when required to do so pursuant to the Rules, the other Party may refer the failure itself to arbitration under the Rules of Arbitration of the International Chamber of Commerce by one or more arbitrators appointed in accordance with the said Rules of Arbitration.

If any Party sends a written notice to the other Party and the CDB expressing its dissatisfaction with a Recommendation or a Decision as provided for in the Rules, or if the CDB does not issue the Recommendation or Decision within the time limit provided for in the Rules, or if the CDB is disbanded pursuant to the Rules, the dispute shall be finally settled under the Rules of Arbitration of the International Chamber of Commerce by one or more arbitrators appointed in accordance with the said Rules of Arbitration.

[*Parties may, if they wish, provide for review by ICC of a CDB's Decisions by inserting the following text in place of the, asterisk above: The CDB shall submit each Decision to ICC for review in accordance with Article 21 of the Rules.]

DISPUTE BOARD RULES OF THE INTERNATIONAL CHAMBER OF COMMERCE

INTRODUCTORY PROVISIONS

Article 1 – Scope of the Rules

Dispute Boards established in accordance with the Dispute Board Rules of the International Chamber of Commerce (the 'Rules') aid the Parties in resolving their business disagreements and disputes. They may provide informal assistance or issue Determinations. Dispute Boards are not arbitral tribunals and their Determinations are not enforceable like arbitral awards. Rather, the Parties contractually agree to be bound by the Determinations under certain specific conditions set forth herein. In application of the Rules, the International Chamber of Commerce ('ICC'), through the ICC Dispute Board Centre ('the Centre'), can provide administrative services to the Parties, which include appointing Dispute Board Members, deciding upon challenges to Dispute Board Members, and reviewing Decisions.

Article 2 – Definitions

In the Rules:

(a) 'Contract' means the agreement of the Parties that contains or is subject to provisions for establishing a Dispute Board under the Rules.
(b) 'Determination' means either a Recommendation or a Decision, issued in writing by the Dispute Board, as described in the Rules.
(c) 'Dispute' means any disagreement arising out of or in connection with the Contract which is referred to a Dispute Board for a Determination under the terms of the Contract and pursuant to the Rules.
(d) 'Dispute Board ('DB') means a Dispute Review Board ('DRB'), a Dispute Adjudication Board ('DAB') or a Combined Dispute Board ('CDB'), composed of one or three Dispute Board members ('DB Members').
(e) 'Party' means a party to the Contract and includes one or more parties, as appropriate.

Article 3 – Agreement to Submit to the Rules

Unless otherwise agreed, the Parties shall establish the DB at the time of entering into the Contract. The Parties shall specify whether the DB shall be a DRB, a DAB or a CDB.

Types of Dispute Boards

Article 4 – Dispute Review Boards (DRBs)

1. DRBs issue Recommendations with respect to Disputes.
2. Upon receipt of a Recommendation, the Parties may comply with it voluntarily but are not required to do so.
3. If no Party has sent a written notice to the other Party and the DRB expressing its dissatisfaction with a Recommendation within 30 days of receiving it, the Recommendation shall become binding on the Parties. The Parties shall thereafter comply with such Recommendation without delay, and they agree not to contest it insofar as such agreement can validly be made.
4. If any Party fails to comply with a Recommendation when required to do so pursuant to this Article 4, the other Party may refer the failure itself to arbitration, if the Parties have so agreed, or, if not, to any court of competent jurisdiction.
5. Any Party that is dissatisfied with a Recommendation shall, within 30 days of receiving it, send a written notice expressing its dissatisfaction to the other

Party and the DRB. For information purposes, such notice may specify the reasons for such Party's dissatisfaction.
6. If any Party submits such a written notice expressing its dissatisfaction with a Recommendation, or if the DRB does not issue its Recommendation within the time limit prescribed in Article 20, or if the DRB is disbanded pursuant to the Rules before a Recommendation regarding a Dispute has been issued, the Dispute in question shall be finally settled by arbitration, if the parties have so agreed, or, if not, by any court of competent jurisdiction.

Article 5 – Dispute Adjudication Boards (DABs)

1. DABs issue Decisions with respect to Disputes.
2. A Decision is binding on the Parties upon its receipt. The Parties shall comply with it without delay, notwithstanding any expression of dissatisfaction pursuant to this Article 5.
3. If no Party has sent a written notice to the other Party and the DAB expressing its dissatisfaction with the Decision within 30 days of receiving it, the Decision shall remain binding on the Parties. The Parties shall continue to comply with the Decision, and they agree not to contest it insofar as such agreement can validly be made.
4. If any Party fails to comply with a Decision when required to do so pursuant to this Article 5, the other Party may refer the failure itself to arbitration, if the Parties have so agreed, or, if not, to any court of competent jurisdiction.
5. Any Party that is dissatisfied with a Decision shall, within 30 days of receiving it, send a written notice expressing its dissatisfaction to the other Party and the DAB. For information purposes, such notice may specify the reasons for such Party's dissatisfaction.
6. If any Party submits such a written notice expressing its dissatisfaction with a Decision, or if the DAB does not issue its Decision within the time limit prescribed in Article 20, or if the DAB is disbanded pursuant to the Rules before a Decision regarding a Dispute has been issued, the Dispute in question shall be finally settled by arbitration, if the Parties have so agreed, or, if not, by any court of competent jurisdiction. Until the Dispute is finally settled by arbitration or otherwise, or unless the arbitral tribunal or the court decides otherwise, the Parties remain bound to comply with the Decision.

Article 6 – Combined Dispute Boards (CDBs)

1. CDBs issue Recommendations with respect to Disputes, pursuant to Article 4, but they may issue Decisions, pursuant to Article 5, as provided in paragraphs 2 and 3 of this Article 6.
2. If any Party requests a Decision with respect to a given Dispute and no other Party objects thereto, the CDB shall issue a Decision.

3. If any Party requests a Decision and another Party objects thereto, the CDB shall make a final decision as to whether it will issue a Recommendation or a Decision. In so deciding, the CDB shall consider, without being limited to, the following factors:
   - whether, due to the urgency of the situation or other relevant considerations, a Decision would facilitate the performance of the Contract or prevent substantial loss or harm to any Party;
   - whether a Decision would prevent disruption of the Contract; and
   - whether a Decision is necessary to preserve evidence.
4. Any request for a Decision by the Party referring a Dispute to the CDB shall be made in the Statement of Case under Article 17. Any such request by another Party should be made in writing no later than in its Response under Article 18.

ESTABLISHMENT OF THE DISPUTE BOARD

Article 7 – Appointment of the DB Members

1. The DB shall be established in accordance with the provisions of the Contract or, where the Contract is silent, in accordance with the Rules.
2. Where the Parties have agreed to establish a DB in accordance with the Rules but have not agreed on the number of DB Members, the DB shall be composed of three members.
3. Where the Parties have agreed that the DB shall have a sole DB Member, they shall jointly appoint the sole DB Member. If the Parties fail to appoint the sole DB Member within 30 days after signing the Contract or within 30 days after the commencement of any performance under the Contract, whichever occurs earlier, or within any other time period agreed upon by the Parties, the sole DB Member shall be appointed by the Centre upon the request of any Party.
4. When the DB is composed of three DB Members, the Parties shall jointly appoint the first two DB Members. If the Parties fail to appoint one or both DB Members within 30 days after signing the Contract or within 30 days after the commencement of any performance under the Contract, whichever occurs earlier, or within any other time period agreed upon by the Parties, both DB Members shall be appointed by the Centre upon the request of any Party.
5. The third DB Member shall be proposed to the Parties by the two DB Members within 30 days following the appointment of the second DB Member. If the Parties do not appoint the proposed third DB Member within 15 days from their receipt of the proposal, or if the two DB Members fail to propose the third DB Member, the third DB Member shall be appointed by the Centre upon the request of any Party. The third DB Member shall act as chairman of the DB unless all DB Members agree upon another chairman with the consent of the Parties.

6. When a DB Member has to be replaced due to death, resignation or termination, the new DB Member shall be appointed in the same manner as the DB Member being replaced, unless otherwise agreed by the Parties. All actions taken by the DB prior to the replacement of a DB Member shall remain valid. When the DB is composed of three DB Members and one of the DB Members is to be replaced, the other two shall continue to be DB Members. Prior to the replacement of the DB Member, the two remaining DB Members shall not hold hearings or issue Determinations without the agreement of all of the Parties.
7. The appointment of any DB Member shall be made by the Centre upon the request of any Party in the event that the Centre is satisfied that there is a sufficient basis for doing so.
8. When appointing a DB Member, the Centre shall consider the prospective DB Member's qualifications relevant to the circumstances, availability nationality and relevant language skills, as well as any observations, comments or requests made by the Parties.

OBLIGATIONS OF THE DISPUTE BOARD MEMBERS

Article 8 – Independence

1. Every DB Member must be and remain independent of the Parties.
2. Every prospective DB Member shall sign a statement of independence and disclose in writing to the Parties, to the other DB Members, and to the Centre, if such DB Member is to be appointed by the Centre, any facts or circumstances which might be of such a nature as to call into question the DB Member's independence in the eyes of the Parties.
3. A DB Member shall immediately disclose in writing to the Parties and the other DB Members any facts or circumstances of a similar nature which may arise in the course of such DB Member's tenure.
4. Should any Party wish to challenge a DB Member on the basis of an alleged lack of independence or otherwise, it may, within 15 days of learning of the facts upon which the challenge is based, submit to the Centre a request for a decision upon the challenge including a written statement of such facts. The Centre will finally decide the challenge after having given the challenged DB Member as well as any other DB Members and the other Party an opportunity to comment on the challenge.
5. If a DB Member is successfully challenged, that DB member's agreement with the Parties shall be terminated forthwith. The resulting vacancy shall be filled following the procedure used to appoint the challenged DB Member, unless otherwise agreed by the Parties.

Article 9 – Work of the DB and Confidentiality

1. By accepting to serve, DB Members undertake to carry out their responsibilities in accordance with the Rules.
2. Unless otherwise agreed by the Parties or otherwise required by applicable law, any information obtained by a DB Member during the course of the DB's activities shall be used by the DB Member only for the purposes of the DB's activities and shall be treated by the DB Member as confidential.
3. Unless otherwise agreed by the Parties, a DB Member shall not act in any judicial, arbitration or similar proceedings relating to any Dispute, whether as a judge, as an arbitrator, as an expert, or as a representative or advisor of a Party.

Article 10 – DB Member Agreement

1. Before commencing DB activities, every DB Member shall sign with all of the Parties a DB Member Agreement. If there are three DB Members, each DB Member Agreement shall have substantive terms that are identical to the other DB Member Agreements, unless otherwise agreed by the Parties and the DB Member concerned.
2. The Parties may at any time, without cause and with immediate effect, jointly terminate the DB Member Agreement of any DB Member but shall pay the Monthly Retainer Fee to such DB member for a minimum of three months following the termination, unless otherwise agreed by the Parties and the DB Member concerned.
3. Every DB Member may terminate the DB Member Agreement at any time by giving a minimum of three months' written notice to the Parties, unless otherwise agreed by the Parties and the DB Member concerned.

OBLIGATION TO COOPERATE

Article 11 – Providing of Information

1. The Parties shall fully cooperate with the DB and communicate information to it in a timely manner. In particular, the Parties and the DB shall cooperate to ensure that, as soon as possible after the DB is constituted, the DB becomes fully informed about the Contract and its performance by the Parties.
2. The Parties shall ensure that the DB is kept informed of the performance of the Contract and of any disagreements arising in the course thereof by such means as progress reports, meetings and, if relevant to the Contract, site visits.

ICC Dispute Board Rules, Clauses and Model Member Agreement

3. The DB shall, after consultation with the Parties, inform the Parties in writing of the nature, format and frequency of any progress reports that the Parties shall send to the DB.
4. If requested by the DB, the Parties, during meetings and site visits, shall provide the DB with adequate working space, accommodation, means of communication, typing facilities and all necessary office and information technology equipment allowing the DB to fulfil its functions.

Article 12 – Meetings and Site Visits

1. At the beginning of its activities the DB shall, in consultation with the Parties, establish a schedule of meetings and, if relevant to the Contract, site visits. The frequency of scheduled meetings and site visits shall be sufficient to keep the DB informed of the performance of the Contract and of any disagreements. Unless otherwise agreed by the Parties and the DB, when site visits are relevant to the Contract, there shall be a minimum of three such visits per year. The Parties and the DB shall attend all such meetings and site visits. In the event that a Party fails to attend, the DB may nevertheless decide to proceed. In the event that a DB Member fails to attend, the DB may proceed if the parties so agree or if the DB so decides.
2. Site visits occur at the site or sites where the Contract is being performed. Meetings can be held at any location agreed by the Parties and the DB. If they do not agree on where to hold a meeting, the location shall be decided by the DB after consultation with the Parties.
3. During scheduled meetings and site visits the DB shall review the performance of the Contract with the parties and may provide informal assistance, pursuant to Article 16, with respect to any disagreements.
4. Any Party may request an urgent meeting or site visit outside the scheduled meetings and site visits. The DB Members shall accommodate such a request at the earliest possible time and shall make best efforts to make themselves available for such urgent meetings or site visits within 30 days of the request.
5. After every meeting and site visit, the DB shall prepare a written summary of such meeting or site visit including a list of those present.

Article 13 – Written Notifications or Communications; Time Limits

1. All written notifications or communications, including any enclosures or attachments, from a Party to the DB or from the DB to the Parties shall be communicated simultaneously to all Parties and DB Members at the address on record for each DB Member and Party.
2. Written notifications or communications shall be sent in the manner agreed between the Parties and the DB or in any manner that provides the sender with proof of the sending thereof.

3. A notification or communication shall be deemed to have been made on the date that it was received by the intended recipient or by its representative or would have been received if made in accordance with this Article 13.
4. Periods of time specified in or fixed under the Rules shall start to run on the day following the date a notification or communication is deemed to have been made in accordance with the preceding paragraph. When the day next following such date is an official holiday or non-business day in the country in which the notification or communication is deemed to have been made, the period of time shall commence on the first following business day. Official holidays or non-business days are included in the calculation of the period of time. If the last day of the relevant period of the time granted is an official holiday or non-business day in the country where the notification or communication is deemed to have been made, the period of time shall expire at the end of the first following business day.

OPERATION OF THE DISPUTE BOARD

Article 14 – Beginning and End of the DB's Activities

1. The DB shall begin its activities after every DB Member and the Parties have signed the DB Member Agreement(s).
2. Unless otherwise agreed by the Parties, the DB shall end its activities upon receiving notice from the Parties of their joint decision to disband the DB.
3. Any dispute which may arise after the DB has been disbanded shall be finally settled by arbitration, if the Parties have so agreed, or, if not, by any court of competent jurisdiction.

Article 15 – Powers of the DB

1. The proceedings before the DB shall be governed by the Rules and, where the Rules are silent, by any rules which the Parties or, failing them, the DB may settle on. In particular, in the absence of an agreement of the Parties with respect thereto, the DB shall have the power, inter alia, to:
   – determine the language or languages of the proceedings before the DB, due regard being given to all relevant circumstances, including the language of the Contract;
   – require the Parties to produce any documents that the DB deems necessary in order to issue a Determination;
   – call meetings, site visits and hearings;
   – decide on all procedural matters arising during any meeting, site visit or hearing;
   – question the Parties, their representatives and any witnesses they may call, in the sequence it chooses;

- issue a Determination even if a Party fails to comply with a request of the DB;
- take any measures necessary for it to fulfil its function as a DB.

2. Decisions of the DB regarding the rules governing the proceedings shall be taken by the sole DB Member or, when there are three DB Members, by majority vote. If there is no majority, the Decision shall be made by the chairman of the DB alone.
3. The DB may take measures for protecting trade secrets and confidential information.
4. If the Contract has more than two Parties, the application of the Rules may be adapted, as appropriate, to apply to the multiparty situation, by agreement of all of the Parties or, failing such agreement, by the DB.

PROCEDURES BEFORE THE DISPUTE BOARD

Article 16 – Informal Assistance with Disagreements

1. On its own initiative or upon the request of any Party and in either case with the agreement of all of the Parties, the DB may informally assist the Parties in resolving any disagreements that may arise during the performance of the Contract. Such informal assistance may occur during any meeting or site visit. A Party proposing the informal assistance of the DB shall endeavour to inform the DB and the other Party thereof well in advance of the meeting or site visit during which such informal assistance would occur.
2. The informal assistance of the DB may take the form of a conversation among the DB and the Parties; separate meetings between the DB and any Party with the prior agreement of the Parties; informal views given by the DB to the Parties; a written note from the DB to the Parties; or any other form of assistance which may help the Parties resolve the disagreement.
3. The DB, if called upon to make a Determination concerning a disagreement with respect to which it has provided informal assistance, shall not be bound by any views, either oral or in writing, which it may have given in the course of its informal assistance.

Article 17 – Formal Referral of Disputes for a Determination; Statement of Case

1. Any Party shall refer a Dispute to the DB by submitting a written statement of its case (the 'Statement of Case') to the other Party and the DB. The Statement of Case shall include:
    - a clear and concise description of the nature and circumstances of the Dispute;

- a list of the issues submitted to the DB for a Determination and a presentation of the referring Party's position thereon;
- any support for the referring Party's position such as documents, drawings, schedules and correspondence;
- a statement of what the referring Party requests the DB to determine; and
- in the case of a CDB, if the referring Party wishes the CDB to issue a Decision, its request for a Decision and the reasons why it believes that the CDB should issue a Decision rather than a Recommendation.

2. The date on which the Statement of Case is received by the sole DB Member or the chairman of the DB, as the case maybe, shall, for all purposes, be deemed to be the date of the commencement of the referral (the 'Date of Commencement').
3. The Parties remain free to settle the Dispute, with or without the assistance of the DB, at any time.

Article 18 – Response and Additional Documentation

1. Unless the Parties agree otherwise or the DB orders otherwise, the responding Party shall respond to the Statement of Case in writing (the 'Response') within 30 days of receiving the Statement of Case. The Response shall include:
   - a clear and concise presentation of the responding Party's position with respect to the Dispute;
   - any support for its position such as documents, drawings, schedules and correspondence;
   - a statement of what the responding Party requests the DB to determine;
   - in the case of a CDB, a response to any request for a Decision made by the referring Party, or if the referring Party has not made such a request, any request for a Decision by the responding Party, including the reasons why it believes that the CDB should issue the type of Determination it desires.
2. The DB may at any time request a Party to submit additional written statements or documentation to assist the DB in preparing its Determination. Each such request shall be communicated in writing by the DB to the Parties.

Article 19 – Organization and Conduct of Hearings

1. A hearing regarding a Dispute shall be held unless the Parties and the DB agree otherwise.
2. Unless the DB orders otherwise, hearings shall be held within 15 days of the date on which the Sole DB Member or the chairman of the DB, as the case may be, receives the Response.
3. Hearings shall be held in the presence of all DB Members unless the DB decides, in the circumstances and after consultation with the Parties, that it is appropriate to hold the hearing in the absence of a DB Member; provided,

however, that prior to the replacement of a DB member a hearing may be held with the two remaining DB members only with the agreement of all of the Parties pursuant to Article 7(6).
4. If any of the Parties refuses or fails to take part in the DB procedure or any stage thereof, the DB shall proceed notwithstanding such refusal or failure.
5. The DB shall be in full charge of the hearings.
6. The DB shall act fairly and impartially and ensure that each Party has a reasonable opportunity to present its case.
7. The Parties shall appear in person or through duly authorized representatives who are in charge of the performance of the Contract. In addition, they may be assisted by advisors.
8. Unless the DB decides otherwise, the hearing shall proceed as follows:
    - presentation of the case, first by the referring Party and then by the responding Party;
    - identification by the DB to the Parties of any matters that need further clarification;
    - clarification by the Parties concerning the matters identified by the DB;
    - responses by each Party to clarifications made by the other Party, to the extent that new issues have been raised in such clarifications.
9. The DB may request the Parties to provide written summaries of their presentations.
10. The DB may deliberate at any location it considers appropriate before issuing its Determination.

Determinations of the Dispute Board

Article 20 – Time Limit for Rendering a Determination

1. The DB shall issue its Determination promptly and, in any event, within 90 days of the Date of Commencement as defined in Article 17(2). However, the Parties may agree to extend the time limit. In deciding whether to do so, the Parties shall consult with the DB and shall take into account the nature and complexity of the Dispute and other relevant circumstances.
2. When the Parties have agreed to submit Decisions to ICC for review, the time limit for issuing a Decision shall be extended by the time required for the Centre to review the Decision. The Centre shall complete its review within 30 days of its receipt of the Decision or of the payment of the administrative fee referred to in Article 3 of the Appendix, whichever occurs later. However, if additional time for such review is required, the Centre shall notify the DB and the Parties thereof in writing before the expiration of the 30 days, specifying the new date by which the Centre's review shall be completed.

Article 21 – Review of Decisions by the Centre

Where the Parties have provided for review by ICC of the Decisions of a DAB or CDB, the DB shall submit the Decision in draft form to the Centre before it is signed. Each Decision must be accompanied by the registration fee referred to in Article 3 of the Appendix. The Centre may lay down modifications only as to the form of the Decision. No such Decision shall be signed by the DB Members or communicated to the Parties prior to the Centre's approval of such Decision.

Article 22 – Contents of a Determination

Determinations shall indicate the date on which they are issued and shall state the findings of the DB as well as the reasons upon which they are based. Determinations may also include, without limitation and not necessarily in the following order:

- a summary of the Dispute, the respective positions of the Parties and the Determination requested;
- a summary of the relevant provisions of the Contract;
- a chronology of relevant events;
- a summary of the procedure followed by the DB; and
- a listing of the submissions and documents provided by the Parties in the course of the procedure.

Article 23 – Making of the Determination

When the DB is composed of three DB Members, the DB shall make every effort to achieve unanimity. If this cannot be achieved, a Determination is given by a majority decision. If there is no majority, the Determination shall be made by the chairman of the DB alone. Any DB Member who disagrees with the Determination shall give the reasons for such disagreement in a separate written report that shall not form part of the Determination but shall be communicated to the Parties. Any failure of a DB Member to give such reasons shall not prevent the issuance or the effectiveness of the Determination.

Article 24 – Correction and Interpretation of Determinations

1. On its own initiative, the DB may correct a clerical, computational or typographical error, or any errors of a similar nature, contained in a Determination, provided such correction is submitted to the Parties within 30 days of the date of such Determination.
2. Any Party may apply to the DB for the correction of an error of the kind referred to in Article 24(1), or for the interpretation of a Determination. Such application must be made to the DB within 30 days of the receipt of the Determination by such Party. After receipt of the application by the sole DB

Member or the chairman of the DB, as the case may be, the DB shall grant the other Party a short time limit from the receipt of the application by that Party, to submit any comments thereon. Any correction or interpretation of the DB shall be issued within 30 days following the expiration of the time limit for the receipt of any comments from the other Party. However, the Parties may agree to extend the time limit for the issuance of any correction or interpretation.

3. Should the DB issue a correction or interpretation of the Determination, all time limits associated with the Determination shall recommence to run upon receipt by the Parties of the correction or interpretation of the Determination.

Article 25 – Admissibility of Determinations in Subsequent Proceedings

Unless otherwise agreed by the Parties, any Determination shall be admissible in any judicial or arbitral proceedings in which all of the parties thereto were Parties to the DB proceedings in which the Determination was issued.

COMPENSATION OF THE DISPUTE BOARD MEMBERS AND ICC

Article 26 – General Considerations

1. All fees and expenses of the DB Members shall be shared equally by the Parties.
2. Unless otherwise agreed by the Parties, when there are three DB members all DB Members shall be treated equally and shall receive the same Monthly Retainer Fee and the same Daily Fee for work performed as a DB Member.
3. Unless otherwise provided in the DB Member Agreement(s), the fees shall be fixed for the first 24 months following the signature of the DB Member Agreement(s) and thereafter shall be adjusted on each anniversary of the DB Member Agreement(s) in accordance with the terms thereof.

Article 27 – Monthly Retainer Fee

1. Unless otherwise provided in the DB Member Agreement(s), each DB Member shall receive a Monthly Retainer Fee as set out in the DB Member Agreement(s) covering the following:
   - being available to attend all DB meetings with the Parties and site visits;
   - being available to attend internal DB meetings;
   - becoming and remaining conversant with the Contract and the progress of its performance;

- the study of progress reports and correspondence submitted by the Parties in the course of the DB's functions; and
- office overhead expenses in the DB Member's place of residence.

2. Unless otherwise agreed in the DB Member Agreement(s), the Monthly Retainer Fee shall be equal to three times the Daily Fee set out in the DB Member Agreement(s) and shall be payable from the date of signature of the DB Member Agreement(s) until termination of the DB Member Agreement(s).

Article 28 – Daily Fee

Unless otherwise agreed in the DB Member Agreement(s), each DB Member shall receive a Daily Fee as set out in the DB Member Agreement(s) covering the time spent for the following activities:

- meetings and site visits;
- hearings;
- travel time;
- internal meetings of the DB;
- study of documents submitted by Parties during procedures before the DB;
- preparation of a DB Determination; and
- activities in coordinating and organizing the operation of the DB.

Article 29 – Travel Costs and other Expenses

1. Unless otherwise provided in the DB Member Agreement(s), air travel expenses shall be reimbursed at unrestricted business class rates between a DB Member's home and the travel destination.
2. Unless otherwise provided in the DB Member Agreement (s), expenses, wherever incurred in DB work, for local transportation, hotels and meals, long distance phone, fax, courier charges, photocopying, postage, visa charges, etc., shall be reimbursed at cost.

Article 30 – Taxes and Charges

1. No taxes and charges, except for value added tax (VAT), levied in connection with the services rendered by a DB Member by the country of the residence or nationality of the DB Member shall be reimbursed by the Parties.
2. All taxes and charges levied in connection with such services by any country other than the DB Member's country of residence or nationality, as well as VAT wherever levied, shall be reimbursed by the Parties.

Article 31 – Payment Arrangements

1. Unless otherwise agreed, invoices shall be submitted by each DB Member to each Party for payment as follows:
   – Monthly Retainer Fees shall be invoiced and paid on a quarterly basis in advance for the next three-month period.
   – Daily Fees and travel expenses shall be invoiced and paid after each meeting, site visit, hearing or Determination.
2. DB Member invoices shall be paid within 30 days after receipt.
3. Failure of any Party to pay its share of fees and expenses within 30 days of receiving a DB Member's invoice shall entitle the DB Member, in addition to any other rights, to suspend work 15 days after providing a notice of suspension to the Parties and any other DB Members, such suspension to remain in effect until receipt of full payment of all outstanding amounts plus simple interest at one-year LIBOR plus 2 per cent, or the 12-month prime interest rate in the currency agreed between the Parties and the DB Members.
4. In the event that a Party fails to pay its share of the fees and expenses of a DB Member when due, any other Party, without waiving its rights, may pay the outstanding amount. The Party making such payment, in addition to any other rights, shall be entitled to reimbursement from the non-paying Party of all such sums paid, plus simple interest at one-year LIBOR plus 2 per cent, or the 12-month prime interest rate in the currency agreed between the Parties and the DB Members.
5. Upon signing the DB Member Agreement, the Parties shall provide the DB Member with the form of the invoice to be sent by DB Members, including the invoicing address, number of copies of invoices required and VAT number, if applicable.

Article 32 – Administrative Expenses of ICC

1. ICC's administrative expenses include an amount for each appointment of a DB Member, an amount for each decision upon a challenge of a DB Member and, when the Parties have agreed to submit Decisions of a DAB or a CDB to ICC for review, an amount for each such review.
2. For each request for appointment of a DB Member, ICC shall receive the non-refundable amount specified in Article 1 of the Appendix. This amount shall represent the total cost for the appointment of one DB Member by the Centre. The Centre shall not proceed with the appointment unless the requisite payment has been received. The cost of each appointment by the Centre shall be shared equally by the Parties.
3. For each decision upon a challenge of a DB Member, the Centre shall fix administrative expenses in an amount not exceeding the maximum sum specified in Article 2 of the Appendix. This amount shall represent the total cost for the decision upon one challenge of a DB Member. The Centre shall

not proceed with the rendering of its decision and the making of the challenge shall have no effect unless the said amount has been received. The cost of each decision by the Centre shall be borne by the Party making the challenge.
4. Where the Parties have provided for the review by ICC of a DAB's or a CDB's Decisions, the Centre shall fix administrative expenses for the review of each Decision in an amount not exceeding the maximum sum specified in Article 3 of the Appendix. This amount shall represent the total cost for the review of one Decision by ICC. The Centre shall not approve a Decision unless the said amount has been received. The cost of reviewing each Decision shall be shared equally by the Parties.
5. If a Party fails to pay its share of the administrative expenses of ICC, the other Party shall be free to pay the entire amount of such administrative expenses.

General Rules

Article 33 – Exclusion of Liability

Neither the DB Members, nor the Centre, nor ICC and its employees, nor the ICC national committees shall be liable to any person for any act or omission in connection with the DB proceedings.

Article 34 – Application of the Rules

In all matters not expressly provided for in the Rules, the DB shall act in the spirit of the Rules and shall make every effort to make sure that Determinations are issued in accordance with the Rules.

APPENDIX – SCHEDULE OF COSTS

Article 1

The non-refundable amount for the request for appointment of a DB Member referred to in Article 32(2) of the Rules is US$ 2,500. No request for appointment of a DB Member shall be processed unless accompanied by the requisite payment.

Article 2

Each request for a decision upon a challenge of a DB member must be accompanied by a registration fee of USD 2,500. No request for a decision upon a challenge of a DB Member shall be processed unless accompanied by the registration fee. Such payment is non-refundable and shall be credited to the administrative expenses for a decision upon a challenge. The Centre shall fix said administrative expenses in an amount not exceeding the maximum sum of USD 10,000.

Article 3

Each Decision of a DAB or a CDB submitted to ICC for review must be accompanied by a registration fee of USD 2,500. No Decision shall be reviewed unless accompanied by the registration fee. Such payment is non-refundable and shall be credited to the administrative expenses for the review of each Decision. The Centre shall fix said administrative expenses in an amount not exceeding the maximum sum of USD 10,000.

MODEL DISPUTE BOARD MEMBER AGREEMENT

This Agreement is entered into between:

DB Member [*full name, title and address*],

hereinafter the 'Dispute Board Member' or 'DB Member'

and

Party 1: [*full name and address*]

Party 2: [*full name and address*],

hereinafter collectively referred to as the Parties.

Whereas:

The Parties have entered into a contract dated…..(the 'Contract') for [*scope of work and/or name of project*], which is to be performed in [*city and country of performance*];

The Contract provides that the parties must refer their disputes to a [*DRB/DAB/CDB*] under the ICC Dispute Board Rules (the 'Rules'); and

The undersigned individual has been appointed to serve as a DB Member.

The DB Member and the Parties therefore agree as follows:

1. UNDERTAKING

The DB Member shall act as [sole DB Member/chairman of the DB/DB Member] and hereby accepts to perform these duties in accordance with the terms of the Contract, the Rules and the terms of this Agreement. The DB Member confirms that he/she is and shall remain independent of the Parties

2. COMPOSITION OF THE DB AND CONTACT DETAILS

- First alternative: The sole DB Member can be contacted as follows: [*name, address, telephone, fax and e-mail details*]
- Second alternative: The Members of the DB are those listed below and can be contacted as follows:

Chairman: [*name, address, telephone, fax and e-mail details*]

DB Member: [*name, address, telephone, fax and e-mail details*]

DB Member: [*name, address, telephone, fax and e-mail details*]

The Parties to the Contract are those indicated above with the following contact details:

Party 1: [*name, person responsible for the Contract, address, telephone, fax and e-mail details*]

Party 2: [*name, person responsible for the Contract, address, telephone, fax and e-mail details*]

Any changes in these contact details shall be immediately communicated to all concerned.

3. QUALIFICATIONS

With respect to any DB Member appointed by the Parties, the undersigned Parties recognize that such DB Member has the necessary professional qualifications and language ability to undertake the duties of a DB Member.

4. FEES

The Monthly Retainer Fee shall be [*specify currency and full amount*], i.e. [*specify multiple*] times the Daily Fee.

The Daily Fee shall be [*specify currency and full amount*] based upon a [*specify number of hours*] -hour day.

These fees shall be fixed for the first 24 months after the signing of the DB Member Agreement and thereafter shall be adjusted automatically on each anniversary of the DB Member Agreement using the following index:......

Expenses of the DB Member, as described in Article 29(2) of the Rules, shall be reimbursed [*at cost/on the basis of a fixed per diem of......*].

5. PAYMENT OF FEES AND EXPENSES

- First alternative: All fees and expenses shall be invoiced to [*Party X*] with a copy to [*Party Y*] and shall be paid to the DB Member by [*Party X*]. [*Party Y*]

shall reimburse half of the fees and expenses to [*Party X*] so that they are borne equally by the Parties.
- Second alternative: All fees and expenses shall be invoiced to and paid by each of the Parties in equal shares.

All payments to the DB Member shall be made without deductions or restrictions to the following account: [*name of bank, account no., SWIFT code, etc.*]. The transfer charges shall be borne by the party making the transfer.

All payments shall be made within 30 days of receipt by a Party of the invoice from the DB Member.

6.   DURATION AND TERMINATION OF THE AGREEMENT

Subject to the provisions of this Article 6, the DB Members agree to serve for the duration of the DB.

The Parties may jointly terminate this Agreement or terminate the whole DB at any time by giving [*specify number*] months' written notice to the DB Member or the whole DB.

The DB Member may resign from the Dispute Board at any time by giving [*specify number*] months' written notice to the Parties.

7.   INDEMNITY

The Parties will jointly and severally indemnify and hold harmless every DB Member from any claims of third parties for anything done or omitted in the discharge or purported discharge of the DB Member's activities, unless the act or omission is shown to have been in bad faith.

8.   DISPUTES AND APPLICABLE LAW

All disputes arising out of or in connection with this Agreement shall be finally settled under the Rules of Arbitration of the International Chamber of Commerce by one arbitrator appointed in accordance with the said Rules of Arbitration. This Agreement shall be governed by [*specify applicable law*]. The place of arbitration shall be [*name of city/country*]. The language of the arbitration shall be [*specify language*].

This Agreement is entered into on [*specify date*] at [*specify place*].

DB Member	Party 1	Party 2
[*signature*]	[*signature*]	[*signature*]

Annex 7
Convention on the Recognition and Enforcement of Foreign Arbitral Awards 1958 (The New York Convention)

UNITED NATIONS CONFERENCE ON INTERNATIONAL COMMERCIAL ARBITRATION

CONVENTION ON THE RECOGNITION AND ENFORCEMENT OF FOREIGN ARBITRAL AWARDS

Article I

1. This Convention shall apply to the recognition and enforcement of arbitral awards made in the territory of a State other than the State where the recognition and enforcement of such awards are sought, and arising out of differences between persons, whether physical or legal. It shall also apply to arbitral awards not considered as domestic awards in the State where their recognition and enforcement are sought.
2. The term 'arbitral awards' shall include not only awards made by arbitrators appointed for each case but also those made by permanent arbitral bodies to which the parties have submitted.
3. When signing, ratifying or acceding to this Convention, or notifying extension under Article X hereof, any State may on the basis of reciprocity declare that it will apply the Convention to the recognition and enforcement of awards made only in the territory of another Contracting State. It may also declare that it will apply the Convention only to differences arising out of legal relationships, whether contractual or not, which are considered as commercial under the national law of the State making such declaration.

Article II

1. Each Contracting State shall recognize an agreement in writing under which the parties undertake to submit to arbitration all or any differences which have

arisen or which may arise between them in respect of a defined legal relationship, whether contractual or not, concerning a subject matter capable of settlement by arbitration.
2. The term 'agreement in writing' shall include an arbitral clause in a contract or an arbitration agreement, signed by the parties or contained in an exchange of letters or telegrams.
3. The court of a Contracting State, when seized of an action in a matter in respect of which the parties have made an agreement within the meaning of this article, shall, at the request of one of the parties, refer the parties to arbitration, unless it finds that the said agreement is null and void, inoperative or incapable of being performed.

Article III

Each Contracting State shall recognize arbitral awards as binding and enforce them in accordance with the rules of procedure of the territory where the award is relied upon, under the conditions laid down in the following articles. There shall not be imposed substantially more onerous conditions or higher fees or charges on the recognition or enforcement of arbitral awards to which this Convention applies than are imposed on the recognition or enforcement of domestic arbitral awards.

Article IV

1. To obtain the recognition and enforcement mentioned in the preceding Article, the party applying for recognition and enforcement shall, at the time of the application, supply:
   (a) The duly authenticated original award or a duly certified copy thereof;
   (b) The original agreement referred to in Article II or a duly certified copy thereof.
2. If the said award or agreement is not made in an official language of the country in which the award is relied upon, the party applying for recognition and enforcement of the award shall produce a translation of these documents into such language. The translation shall be certified by an official or sworn translator or by a diplomatic or consular agent.

Article V

1. Recognition and enforcement of the award may be refused, at the request of the party against whom it is invoked, only if that party furnishes to the competent authority where the recognition and enforcement is sought, proof that:
   (a) The parties to the agreement referred to in Article II were, under the law applicable to them, under some incapacity, or the said agreement is not valid under the law to which the parties have subjected it or, failing any

indication thereon, under the law of the country where the award was made; or

(b) The party against whom the award is invoked was not given proper notice of the appointment of the arbitrator or of the arbitration proceedings or was otherwise unable to present his case; or

(c) The award deals with a difference not contemplated by or not falling within the terms of the submission to arbitration, or it contains decisions on matters beyond the scope of the submission to arbitration, provided that, if the decisions on matters submitted to arbitration can be separated from those not so submitted, that part of the award which contains decisions on matters submitted to arbitration may be recognized and enforced; or

(d) The composition of the arbitral authority or the arbitral procedure was not in accordance with the agreement of the parties, or, failing such agreement, was not in accordance with the law of the country where the arbitration took place; or

(e) The award has not yet become binding on the parties, or has been set aside or suspended by a competent authority of the country in which, or under the law of which, that award was made.

2. Recognition and enforcement of an arbitral award may also be refused if the competent authority in the country where recognition and enforcement is sought finds that:

(a) The subject matter of the difference is not capable of settlement by arbitration under the law of that country; or

(b) The recognition or enforcement of the award would be contrary to the public policy of that country.

Article VI

If an application for the setting aside or suspension of the award has been made to a competent authority referred to in Article V (1) (e), the authority before which the award is sought to be relied upon may, if it considers it proper, adjourn the decision on the enforcement of the award and may also, on the application of the party claiming enforcement of the award, order the other party to give suitable security.

Article VII

1. The provisions of the present Convention shall not affect the validity of multilateral or bilateral agreements concerning the recognition and enforcement of arbitral awards entered into by the Contracting States nor deprive any interested party of any right he may have to avail himself of an arbitral award

in the manner and to the extent allowed by the law or the treaties of the country where such award is sought to be relied upon.
2. The Geneva Protocol on Arbitration Clauses of 1923 and the Geneva Convention on the Execution of Foreign Arbitral Awards of 1927 shall cease to have effect between Contracting States on their becoming bound and to the extent that they become bound, by this Convention.

Article VIII

1. This Convention shall be open until 31 December 1958 for signature on behalf of any Member of the United Nations and also on behalf of any other State which is or hereafter becomes a member of any specialized agency of the United Nations, or which is or hereafter becomes a party to the Statute of the International Court of Justice, or any other State to which an invitation has been addressed by the General Assembly of the United Nations.
2. This Convention shall be ratified and the instrument of ratification shall be deposited with the Secretary-General of the United Nations.

Article IX

1. This Convention shall be open for accession to all States referred to in Article VIII.
2. Accession shall be effected by the deposit of an instrument of accession with the Secretary-General of the United Nations.

Article X

1. Any State may, at the time of signature, ratification or accession, declare that this Convention shall extend to all or any of the territories for the international relations of which it is responsible. Such a declaration shall take effect when the Convention enters into force for the State concerned.
2. At any time thereafter any such extension shall be made by notification addressed to the Secretary-General of the United Nations and shall take effect as from the ninetieth day after the day of receipt by the Secretary-General of the United Nations of this notification, or as from the date of entry into force of the Convention for the State concerned, whichever is the later.
3. With respect to those territories to which this Convention is not extended at the time of signature, ratification or accession, each State concerned shall consider the possibility of taking the necessary steps in order to extend the application of this Convention to such territories, subject, where necessary for constitutional reasons, to the consent of the Governments of such territories.

Article XI

In the case of a federal or non-unitary State, the following provisions shall apply:

(a) With respect to those articles of this Convention that come within the legislative jurisdiction of the federal authority, the obligations of the federal Government shall to this extent be the same as those of Contracting States which are not federal States;
(b) With respect to those articles of this Convention that come within the legislative jurisdiction of constituent states or provinces which are not, under the constitutional system of the federation, bound to take legislative action, the federal Government shall bring such articles with a favourable recommendation to the notice of the appropriate authorities of constituent states or provinces at the earliest possible moment;
(c) A federal State Party to this Convention shall, at the request of any other Contracting State transmitted through the Secretary-General of the United Nations, supply a statement of the law and practice of the federation and its constituent units in regard to any particular provision of this Convention, showing the extent to which effect has been given to that provision by legislative or other action.

Article XII

1. This Convention shall come into force on the 90th day following the date of deposit of the third instrument of ratification or accession.
2. For each State ratifying or acceding to this Convention after the deposit of the third instrument of ratification or accession, this Convention shall enter into force on the 90th day after deposit by such State of its instrument of ratification or accession.

Article XIII

1. Any Contracting State may denounce this Convention by a written notification to the Secretary-General of the United Nations. Denunciation shall take effect one year after the date of receipt of the notification by the Secretary-General.
2. Any State which has made a declaration or notification under Article X may, at any time thereafter, by notification to the Secretary-General of the United Nations, declare that this Convention shall cease to extend to the territory concerned one year after the date of the receipt of the notification by the Secretary-General.
3. This Convention shall continue to be applicable to arbitral awards in respect of which recognition or enforcement proceedings have been instituted before the denunciation takes effect.

Article XIV

A Contracting State shall not be entitled to avail itself of the present Convention against other Contracting States except to the extent that it is itself bound to apply the Convention.

Article XV

The Secretary-General of the United Nations shall notify the States contemplated in article VIII of the following:

(a) Signatures and ratifications in accordance with Article VIII;
(b) Accessions in accordance with Article IX;
(c) Declarations and notifications under Articles I, X and XI;
(d) The date upon which this Convention enters into force in accordance with Article XII;
(e) Denunciations and notifications in accordance with Article XIII.

Article XVI

1. This Convention, of which the Chinese, English, French, Russian and Spanish texts shall be equally authentic, shall be deposited in the archives of the United Nations.
2. The Secretary-General of the United Nations shall transmit a certified copy of this Convention to the States contemplated in Article VIII.

Annex 8
UNCITRAL Model Law on International Commercial Arbitration

UNITED NATIONS COMMISSION ON INTERNATIONAL TRADE LAW (UNCITRAL)

UNCITRAL MODEL LAW ON INTERNATIONAL COMMERCIAL ARBITRATION

(United Nations document A/40/17, Annex I)
[As adopted by the United Nations Commission on International Trade Law on 21 June 1985]

CHAPTER I – GENERAL PROVISIONS

Article 1 – Scope of Application[1]

(1) This Law applies to international commercial[2] arbitration, subject to any agreement in force between this State and any other State or States.
(2) The provisions of this Law, except Articles 8, 9, 35 and 36, apply only if the place of arbitration is in the territory of this State.

---

1 Article headings are for reference purposes only and are not to be used for purposes of interpretation.
2 The term 'commercial' should be given a wide interpretation so as to cover matters arising from all relationships of a commercial nature, whether contractual or not. Relationships of a commercial nature include, but are not limited to, the following transactions: any trade transaction for the supply or exchange of goods or services; distribution agreement; commercial representation or agency; factoring; leasing; construction of works; consulting; engineering; licensing; investment; financing; banking, insurance; exploitation agreement or concession; joint venture and other forms of industrial or business co-operation; carriage of goods or passengers by air, sea, rail or road.

(3) An arbitration is international if:
   (a) the parties to an arbitration agreement have, at the time of the conclusion of that agreement, their places of business in different States; or
   (b) one of the following places is situated outside the State in which the parties have their places of business:
      (i) the place of arbitration if determined in, or pursuant to, the arbitration agreement;
      (ii) any place where a substantial part of the obligations of the commercial relationship is to be performed or the place with which the subject-matter of the dispute is most closely connected; or
   (c) the parties have expressly agreed that the subject-matter of the arbitration agreement relates to more than one country.
(4) For the purposes of paragraph (3) of this article:
   (a) if a party has more than one place of business, the place of business is that which has the closest relationship to the arbitration agreement;
   (b) if a party does not have a place of business, reference is to be made to his habitual residence.
(5) This Law shall not affect any other law of this State by virtue of which certain disputes may not be submitted to arbitration or may be submitted to arbitration only according to provisions other than those of this Law.

Article 2 – Definitions and Rules of Interpretation

For the purposes of this Law:

(a) 'arbitration' means any arbitration whether or not administered by a permanent arbitral institution;
(b) 'arbitral tribunal' means a sole arbitrator or a panel of arbitrators;
(c) 'court' means a body or organ of the judicial system of a State;
(d) where a provision of this Law, except Article 28, leaves the parties free to determine a certain issue, such freedom includes the right of the parties to authorize a third party, including an institution, to make that determination;
(e) where a provision of this Law refers to the fact that the parties have agreed or that they may agree or in any other way refers to an agreement of the parties, such agreement includes any arbitration rules referred to in that agreement;
(f) where a provision of this Law, other than in Articles 25(a) and 32(2)(a), refers to a claim, it also applies to a counter-claim, and where it refers to a defence, it also applies to a defence to such counter-claim.

Article 3 – Receipt of Written Communications

(1) Unless otherwise agreed by the parties:
   (a) any written communication is deemed to have been received if it is delivered to the addressee personally or if it is delivered at his place of

business, habitual residence or mailing address; if none of these can be found after making a reasonable inquiry, a written communication is deemed to have been received if it is sent to the addressee's last-known place of business, habitual residence or mailing address by registered letter or any other means which provides a record of the attempt to deliver it;
  (b) the communication is deemed to have been received on the day it is so delivered.
(2) The provisions of this article do not apply to communications in court proceedings.

Article 4 – Waiver of Right to Object

A party who knows that any provision of this Law from which the parties may derogate or any requirement under the arbitration agreement has not been complied with and yet proceeds with the arbitration without stating his objection to such non-compliance without undue delay or, if a time-limit is provided therefor, within such period of time, shall be deemed to have waived his right to object.

Article 5 – Extent of Court Intervention

In matters governed by this Law, no court shall intervene except where so provided in this Law.

Article 6 – Court or Other Authority for Certain Functions of Arbitration Assistance and Supervision

The functions referred to in Articles 11(3), 11(4), 13(3), 14, 16(3) and 34(2) shall be performed by … [Each State enacting this model law specifies the court, courts or, where referred to therein, other authority competent to perform these functions.]

CHAPTER II – ARBITRATION AGREEMENT

Article 7 – Definition and Form of Arbitration Agreement

(1) 'Arbitration agreement' is an agreement by the parties to submit to arbitration all or certain disputes which have arisen or which may arise between them in respect of a defined legal relationship, whether contractual or not. An arbitration agreement may be in the form of an arbitration clause in a contract or in the form of a separate agreement.
(2) The arbitration agreement shall be in writing. An agreement is in writing if it is contained in a document signed by the parties or in an exchange of letters, telex, telegrams or other means of telecommunication which provide a record

of the agreement, or in an exchange of statements of claim and defence in which the existence of an agreement is alleged by one party and not denied by another. The reference in a contract to a document containing an arbitration clause constitutes an arbitration agreement provided that the contract is in writing and the reference is such as to make that clause part of the contract.

Article 8 – Arbitration Agreement and Substantive Claim before Court

(1) A court before which an action is brought in a matter which is the subject of an arbitration agreement shall, if a party so requests not later than when submitting his first statement on the substance of the dispute, refer the parties to arbitration unless it finds that the agreement is null and void, inoperative or incapable of being performed.
(2) Where an action referred to in paragraph (1) of this article has been brought, arbitral proceedings may nevertheless be commenced or continued, and an award may be made, while the issue is pending before the court.

Article 9 – Arbitration Agreement and Interim Measures by Court

It is not incompatible with an arbitration agreement for a party to request, before or during arbitral proceedings, from a court an interim measure of protection and for a court to grant such measure.

CHAPTER III – COMPOSITION OF ARBITRAL TRIBUNAL

Article 10 – Number of Arbitrators

(1) The parties are free to determine the number of arbitrators.
(2) Failing such determination, the number of arbitrators shall be three.

Article 11 – Appointment of Arbitrators

(1) No person shall be precluded by reason of his nationality from acting as an arbitrator, unless otherwise agreed by the parties.
(2) The parties are free to agree on a procedure of appointing the arbitrator or arbitrators, subject to the provisions of paragraphs (4) and (5) of this article.
(3) Failing such agreement,
    (a) in an arbitration with three arbitrators, each party shall appoint one arbitrator, and the two arbitrators thus appointed shall appoint the third arbitrator; if a party fails to appoint the arbitrator within 30 days of receipt of a request to do so from the other party, or if the two arbitrators fail to agree on the third arbitrator within 30 days of their appointment, the

appointment shall be made, upon request of a party, by the court or other authority specified in Article 6;
  (b) in an arbitration with a sole arbitrator, if the parties are unable to agree on the arbitrator, he shall be appointed, upon request of a party, by the court or other authority specified in Article 6.
(4) Where, under an appointment procedure agreed upon by the parties,
  (a) a party fails to act as required under such procedure, or
  (b) the parties, or two arbitrators, are unable to reach an agreement expected of them under such procedure, or
  (c) a third party, including an institution, fails to perform any function entrusted to it under such procedure,
any party may request the court or other authority specified in Article 6 to take the necessary measure, unless the agreement on the appointment procedure provides other means for securing the appointment.
(5) A decision on a matter entrusted by paragraph (3) or (4) of this article to the court or other authority specified in Article 6 shall be subject to no appeal. The court or other authority, in appointing an arbitrator, shall have due regard to any qualifications required of the arbitrator by the agreement of the parties and to such considerations as are likely to secure the appointment of an independent and impartial arbitrator and, in the case of a sole or third arbitrator, shall take into account as well the advisability of appointing an arbitrator of a nationality other than those of the parties.

Article 12 – Grounds for Challenge

(1) When a person is approached in connection with his possible appointment as an arbitrator, he shall disclose any circumstances likely to give rise to justifiable doubts as to his impartiality or independence. An arbitrator, from the time of his appointment and throughout the arbitral proceedings, shall without delay disclose any such circumstances to the parties unless they have already been informed of them by him.
(2) An arbitrator may be challenged only if circumstances exist that give rise to justifiable doubts as to his impartiality or independence, or if he does not possess qualifications agreed to by the parties. A party may challenge an arbitrator appointed by him, or in whose appointment he has participated, only for reasons of which he becomes aware after the appointment has been made.

Article 13 – Challenge Procedure

(1) The parties are free to agree on a procedure for challenging an arbitrator, subject to the provisions of paragraph (3) of this article.
(2) Failing such agreement, a party who intends to challenge an arbitrator shall, within 15 days after becoming aware of the constitution of the arbitral tribunal or after becoming aware of any circumstance referred to in Article 12(2), send

a written statement of the reasons for the challenge to the arbitral tribunal. Unless the challenged arbitrator withdraws from his office or the other party agrees to the challenge, the arbitral tribunal shall decide on the challenge.

(3) If a challenge under any procedure agreed upon by the parties or under the procedure of paragraph (2) of this article is not successful, the challenging party may request, within 30 days after having received notice of the decision rejecting the challenge, the court or other authority specified in Article 6 to decide on the challenge, which decision shall be subject to no appeal; while such a request is pending, the arbitral tribunal, including the challenged arbitrator, may continue the arbitral proceedings and make an award.

Article 14 – Failure or Impossibility to Act

(1) If an arbitrator becomes de jure or de facto unable to perform his functions or for other reasons fails to act without undue delay, his mandate terminates if he withdraws from his office or if the parties agree on the termination. Otherwise, if a controversy remains concerning any of these grounds, any party may request the court or other authority specified in Article 6 to decide on the termination of the mandate, which decision shall be subject to no appeal.

(2) If, under this article or Article 13(2), an arbitrator withdraws from his office or a party agrees to the termination of the mandate of an arbitrator, this does not imply acceptance of the validity of any ground referred to in this article or Article 12(2).

Article 15 – Appointment of Substitute Arbitrator

Where the mandate of an arbitrator terminates under Article 13 or 14 or because of his withdrawal from office for any other reason or because of the revocation of his mandate by agreement of the parties or in any other case of termination of his mandate, a substitute arbitrator shall be appointed according to the rules that were applicable to the appointment of the arbitrator being replaced.

CHAPTER IV – JURISDICTION OF ARBITRAL TRIBUNAL

Article 16 – Competence of Arbitral Tribunal to Rule on Its Jurisdiction

(1) The arbitral tribunal may rule on its own jurisdiction, including any objections with respect to the existence or validity of the arbitration agreement. For that purpose, an arbitration clause which forms part of a contract shall be treated as an agreement independent of the other terms of the contract. A decision by the arbitral tribunal that the contract is null and void shall not entail ipso jure the invalidity of the arbitration clause.

(2) A plea that the arbitral tribunal does not have jurisdiction shall be raised not later than the submission of the statement of defence. A party is not precluded from raising such a plea by the fact that he has appointed, or participated in the appointment of, an arbitrator. A plea that the arbitral tribunal is exceeding the scope of its authority shall be raised as soon as the matter alleged to be beyond the scope of its authority is raised during the arbitral proceedings. The arbitral tribunal may, in either case, admit a later plea if it considers the delay justified.

(3) The arbitral tribunal may rule on a plea referred to in paragraph (2) of this article either as a preliminary question or in an award on the merits. If the arbitral tribunal rules as a preliminary question that it has jurisdiction, any party may request, within 30 days after having received notice of that ruling, the court specified in Article 6 to decide the matter, which decision shall be subject to no appeal; while such a request is pending, the arbitral tribunal may continue the arbitral proceedings and make an award.

Article 17 – Power of Arbitral Tribunal to Order Interim Measures

Unless otherwise agreed by the parties, the arbitral tribunal may, at the request of a party, order any party to take such interim measure of protection as the arbitral tribunal may consider necessary in respect of the subject-matter of the dispute. The arbitral tribunal may require any party to provide appropriate security in connection with such measure.

CHAPTER V – CONDUCT OF ARBITRAL PROCEEDINGS

Article 18 – Equal Treatment of Parties

The parties shall be treated with equality and each party shall be given a full opportunity of presenting his case.

Article 19 – Determination of Rules of Procedure

(1) Subject to the provisions of this Law, the parties are free to agree on the procedure to be followed by the arbitral tribunal in conducting the proceedings.
(2) Failing such agreement, the arbitral tribunal may, subject to the provisions of this Law, conduct the arbitration in such manner as it considers appropriate. The power conferred upon the arbitral tribunal includes the power to determine the admissibility, relevance, materiality and weight of any evidence.

Article 20 – Place of Arbitration

(1) The parties are free to agree on the place of arbitration. Failing such agreement, the place of arbitration shall be determined by the arbitral tribunal having regard to the circumstances of the case, including the convenience of the parties.
(2) Notwithstanding the provisions of paragraph (1) of this article, the arbitral tribunal may, unless otherwise agreed by the parties, meet at any place it considers appropriate for consultation among its members, for hearing witnesses, experts or the parties, or for inspection of goods, other property or documents.

Article 21 – Commencement of Arbitral Proceedings

Unless otherwise agreed by the parties, the arbitral proceedings in respect of a particular dispute commence on the date on which a request for that dispute to be referred to arbitration is received by the respondent.

Article 22 – Language

(1) The parties are free to agree on the language or languages to be used in the arbitral proceedings. Failing such agreement, the arbitral tribunal shall determine the language or languages to be used in the proceedings. This agreement or determination, unless otherwise specified therein, shall apply to any written statement by a party, any hearing and any award, decision or other communication by the arbitral tribunal.
(2) The arbitral tribunal may order that any documentary evidence shall be accompanied by a translation into the language or languages agreed upon by the parties or determined by the arbitral tribunal.

Article 23 – Statements of Claim and Defence

(1) Within the period of time agreed by the parties or determined by the arbitral tribunal, the claimant shall state the facts supporting his claim, the points at issue and the relief or remedy sought, and the respondent shall state his defence in respect of these particulars, unless the parties have otherwise agreed as to the required elements of such statements. The parties may submit with their statements all documents they consider to be relevant or may add a reference to the documents or other evidence they will submit.
(2) Unless otherwise agreed by the parties, either party may amend or supplement his claim or defence during the course of the arbitral proceedings, unless the arbitral tribunal considers it inappropriate to allow such amendment having regard to the delay in making it.

Article 24 – Hearings and Written Proceedings

(1) Subject to any contrary agreement by the parties, the arbitral tribunal shall decide whether to hold oral hearings for the presentation of evidence or for oral argument, or whether the proceedings shall be conducted on the basis of documents and other materials. However, unless the parties have agreed that no hearings shall be held, the arbitral tribunal shall hold such hearings at an appropriate stage of the proceedings, if so requested by a party.
(2) The parties shall be given sufficient advance notice of any hearing and of any meeting of the arbitral tribunal for the purposes of inspection of goods, other property or documents.
(3) All statements, documents or other information supplied to the arbitral tribunal by one party shall be communicated to the other party. Also any expert report or evidentiary document on which the arbitral tribunal may rely in making its decision shall be communicated to the parties.

Article 25 – Default of a Party

Unless otherwise agreed by the parties, if, without showing sufficient cause,

(a) the claimant fails to communicate his statement of claim in accordance with Article 23(1), the arbitral tribunal shall terminate the proceedings;
(b) the respondent fails to communicate his statement of defence in accordance with Article 23(1), the arbitral tribunal shall continue the proceedings without treating such failure in itself as an admission of the claimant's allegations;
(c) any party fails to appear at a hearing or to produce documentary evidence, the arbitral tribunal may continue the proceedings and make the award on the evidence before it.

Article 26 – Expert Appointed by Arbitral Tribunal

(1) Unless otherwise agreed by the parties, the arbitral tribunal
   (a) may appoint one or more experts to report to it on specific issues to be determined by the arbitral tribunal;
   (b) may require a party to give the expert any relevant information or to produce, or to provide access to, any relevant documents, goods or other property for his inspection.
(2) Unless otherwise agreed by the parties, if a party so requests or if the arbitral tribunal considers it necessary, the expert shall, after delivery of his written or oral report, participate in a hearing where the parties have the opportunity to put questions to him and to present expert witnesses in order to testify on the points at issue.

Article 27 – Court Assistance in Taking Evidence

The arbitral tribunal or a party with the approval of the arbitral tribunal may request from a competent court of this State assistance in taking evidence. The court may execute the request within its competence and according to its rules on taking evidence.

CHAPTER VI – MAKING OF AWARD AND TERMINATION OF PROCEEDINGS

Article 28 – Rules Applicable to Substance of Dispute

(1) The arbitral tribunal shall decide the dispute in accordance with such rules of law as are chosen by the parties as applicable to the substance of the dispute. Any designation of the law or legal system of a given State shall be construed, unless otherwise expressed, as directly referring to the substantive law of that State and not to its conflict of laws rules.
(2) Failing any designation by the parties, the arbitral tribunal shall apply the law determined by the conflict of laws rules which it considers applicable.
(3) The arbitral tribunal shall decide ex aequo et bono or as amiable compositeur only if the parties have expressly authorized it to do so.
(4) In all cases, the arbitral tribunal shall decide in accordance with the terms of the contract and shall take into account the usages of the trade applicable to the transaction.

Article 29 – Decision Making by Panel of Arbitrators

In arbitral proceedings with more than one arbitrator, any decision of the arbitral tribunal shall be made, unless otherwise agreed by the parties, by a majority of all its members. However, questions of procedure may be decided by a presiding arbitrator, if so authorized by the parties or all members of the arbitral tribunal.

Article 30 – Settlement

(1) If, during arbitral proceedings, the parties settle the dispute, the arbitral tribunal shall terminate the proceedings and, if requested by the parties and not objected to by the arbitral tribunal, record the settlement in the form of an arbitral award on agreed terms.
(2) An award on agreed terms shall be made in accordance with the provisions of Article 31 and shall state that it is an award. Such an award has the same status and effect as any other award on the merits of the case.

Article 31 – Form and Contents of Award

(1) The award shall be made in writing and shall be signed by the arbitrator or arbitrators. In arbitral proceedings with more than one arbitrator, the signatures of the majority of all members of the arbitral tribunal shall suffice, provided that the reason for any omitted signature is stated.
(2) The award shall state the reasons upon which it is based, unless the parties have agreed that no reasons are to be given or the award is an award on agreed terms under Article 30.
(3) The award shall state its date and the place of arbitration as determined in accordance with Article 20(1). The award shall be deemed to have been made at that place.
(4) After the award is made, a copy signed by the arbitrators in accordance with paragraph (1) of this article shall be delivered to each party.

Article 32 – Termination of Proceedings

(1) The arbitral proceedings are terminated by the final award or by an order of the arbitral tribunal in accordance with paragraph (2) of this article.
(2) The arbitral tribunal shall issue an order for the termination of the arbitral proceedings when:
   (a) the claimant withdraws his claim, unless the respondent objects thereto and the arbitral tribunal recognizes a legitimate interest on his part in obtaining a final settlement of the dispute;
   (b) the parties agree on the termination of the proceedings;
   (c) the arbitral tribunal finds that the continuation of the proceedings has for any other reason become unnecessary or impossible.
(3) The mandate of the arbitral tribunal terminates with the termination of the arbitral proceedings, subject to the provisions of Articles 33 and 34(4).

Article 33 – Correction and Interpretation of Award; Additional Award

(1) Within 30 days of receipt of the award, unless another period of time has been agreed upon by the parties:
   (a) a party, with notice to the other party, may request the arbitral tribunal to correct in the award any errors in computation, any clerical or typographical errors or any errors of similar nature;
   (b) if so agreed by the parties, a party, with notice to the other party, may request the arbitral tribunal to give an interpretation of a specific point or part of the award.
   If the arbitral tribunal considers the request to be justified, it shall make the correction or give the interpretation within 30 days of receipt of the request. The interpretation shall form part of the award.

(2) The arbitral tribunal may correct any error of the type referred to in paragraph (1)(a) of this article on its own initiative within 30 days of the date of the award.
(3) Unless otherwise agreed by the parties, a party, with notice to the other party, may request, within 30 days of receipt of the award, the arbitral tribunal to make an additional award as to claims presented in the arbitral proceedings but omitted from the award. If the arbitral tribunal considers the request to be justified, it shall make the additional award within 60 days.
(4) The arbitral tribunal may extend, if necessary, the period of time within which it shall make a correction, interpretation or an additional award under paragraph (1) or (3) of this article.
(5) The provisions of Article 31 shall apply to a correction or interpretation of the award or to an additional award.

CHAPTER VII – RECOURSE AGAINST AWARD

Article 34 – Application for Setting Aside as Exclusive Recourse Against Arbitral Award

(1) Recourse to a court against an arbitral award may be made only by an application for setting aside in accordance with paragraphs (2) and (3) of this article.
(2) An arbitral award may be set aside by the court specified in Article 6 only if:
   (a) the party making the application furnishes proof that:
      (i) a party to the arbitration agreement referred to in Article 7 was under some incapacity; or the said agreement is not valid under the law to which the parties have subjected it or, failing any indication thereon, under the law of this State; or
     (ii) the party making the application was not given proper notice of the appointment of an arbitrator or of the arbitral proceedings or was otherwise unable to present his case; or
    (iii) the award deals with a dispute not contemplated by or not falling within the terms of the submission to arbitration, or contains decisions on matters beyond the scope of the submission to arbitration, provided that, if the decisions on matters submitted to arbitration can be separated from those not so submitted, only that part of the award which contains decisions on matters not submitted to arbitration may be set aside; or
    (iv) the composition of the arbitral tribunal or the arbitral procedure was not in accordance with the agreement of the parties, unless such agreement was in conflict with a provision of this Law from which

the parties cannot derogate, or, failing such agreement, was not in accordance with this Law; or
  (b) the court finds that:
    (i) the subject-matter of the dispute is not capable of settlement by arbitration under the law of this State; or
    (ii) the award is in conflict with the public policy of this State.
(3) An application for setting aside may not be made after three months have elapsed from the date on which the party making that application had received the award or, if a request had been made under Article 33, from the date on which that request had been disposed of by the arbitral tribunal.
(4) The court, when asked to set aside an award, may, where appropriate and so requested by a party, suspend the setting aside proceedings for a period of time determined by it in order to give the arbitral tribunal an opportunity to resume the arbitral proceedings or to take such other action as in the arbitral tribunal's opinion will eliminate the grounds for setting aside.

CHAPTER VIII – RECOGNITION AND ENFORCEMENT OF AWARDS

Article 35 – Recognition and Enforcement

(1) An arbitral award, irrespective of the country in which it was made, shall be recognized as binding and, upon application in writing to the competent court, shall be enforced subject to the provisions of this article and of Article 36.
(2) The party relying on an award or applying for its enforcement shall supply the duly authenticated original award or a duly certified copy thereof, and the original arbitration agreement referred to in Article 7 or a duly certified copy thereof. If the award or agreement is not made in an official language of this State, the party shall supply a duly certified translation thereof into such language.[3]

Article 36 – Grounds for Refusing Recognition or Enforcement

(1) Recognition or enforcement of an arbitral award, irrespective of the country in which it was made, may be refused only:
  (a) at the request of the party against whom it is invoked, if that party furnishes to the competent court where recognition or enforcement is sought proof that:
    (i) a party to the arbitration agreement referred to in Article 7 was under some incapacity; or the said agreement is not valid under the law to

---

[3] The conditions set forth in this paragraph are intended to set maximum standards. It would, thus, not be contrary to the harmonization to be achieved by the model law if a State retained even less onerous conditions.

which the parties have subjected it or, failing any indication thereon, under the law of the country where the award was made; or

(ii) the party against whom the award is invoked was not given proper notice of the appointment of an arbitrator or of the arbitral proceedings or was otherwise unable to present his case; or

(iii) the award deals with a dispute not contemplated by or not falling within the terms of the submission to arbitration, or it contains decisions on matters beyond the scope of the submission to arbitration, provided that, if the decisions on matters submitted to arbitration can be separated from those not so submitted, that part of the award which contains decisions on matters submitted to arbitration may be recognized and enforced; or

(iv) the composition of the arbitral tribunal or the arbitral procedure was not in accordance with the agreement of the parties or, failing such agreement, was not in accordance with the law of the country where the arbitration took place; or

(v) the award has not yet become binding on the parties or has been set aside or suspended by a court of the country in which, or under the law of which, that award was made; or

(b) if the court finds that:

(i) the subject-matter of the dispute is not capable of settlement by arbitration under the law of this State; or

(ii) the recognition or enforcement of the award would be contrary to the public policy of this State.

(2) If an application for setting aside or suspension of an award has been made to a court referred to in paragraph (1)(a)(v) of this article, the court where recognition or enforcement is sought may, if it considers it proper, adjourn its decision and may also, on the application of the party claiming recognition or enforcement of the award, order the other party to provide appropriate security.

EXPLANATORY NOTE BY THE UNCITRAL SECRETARIAT ON THE MODEL LAW ON INTERNATIONAL COMMERCIAL ARBITRATION[4]

1. The UNCITRAL Model Law on International Commercial Arbitration was adopted by the United Nations Commission on International Trade Law (UNCITRAL) on 21 June 1985, at the close of the Commission's 18th annual session. The General Assembly, in its Resolution 40/72 of 11 December 1985, recommended 'that all States give due consideration to the Model Law on

---

[4] This note has been prepared by the secretariat of the United Nations Commission on International Trade Law (UNICITRAL) for informational purposes only; it is not an official commentary on the Model Law. A commentary prepared by the Secretariat on an earlier draft of the Model Law appears in document A/CN.9/264 (reproduced in UNCITRAL Yearbook, vol. XVI – 1985) (United Nations publication, Sales No. E.87.V4).

International Commercial Arbitration, in view of the desirability of uniformity of the law of arbitral procedures and the specific needs of international commercial arbitration practice'.
2. The Model Law constitutes a sound and promising basis for the desired harmonization and improvement of national laws. It covers all stages of the arbitral process from the arbitration agreement to the recognition and enforcement of the arbitral award and reflects a worldwide consensus on the principles and important issues of international arbitration practice. It is acceptable to States of all regions and the different legal or economic systems of the world.
3. The form of a model law was chosen as the vehicle for harmonization and improvement in view of the flexibility it gives to States in preparing new arbitration laws. It is advisable to follow the model as closely as possible since that would be the best contribution to the desired harmonization and in the best interest of the users of international arbitration, who are primarily foreign parties and their lawyers.

A.   BACKGROUND TO THE MODEL LAW

4. The Model Law is designed to meet concerns relating to the current state of national laws on arbitration. The need for improvement and harmonization is based on findings that domestic laws are often inappropriate for international cases and that considerable disparity exists between them.

1.   Inadequacy of Domestic Laws

5. A global survey of national laws on arbitration revealed considerable disparities not only as regards individual provisions and solutions but also in terms of development and refinement. Some laws may be regarded as outdated, sometimes going back to the nineteenth century and often equating the arbitral process with court litigation. Other laws may be said to be fragmentary in that they do not address all relevant issues. Even most of those laws which appear to be up-to-date and comprehensive were drafted with domestic arbitration primarily, if not exclusively, in mind. While this approach is understandable in view of the fact that even today the bulk of cases governed by a general arbitration law would be of a purely domestic nature, the unfortunate consequence is that traditional local concepts are imposed on international cases and the needs of modern practice are often not met.
6. The expectations of the parties as expressed in a chosen set of arbitration rules or a 'one-off' arbitration agreement may be frustrated, especially by a mandatory provision of the applicable law.

Unexpected and undesired restrictions found in national laws relate, for example, to the parties' ability effectively to submit future disputes to arbitration, to their power to select the arbitrator freely, or to their interest in having the arbitral proceedings conducted according to the agreed rules of procedure and with no more court involvement than is appropriate. Frustrations may also ensue from non-mandatory provisions which may impose undesired requirements on unwary parties who did not provide otherwise.

Even the absence of non-mandatory provisions may cause difficulties by not providing answers to the many procedural issues relevant in an arbitration and not always settled in the arbitration agreement.

2. Disparity Between National Laws

7. Problems and undesired consequences, whether emanating from mandatory or non-mandatory provisions or from a lack of pertinent provisions, are aggravated by the fact that national laws on arbitral procedure differ widely. The differences are a frequent source of concern in international arbitration, where at least one of the parties is, and often both parties are, confronted with foreign and unfamiliar provisions and procedures. For such a party it may be expensive, impractical or impossible to obtain a full and precise account of the law applicable to the arbitration.

8. Uncertainty about the local law with the inherent risk of frustration may adversely affect not only the functioning of the arbitral process but already the selection of the place of arbitration. A party may well for those reasons hesitate or refuse to agree to a place which otherwise, for practical reasons, would be appropriate in the case at hand. The choice of places of arbitration would thus be widened and the smooth functioning of the arbitral proceedings would be enhanced if States were to adopt the Model Law which is easily recognizable, meets the specific needs of international commercial arbitration and provides an international standard with solutions acceptable to parties from different States and legal systems.

B. SALIENT FEATURES OF THE MODEL LAW

1. Special Procedural Regime for International Commercial Arbitration

9. The principles and individual solutions adopted in the Model Law aim at reducing or eliminating the above concerns and difficulties. As a response to the inadequacies and disparities of national laws, the Model Law presents a special legal regime geared to international commercial arbitration, without affecting any relevant treaty in force in the State adopting the Model Law. While the need for uniformity exists only in respect of international cases, the

desire of updating and improving the arbitration law may be felt by a State also in respect of non-international cases and could be met by enacting modern legislation based on the Model Law for both categories of cases.

a.   *Substantive and Territorial Scope of Application*

10. The Model Law defines an arbitration as international if 'the parties to an arbitration agreement have, at the time of the conclusion of that agreement, their places of business in different States' (Article 1(3)). The vast majority of situations commonly regarded as international will fall under this criterion. In addition, an arbitration is international if the place of arbitration, the place of contract performance, or the place of the subject-matter of the dispute is situated in a State other than where the parties have their place of business, or if the parties have expressly agreed that the subject-matter of the arbitration agreement relates to more than one country.
11. As regards the term 'commercial', no hard and fast definition could be provided. Article 1 contains a note calling for 'a wide interpretation so as to cover matters arising from all relationships of a commercial nature, whether contractual or not'. The footnote to Article 1 then provides an illustrative list of relationships that are to be considered commercial, thus emphasizing the width of the suggested interpretation and indicating that the determinative test is not based on what the national law may regard as 'commercial'.
12. Another aspect of applicability is what one may call the territorial scope of application. According to Article 1(2), the Model Law as enacted in a given State would apply only if the place of arbitration is in the territory of that State. However, there is an important and reasonable exception. Articles 8(1) and 9 which deal with recognition of arbitration agreements, including their compatibility with interim measures of protection, and Articles 35 and 36 on recognition and enforcement of arbitral awards are given a global scope, i.e. they apply irrespective of whether the place of arbitration is in that State or in another State and, as regards Articles 8 and 9, even if the place of arbitration is not yet determined.
13. The strict territorial criterion, governing the bulk of the provisions of the Model Law, was adopted for the sake of certainty and in view of the following facts. The place of arbitration is used as the exclusive criterion by the great majority of national laws and, where national laws allow parties to choose the procedural law of a State other than that where the arbitration takes place, experience shows that parties in practice rarely make use of that facility. The Model Law, by its liberal contents, further reduces the need for such choice of a 'foreign' law in lieu of the (Model) Law of the place of arbitration, not the least because it grants parties wide freedom in shaping the rules of the arbitral proceedings. This includes the possibility of incorporating into the arbitration agreement procedural provisions of a 'foreign' law, provided there is no conflict with the few mandatory provisions of the Model Law. Furthermore, the strict territorial

criterion is of considerable practical benefit in respect of Articles 11, 13, 14, 16, 27 and 34, which entrust the courts of the respective State with functions of arbitration assistance and supervision.

b. *Delimitation of Court Assistance and Supervision*

14. As evidenced by recent amendments to arbitration laws, there exists a trend in favour of limiting court involvement in international commercial arbitration. This seems justified in view of the fact that the parties to an arbitration agreement make a conscious decision to exclude court jurisdiction and, in particular in commercial cases, prefer expediency and finality to protracted battles in court.
15. In this spirit, the Model Law envisages court involvement in the following instances. A first group comprises appointment, challenge and termination of the mandate of an arbitrator (Articles 11, 13 and 14), jurisdiction of the arbitral tribunal (Article 16) and setting aside of the arbitral award (Article 34). These instances are listed in Article 6 as functions which should be entrusted, for the sake of centralization, specialization and acceleration, to a specially designated court or, as regards Articles 11, 13 and 14, possibly to another authority (e.g. arbitral institution, chamber of commerce). A second group comprises court assistance in taking evidence (Article 27), recognition of the arbitration agreement, including its compatibility with court-ordered interim measures of protection (Articles 8 and 9), and recognition and enforcement of arbitral awards (Articles 35 and 36).
16. Beyond the instances in these two groups, 'no court shall intervene, in matters governed by this Law'. This is stated in the innovative Article 5, which by itself does not take a stand on what is the appropriate role of the courts but guarantees the reader and user that he will find all instances of possible court intervention in this Law, except for matters not regulated by it (e.g., consolidation of arbitral proceedings, contractual relationship between arbitrators and parties or arbitral institutions, or fixing of costs and fees, including deposits). Especially foreign readers and users, who constitute the majority of potential users and may be viewed as the primary addressees of any special law on international commercial arbitration, will appreciate that they do not have to search outside this Law.

2. Arbitration Agreement

17. Chapter II of the Model Law deals with the arbitration agreement, including its recognition by courts. The provisions follow closely Article II of the Convention on the Recognition and Enforcement of Foreign Arbitral Awards (New York, 1958) (hereafter referred to as '1958 New York Convention'), with a number of useful clarifications added.

a. *Definition and Form of Arbitration Agreement*

18. Article 7(1) recognizes the validity and effect of a commitment by the parties to submit to arbitration an existing dispute ('compromis') or a future dispute ('clause compromissoire'). The latter type of agreement is presently not given full effect under certain national laws.
19. While oral arbitration agreements are found in practice and are recognized by some national laws, Article 7(2) follows the 1958 New York Convention in requiring written form. It widens and clarifies the definition of written form of Article II(2) of that Convention by adding 'telex or other means of telecommunication which provide a record of the agreement', by covering the submission-type situation of 'an exchange of statements of claim and defence in which the existence of an agreement is alleged by one party and not denied by another', and by providing that 'the reference in a contract to a document' (e.g. general conditions) 'containing an arbitration clause constitutes an arbitration agreement provided that the contract is in writing and the reference is such as to make that clause part of the contract'.

b. *Arbitration Agreement and the Courts*

20. Articles 8 and 9 deal with two important aspects of the complex issue of the relationship between the arbitration agreement and resort to courts. Modelled on Article II(3) of the 1958 New York Convention, Article 8(1) of the Model Law obliges any court to refer the parties to arbitration if seized with a claim on the same subject-matter unless it finds that the arbitration agreement is null and void, inoperative or incapable of being performed. The referral is dependent on a request which a party may make not later than when submitting his first statement on the substance of the dispute. While this provision, where adopted by a State when it adopts the Model Law, by its nature binds merely the courts of that State, it is not restricted to agreements providing for arbitration in that State and, thus, helps to give universal recognition and effect to international commercial arbitration agreements.
21. Article 9 expresses the principle that any interim measures of protection that may be obtained from courts under their procedural law (e.g. pre-award attachments) are compatible with an arbitration agreement. Like Article 8, this provision is addressed to the courts of a given State, insofar as it determines their granting of interim measures as being compatible with an arbitration agreement, irrespective of the place of arbitration. Insofar as it declares it to be compatible with an arbitration agreement for a party to request such measure from a court, the provision would apply irrespective of whether the request is made to a court of the given State or of any other country. Wherever such request may be made, it may not be relied upon, under the Model Law, as an objection against the existence or effect of an arbitration agreement.

3. Composition of Arbitral Tribunal

22. Chapter III contains a number of detailed provisions on appointment, challenge, termination of mandate and replacement of an arbitrator. The chapter illustrates the approach of the Model Law in eliminating difficulties arising from inappropriate or fragmentary laws or rules. The approach consists, first, of recognizing the freedom of the parties to determine, by reference to an existing set of arbitration rules or by an ad hoc agreement, the procedure to be followed, subject to fundamental requirements of fairness and justice. Secondly, where the parties have not used their freedom to lay down the rules of procedure or a particular issue has not been covered, the Model Law ensures, by providing a set of suppletive rules, that the arbitration may commence and proceed effectively to the resolution of the dispute.

23. Where under any procedure, agreed upon by the parties or based upon the suppletive rules of the Model Law, difficulties arise in the process of appointment, challenge or termination of the mandate of an arbitrator, Articles 11, 13 and 14 provide for assistance by courts or other authorities. In view of the urgency of the matter and in order to reduce the risk and effect of any dilatory tactics, instant resort may be had by a party within a short period of time and the decision is not appealable.

4. Jurisdiction of Arbitral Tribunal

a. *Competence to Rule on Own Jurisdiction*

24. Article 16(1) adopts the two important (not yet generally recognized) principles of 'Kompetenz- Kompetenz' and of separability or autonomy of the arbitration clause. The arbitral tribunal may rule on its own jurisdiction, including any objections with respect to the existence or validity of the arbitration agreement. For that purpose, an arbitration clause shall be treated as an agreement independent of the other terms of the contract, and a decision by the arbitral tribunal that the contract is null and void shall not entail ipso jure the invalidity of the arbitration clause. Detailed provisions in paragraph (2) require that any objections relating to the arbitrators' jurisdiction be made at the earliest possible time.

25. The arbitral tribunal's competence to rule on its own jurisdiction, i.e. on the very foundation of its mandate and power, is, of course, subject to court control. Where the arbitral tribunal rules as a preliminary question that it has jurisdiction, Article 16(3) provides for instant court control in order to avoid unnecessary waste of money and time. However, three procedural safeguards are added to reduce the risk and effect of dilatory tactics: short time-period for resort to court (30 days), court decision is not appealable, and discretion of the arbitral tribunal to continue the proceedings and make an award while the matter is pending with the court. In those less common cases where the

arbitral tribunal combines its decision on jurisdiction with an award on the merits, judicial review on the question of jurisdiction is available in setting aside proceedings under Article 34 or in enforcement proceedings under Article 36.

b. *Power to Order Interim Measures*

26. Unlike some national laws, the Model Law empowers the arbitral tribunal, unless otherwise agreed by the parties, to order any party to take an interim measure of protection in respect of the subject-matter of the dispute, if so requested by a party (Article 17). It may be noted that the article does not deal with enforcement of such measures; any State adopting the Model Law would be free to provide court assistance in this regard.

5. Conduct of Arbitral Proceedings

27. Chapter V provides the legal framework for a fair and effective conduct of the arbitral proceedings. It opens with two provisions expressing basic principles that permeate the arbitral procedure governed by the Model Law. Article 18 lays down fundamental requirements of procedural justice and Article 19 the rights and powers to determine the rules of procedure.

a. *Fundamental Procedural Rights of a Party*

28. Article 18 embodies the basic principle that the parties shall be treated with equality and each party shall be given a full opportunity of presenting his case. Other provisions implement and specify the basic principle in respect of certain fundamental rights of a party. Article 24(1) provides that, unless the parties have validly agreed that no oral hearings for the presentation of evidence or for oral argument be held, the arbitral tribunal shall hold such hearings at an appropriate stage of the proceedings, if so requested by a party. It should be noted that Article 24(1) deals only with the general right of a party to oral hearings (as an alternative to conducting the proceedings on the basis of documents and other materials) and not with the procedural aspects such as the length, number or timing of hearings.

29. Another fundamental right of a party of being heard and being able to present his case relates to evidence by an expert appointed by the arbitral tribunal. Article 26(2) obliges the expert, after having delivered his written or oral report, to participate in a hearing where the parties may put questions to him and present expert witnesses in order to testify on the points at issue, if such a hearing is requested by a party or deemed necessary by the arbitral tribunal. As another provision aimed at ensuring fairness, objectivity and impartiality, Article 24(3) provides that all statements, documents and other information supplied to the arbitral tribunal by one party shall be communicated to the

other party, and that any expert report or evidentiary document on which the arbitral tribunal may rely in making its decision shall be communicated to the parties. In order to enable the parties to be present at any hearing and at any meeting of the arbitral tribunal for inspection purposes, they shall be given sufficient notice in advance (Article 24(2)).

b. *Determination of Rules of Procedure*

30. Article 19 guarantees the parties' freedom to agree on the procedure to be followed by the arbitral tribunal in conducting the proceedings, subject to a few mandatory provisions on procedure, and empowers the arbitral tribunal, failing agreement by the parties, to conduct the arbitration in such a manner as it considers appropriate. The power conferred upon the arbitral tribunal includes the power to determine the admissibility, relevance, materiality and weight of any evidence.
31. Autonomy of the parties to determine the rules of procedure is of special importance in international cases since it allows the parties to select or tailor the rules according to their specific wishes and needs, unimpeded by traditional domestic concepts and without the earlier mentioned risk of frustration. The supplementary discretion of the arbitral tribunal is equally important in that it allows the tribunal to tailor the conduct of the proceedings to the specific features of the case without restraints of the traditional local law, including any domestic rules on evidence. Moreover, it provides a means for solving any procedural questions not regulated in the arbitration agreement or the Model Law.
32. In addition to the general provisions of Article 19, there are some special provisions using the same approach of granting the parties autonomy and, failing agreement, empowering the arbitral tribunal to decide the matter. Examples of particular practical importance in international cases are Article 20 on the place of arbitration and Article 22 on the language of the proceedings.

c. *Default of a Party*

33. Only if due notice was given, may the arbitral proceedings be continued in the absence of a party. This applies, in particular, to the failure of a party to appear at a hearing or to produce documentary evidence without showing sufficient cause for the failure (Article 25(c)). The arbitral tribunal may also continue the proceedings where the respondent fails to communicate his statement of defence, while there is no need for continuing the proceedings if the claimant fails to submit his statement of claim (Article 25(a), (b)).
34. Provisions which empower the arbitral tribunal to carry out its task even if one of the parties does not participate are of considerable practical importance since, as experience shows, it is not uncommon that one of the parties has little interest in co-operating and in expediting matters. They would, thus, give

international commercial arbitration its necessary effectiveness, within the limits of fundamental requirements of procedural justice.

6. Making of Award and Termination of Proceedings

a. Rules Applicable to Substance of Dispute

35. Article 28 deals with the substantive law aspects of arbitration. Under paragraph (1), the arbitral tribunal decides the dispute in accordance with such rules of law as may be agreed by the parties. This provision is significant in two respects. It grants the parties the freedom to choose the applicable substantive law, which is important in view of the fact that a number of national laws do not clearly or fully recognize that right. In addition, by referring to the choice of 'rules of law' instead of 'law', the Model Law gives the parties a wider range of options as regards the designation of the law applicable to the substance of the dispute in that they may, for example, agree on rules of law that have been elaborated by an international forum but have not yet been incorporated into any national legal system. The power of the arbitral tribunal, on the other hand, follows more traditional lines. When the parties have not designated the applicable law, the arbitral tribunal shall apply the law, i.e. the national law, determined by the conflict of laws rules which it considers applicable.

36. According to Article 28(3), the parties may authorize the arbitral tribunal to decide the dispute ex aequo et bono or as amiables compositeurs. This type of arbitration is currently not known or used in all legal systems and there exists no uniform understanding as regards the precise scope of the power of the arbitral tribunal. When parties anticipate an uncertainty in this respect, they may wish to provide a clarification in the arbitration agreement by a more specific authorization to the arbitral tribunal. Paragraph (4) makes clear that in all cases, i.e. including an arbitration ex aequo et bono, the arbitral tribunal must decide in accordance with the terms of the contract and shall take into account the usages of the trade applicable to the transaction.

b. Making of Award and Other Decisions

37. In its rules on the making of the award (Articles 29–31), the Model Law pays special attention to the rather common case that the arbitral tribunal consists of a plurality of arbitrators (in particular, three). It provides that, in such case, any award and other decision shall be made by a majority of the arbitrators, except on questions of procedure, which may be left to a presiding arbitrator. The majority principle applies also to the signing of the award, provided that the reason for any omitted signature is stated.

38. Article 31(3) provides that the award shall state the place of arbitration and that it shall be deemed to have been made at that place. As to this presumption,

it may be noted that the final making of the award constitutes a legal act, which in practice is not necessarily one factual act but may be done in deliberations at various places, by telephone conversation or correspondence; above all, the award need not be signed by the arbitrators at the same place.

39. The arbitral award must be in writing and state its date. It must also state the reasons on which it is based, unless the parties have agreed otherwise or the award is an award on agreed terms, i.e. an award which records the terms of an amicable settlement by the parties. It may be added that the Model Law neither requires nor prohibits 'dissenting opinions'.

7. Recourse Against Award

40. National laws on arbitration, often equating awards with court decisions, provide a variety of means of recourse against arbitral awards, with varying and often long time-periods and with extensive lists of grounds that differ widely in the various legal systems. The Model Law attempts to ameliorate this situation, which is of considerable concern to those involved in international commercial arbitration.

a. *Application for Setting Aside as Exclusive Recourse*

41. The first measure of improvement is to allow only one type of recourse, to the exclusion of any other means of recourse regulated in another procedural law of the State in question. An application for setting aside under Article 34 must be made within three months of receipt of the award. It should be noted that 'recourse' means actively 'attacking' the award; a party is, of course, not precluded from seeking court control by way of defence in enforcement proceedings (Article 36). Furthermore, 'recourse' means resort to a court, i.e. an organ of the judicial system of a State; a party is not precluded from resorting to an arbitral tribunal of second instance if such a possibility has been agreed upon by the parties (as is common in certain commodity trades).

b. *Grounds for Setting Aside*

42. As a further measure of improvement, the Model Law contains an exclusive list of limited grounds on which an award may be set aside. This list is essentially the same as the one in Article 36(1), taken from Article V of the 1958 New York Convention: lack of capacity of parties to conclude arbitration agreement or lack of valid arbitration agreement; lack of notice of appointment of an arbitrator or of the arbitral proceedings or inability of a party to present his case; award deals with matters not covered by submission to arbitration; composition of arbitral tribunal or conduct of arbitral proceedings contrary to effective agreement of parties or, failing agreement, to the Model Law; non-arbitrability of subject matter of dispute and violation of public policy,

which would include serious departures from fundamental notions of procedural justice.
43. Such a parallelism of the grounds for setting aside with those provided in Article V of the 1958 New York Convention for refusal of recognition and enforcement was already adopted in the European Convention on International Commercial Arbitration (Geneva, 1961). Under its Article IX, the decision of a foreign court setting aside an award for a reason other than the ones listed in Article V of the 1958 New York Convention does not constitute a ground for refusing enforcement. The Model Law takes this philosophy one step further by directly limiting the reasons for setting aside.
44. Although the grounds for setting aside are almost identical to those for refusing recognition or enforcement, two practical differences should be noted. Firstly, the grounds relating to public policy, including non-arbitrability, may be different in substance, depending on the State in question (i.e. State of setting aside or State of enforcement). Secondly, and more importantly, the grounds for refusal of recognition or enforcement are valid and effective only in the State (or States) where the winning party seeks recognition and enforcement, while the grounds for setting aside have a different impact: The setting aside of an award at the place of origin prevents enforcement of that award in all other countries by virtue of Article V(1)(e) of the 1958 New York Convention and Article 36(1)(a)(v) of the Model Law.

8. Recognition and Enforcement of Awards

45. The eighth and last chapter of the Model Law deals with recognition and enforcement of awards. Its provisions reflect the significant policy decision that the same rules should apply to arbitral awards whether made in the country of enforcement or abroad, and that those rules should follow closely the 1958 New York Convention.

a. *Towards Uniform Treatment of All Awards Irrespective Of Country of Origin*

46. By treating awards rendered in international commercial arbitration in a uniform manner irrespective of where they were made, the Model Law draws a new demarcation line between 'international' and 'non-international' awards instead of the traditional line between 'foreign' and 'domestic' awards. This new line is based on substantive grounds rather than territorial borders, which are inappropriate in view of the limited importance of the place of arbitration in international cases. The place of arbitration is often chosen for reasons of convenience of the parties and the dispute may have little or no connection with the State where the arbitration takes place. Consequently, the recognition

and enforcement of 'international' awards, whether 'foreign' or 'domestic', should be governed by the same provisions.

47. By modelling the recognition and enforcement rules on the relevant provisions of the 1958 New York Convention, the Model Law supplements, without conflicting with, the regime of recognition and enforcement created by that successful Convention.

b. *Procedural Conditions of Recognition and Enforcement*

48. Under Article 35(1) any arbitral award, irrespective of the country in which it was made, shall be recognized as binding and enforceable, subject to the provisions of Article 35(2) and of Article 36 (which sets forth the grounds on which recognition or enforcement may be refused). Based on the above consideration of the limited importance of the place of arbitration in international cases and the desire of overcoming territorial restrictions, reciprocity is not included as a condition for recognition and enforcement.

49. The Model Law does not lay down procedural details of recognition and enforcement since there is no practical need for unifying them, and since they form an intrinsic part of the national procedural law and practice. The Model Law merely sets certain conditions for obtaining enforcement: application in writing, accompanied by the award and the arbitration agreement (Article 35(2)).

c. *Grounds for Refusing Recognition or Enforcement*

50. As noted earlier, the grounds on which recognition or enforcement may be refused under the Model Law are identical to those listed in Article V of the New York Convention. Only, under the Model Law, they are relevant not merely to foreign awards but to all awards rendered in international commercial arbitration. While some provisions of that Convention, in particular as regards their drafting, may have called for improvement, only the first ground on the list (i.e. 'the parties to the arbitration agreement were, under the law applicable to them, under some incapacity') was modified since it was viewed as containing an incomplete and potentially misleading conflicts rule. Generally, it was deemed desirable to adopt, for the sake of harmony, the same approach and wording as this important Convention.

Further information on the Model Law may be obtained from:

UNCITRAL Secretariat
Vienna International Centre
P.O. Box 500
A-1400 Vienna
Austria

Telephone: (43)(1) 26060-4060 or 4061
Telefax: (43)(1) 26060-5813
E-mail: uncitral@uncitral.org

Annex 9
IBA Rules on the Taking of Evidence in International Commercial Arbitration

[Adopted by a resolution of the IBA Council 1 June 1999]

THE RULES

Preamble

1. These IBA Rules on the Taking of Evidence in International Commercial Arbitration (the 'IBA Rules of Evidence') are intended to govern in an efficient and economical manner the taking of evidence in international commercial arbitrations, particularly those between Parties from different legal traditions. They are designed to supplement the legal provisions and the institutional or ad hoc rules according to which the Parties are conducting their arbitration.
2. Parties and Arbitral Tribunals may adopt the IBA Rules of Evidence, in whole or in part, to govern arbitration proceedings, or they may vary them or use them as guidelines in developing their own procedures. The Rules are not intended to limit the flexibility that is inherent in, and an advantage of, international arbitration, and Parties and Arbitral Tribunals are free to adapt them to the particular circumstances of each arbitration.
3. Each Arbitral Tribunal is encouraged to identify to the Parties, as soon as it considers it to be appropriate, the issues that it may regard as relevant and material to the outcome of the case, including issues where a preliminary determination may be appropriate.
4. The taking of evidence shall be conducted on the principle that each Party shall be entitled to know, reasonably in advance of any Evidentiary Hearing, the evidence on which the other Parties rely.

Article 1 – Definitions

In the IBA Rules of Evidence:

- *'Arbitral Tribunal'* means a sole arbitrator or a panel of arbitrators validly deciding by majority or otherwise;
- *'Claimant'* means the Party or Parties who commenced the arbitration and any Party who, through joinder or otherwise, becomes aligned with such Party or Parties;
- *'Document'* means a writing of any kind, whether recorded on paper, electronic means, audio or visual recordings or any other mechanical or electronic means of storing or recording information;
- *'Evidentiary Hearing"* means any hearing, whether or not held on consecutive days, at which the Arbitral Tribunal receives oral evidence;
- *'Expert Report'* means a written statement by a Tribunal-Appointed Expert or a Party-Appointed Expert submitted pursuant to the IBA Rules of Evidence;
- *'General Rule"* mean the institutional or ad hoc rules according to which the Parties are conducting their arbitration;
- *'Party'* means a party to the arbitration;
- *'Party-Appointed Expert'* means an expert witness presented by a Party;
- *'Request to Produce'* means a request by a Party for a procedural order by which the Arbitral Tribunal would direct another Party to produce documents;
- *'Respondent'* means the Party or Parties against whom the Claimant made its claim, and any Party who, through joinder or otherwise, becomes aligned with such Party or Parties, and includes a Respondent making a counter-claim;
- *'Tribunal-Appointed Expert'* means a person or organization appointed by the Arbitral Tribunal in order to report to it on specific issues determined by the Arbitral Tribunal.

Article 2 – Scope of Application

1. Whenever the Parties have agreed or the Arbitral Tribunal has determined to apply the IBA Rules of Evidence, the Rules shall govern the taking of evidence, except to the extent that any specific provision of them may be found to be in conflict with any mandatory provision of law determined to be applicable to the case by the Parties or by the Arbitral Tribunal.
2. In case of conflict between any provisions of the IBA Rules of Evidence and the General Rules, the Arbitral Tribunal shall apply the IBA Rules of Evidence in the manner that it determines best in order to accomplish the purposes of both the General Rules and the IBA Rules of Evidence, unless the Parties agree to the contrary.

3. In the event of any dispute regarding the meaning of the IBA Rules of Evidence, the Arbitral Tribunal shall interpret them according to their purpose and in the manner most appropriate for the particular arbitration.
4. Insofar as the IBA Rules of Evidence and the General Rules are silent on any matter concerning the taking of evidence and the Parties have not agreed otherwise, the Arbitral Tribunal may conduct the taking of evidence as it deems appropriate, in accordance with the general principles of the IBA Rules of Evidence.

Article 3 – Documents

1. Within the time ordered by the Arbitral Tribunal, each Party shall submit to the Arbitral Tribunal and to the other Parties all documents available to it on which it relies, including public documents and those in the public domain, except for any documents that have already been submitted by another Party.
2. Within the time ordered by the Arbitral Tribunal, any Party may submit to the Arbitral Tribunal a Request to Produce.
3. A Request to Produce shall contain:
    (a) (i) a description of a requested document sufficient to identify it, or (ii) a description in sufficient detail (including subject matter) of a narrow and specific requested category of documents that are reasonably believed to exist;
    (b) a description of how the documents requested are relevant and material to the outcome of the case; and
    (c) a statement that the documents requested are not in the possession, custody or control of the requesting Party, and of the reason why that Party assumes the documents requested to be in the possession, custody or control of the other Party.
4. Within the time ordered by the Arbitral Tribunal, the Party to whom the Request to Produce is addressed shall produce to the Arbitral Tribunal and to the other Parties all the documents requested in its possession, custody or control as to which no objection is made.
5. If the Party to whom the Request to Produce is addressed has objections to some or all of the documents requested, it shall state them in writing to the Arbitral Tribunal within the time ordered by the Arbitral Tribunal. The reasons for such objections shall be any of those set forth in Article 9.2.
6. The Arbitral Tribunal shall, in consultation with the Parties and in timely fashion, consider the Request to Produce and the objections. The Arbitral Tribunal may order the Party to whom such Request is addressed to produce to the Arbitral Tribunal and to the other Parties those requested documents in its possession, custody or control as to which the Arbitral Tribunal determines that (i) the issues that the requesting Party wishes to prove are relevant and material to the outcome of the case, and (ii) none of the reasons for objection set forth in Article 9.2 apply.

7. In exceptional circumstances, if the propriety of an objection can only be determined by review of the document, the Arbitral Tribunal may determine that it should not review the document. In that event, the Arbitral Tribunal may, after consultation with the Parties, appoint an independent and impartial expert, bound to confidentiality, to review any such document and to report on the objection. To the extent that the objection is upheld by the Arbitral Tribunal, the expert shall not disclose to the Arbitral Tribunal and to the other Parties the contents of the document reviewed.
8. If a Party wishes to obtain the production of documents from a person or organization who is not a Party to the arbitration and from whom the Party cannot obtain the documents on its own, the Party may, within the time ordered by the Arbitral Tribunal, ask it to take whatever steps are legally available to obtain the requested documents. The Party shall identify the documents in sufficient detail and state why such documents are relevant and material to the outcome of the case. The Arbitral Tribunal shall decide on this request and shall take the necessary steps if in its discretion it determines that the documents would be relevant and material.
9. The Arbitral Tribunal, at any time before the arbitration is concluded, may request a Party to produce to the Arbitral Tribunal and to the other Parties any documents that it believes to be relevant and material to the outcome of the case. A Party may object to such a request based on any of the reasons set forth in Article 9.2. If a Party raises such an objection, the Arbitral Tribunal shall decide whether to order the production of such documents based upon the considerations set forth in Article 3.6 and, if the Arbitral Tribunal considers it appropriate, through the use of the procedures set forth in Article 3.7.
10. Within the time ordered by the Arbitral Tribunal, the Parties may submit to the Arbitral Tribunal and to the other Parties any additional documents which they believe have become relevant and material as a consequence of the issues raised in documents, Witness Statements or Expert Reports submitted or produced by another Party or in other submissions of the Parties.
11. If copies are submitted or produced, they must conform fully to the originals. At the request of the Arbitral Tribunal, any original must be presented for inspection.
12. All documents produced by a Party pursuant to the IBA Rules of Evidence (or by a non-Party pursuant to Article 3.8) shall be kept confidential by the Arbitral Tribunal and by the other Parties, and they shall be used only in connection with the arbitration. The Arbitral Tribunal may issue orders to set forth the terms of this confidentiality. This requirement is without prejudice to all other obligations of confidentiality in arbitration.

Article 4 – Witnesses of Fact

1. Within the time ordered by the Arbitral Tribunal, each Party shall identify the witnesses on whose testimony it relies and the subject matter of that testimony.

2. Any person may present evidence as a witness, including a Party or a Party's officer, employee or other representative.
3. It shall not be improper for a Party, its officers, employees, legal advisors or other representatives to interview its witnesses or potential witnesses.
4. The Arbitral Tribunal may order each Party to submit within a specified time to the Arbitral Tribunal and to the other Parties a written statement by each witness on whose testimony it relies, except for those witnesses whose testimony is sought pursuant to Article 4.10 (the 'Witness Statement'). If Evidentiary Hearings are organized on separate issues (such as liability and damages), the Arbitral Tribunal or the Parties by agreement may schedule the submission of Witness Statements separately for each Evidentiary Hearing.
5. Each Witness Statement shall contain:
   (a) the full name and address of the witness, his or her present and past relationship (if any) with any of the Parties, and a description of his or her background, qualifications, training and experience, if such a description may be relevant and material to the dispute or to the contents of the statement;
   (b) a full and detailed description of the facts, and the source of the witness's information as to those facts, sufficient to serve as that witness's evidence in the matter in dispute;
   (c) an affirmation of the truth of the statement; and
   (d) the signature of the witness and its date and place.
6. If Witness Statements are submitted, any Party may, within the time ordered by the Arbitral Tribunal, submit to the Arbitral Tribunal and to the other Parties revised or additional Witness Statements, including statements from persons not previously named as witnesses, so long as any such revisions or additions only respond to matters contained in another Party's Witness Statement or Expert Report and such matters have not been previously presented in the arbitration.
7. Each witness who has submitted a Witness Statement shall appear for testimony at an Evidentiary Hearing, unless the Parties agree otherwise.
8. If a witness who has submitted a Witness Statement does not appear without a valid reason for testimony at an Evidentiary Hearing, except by agreement of the Parties, the Arbitral Tribunal shall disregard that Witness Statement unless, in exceptional circumstances, the Arbitral Tribunal determines otherwise.
9. If the Parties agree that a witness who has submitted a Witness Statement does not need to appear for testimony at an Evidentiary Hearing, such an agreement shall not be considered to reflect an agreement as to the correctness of the content of the Witness Statement.
10. If a Party wishes to present evidence from a person who will not appear voluntarily at its request, the Party may, within the time ordered by the Arbitral Tribunal, ask it to take whatever steps are legally available to obtain the testimony of that person. The Party shall identify the intended witness, shall

describe the subjects on which the witness's testimony is sought and shall state why such subjects are relevant and material to the outcome of the case. The Arbitral Tribunal shall decide on this request and shall take the necessary steps if in its discretion it determines that the testimony of that witness would be relevant and material.
11. The Arbitral Tribunal may, at any time before the arbitration is concluded, order any Party to provide, or to use its best efforts to provide, the appearance for testimony at an Evidentiary Hearing of any person, including one whose testimony has not yet been offered.

Article 5 – Party-Appointed Experts

1. A Party may rely on a Party-Appointed Expert as a means of evidence on specific issues. Within the time ordered by the Arbitral Tribunal, a Party-Appointed Expert shall submit an Expert Report.
2. The Expert Report shall contain:
   (a) the full name and address of the Party-Appointed Expert, his or her present and past relationship (if any) with any of the Parties, and a description of his or her background, qualifications, training and experience;
   (b) a statement of the facts on which he or she is basing his or her expert opinions and conclusions;
   (c) his or her expert opinions and conclusions, including a description of the method, evidence and information used in arriving at the conclusions;
   (d) an affirmation of the truth of the Expert Report; and
   (e) the signature of the Party-Appointed Expert and its date and place.
3. The Arbitral Tribunal in its discretion may order that any Party-Appointed Experts who have submitted Expert Reports on the same or related issues meet and confer on such issues. At such meeting, the Party-Appointed Experts shall attempt to reach agreement on those issues as to which they had differences of opinion in their Expert Reports, and they shall record in writing any such issues on which they reach agreement.
4. Each Party-Appointed Expert shall appear for testimony at an Evidentiary Hearing, unless the Parties agree otherwise and the Arbitral Tribunal accepts this agreement.
5. If a Party-Appointed Expert does not appear without a valid reason for testimony at an Evidentiary Hearing, except by agreement of the Parties accepted by the Arbitral Tribunal, the Arbitral Tribunal shall disregard his or her Expert Report unless, in exceptional circumstances, the Arbitral Tribunal determines otherwise.
6. If the Parties agree that a Party-Appointed Expert does not need to appear for testimony at an Evidentiary Hearing, such an agreement shall not be considered to reflect an agreement as to the correctness of the content of the Expert Report.

Article 6 – Tribunal-Appointed Experts

1. The Arbitral Tribunal, after having consulted with the Parties, may appoint one or more independent Tribunal-Appointed Experts to report to it on specific issues designated by the Arbitral Tribunal. The Arbitral Tribunal shall establish the terms of reference for any Tribunal-Appointed Expert report after having consulted with the Parties. A copy of the final terms of reference shall be sent by the Arbitral Tribunal to the Parties.
2. The Tribunal-Appointed Expert shall, before accepting appointment, submit to the Arbitral Tribunal and to the Parties a statement of his or her independence from the Parties and the Arbitral Tribunal. Within the time ordered by the Arbitral Tribunal, the Parties shall inform the Arbitral Tribunal whether they have any objections to the Tribunal-Appointed Expert's independence. The Arbitral Tribunal shall decide promptly whether to accept any such objection.
3. Subject to the provisions of Article 9.2, the Tribunal-Appointed Expert may request a Party to provide any relevant and material information or to provide access to any relevant documents, goods, samples, property or site for inspection.
    The authority of a Tribunal-Appointed Expert to request such information or access shall be the same as the authority of the Arbitral Tribunal. The Parties and their representatives shall have the right to receive any such information and to attend any such inspection. Any disagreement between a Tribunal-Appointed Expert and a Party as to the relevance, materiality or appropriateness of such a request shall be decided by the Arbitral Tribunal, in the manner provided in Articles 3.5 through 3.7. The Tribunal-Appointed Expert shall record in the report any non-compliance by a Party with an appropriate request or decision by the Arbitral Tribunal and shall describe its effects on the determination of the specific issue.
4. The Tribunal-Appointed Expert shall report in writing to the Arbitral Tribunal. The Tribunal-Appointed Expert shall describe in the report the method, evidence and information used in arriving at the conclusions.
5. The Arbitral Tribunal shall send a copy of such Expert Report to the Parties. The Parties may examine any document that the Tribunal-Appointed Expert has examined and any correspondence between the Arbitral Tribunal and the Tribunal-Appointed Expert. Within the time ordered by the Arbitral Tribunal, any Party shall have the opportunity to respond to the report in a submission by the Party or through an Expert Report by a Party-Appointed Expert. The Arbitral Tribunal shall send the submission or Expert Report to the Tribunal-Appointed Expert and to the other Parties.
6. At the request of a Party or of the Arbitral Tribunal, the Tribunal-Appointed Expert shall be present at an Evidentiary Hearing. The Arbitral Tribunal may question the Tribunal-Appointed Expert, and he or she may be questioned by the Parties or by any Party-Appointed Expert on issues raised in the Parties'

submissions or in the Expert Reports made by the Party-Appointed Experts pursuant to Article 6.5.
7. Any Expert Report made by a Tribunal-Appointed Expert and its conclusions shall be assessed by the Arbitral Tribunal with due regard to all circumstances of the case.
8. The fees and expenses of a Tribunal-Appointed Expert, to be funded in a manner determined by the Arbitral Tribunal, shall form part of the costs of the arbitration.

Article 7 – On Site Inspection

Subject to the provisions of Article 9.2, the Arbitral Tribunal may, at the request of a Party or on its own motion, inspect or require the inspection by a Tribunal-Appointed Expert of any site, property, machinery or any other goods or process, or documents, as it deems appropriate. The Arbitral Tribunal shall, in consultation with the Parties, determine the timing and arrangement for the inspection. The Parties and their representatives shall have the right to attend any such inspection.

Article 8 – Evidentiary Hearing

1. The Arbitral Tribunal shall at all times have complete control over the Evidentiary Hearing. The Arbitral Tribunal may limit or exclude any question to, answer by or appearance of a witness (which term includes, for the purposes of this Article, witnesses of fact and any Experts), if it considers such question, answer or appearance to be irrelevant, immaterial, burdensome, duplicative or covered by a reason for objection set forth in Article 9.2. Questions to a witness during direct and redirect testimony may not be unreasonably leading.
2. The Claimant shall ordinarily first present the testimony of its witnesses, followed by the Respondent presenting testimony of its witnesses, and then by the presentation by Claimant of rebuttal witnesses, if any. Following direct testimony, any other Party may question such witness, in an order to be determined by the Arbitral Tribunal. The Party who initially presented the witness shall subsequently have the opportunity to ask additional questions on the matters raised in the other Parties' questioning. The Arbitral Tribunal, upon request of a Party or on its own motion, may vary this order of proceeding, including the arrangement of testimony by particular issues or in such a manner that witnesses presented by different Parties be questioned at the same time and in confrontation with each other. The Arbitral Tribunal may ask questions to a witness at any time.
3. Any witness providing testimony shall first affirm, in a manner determined appropriate by the Arbitral Tribunal, that he or she is telling the truth. If the witness has submitted a Witness Statement or an Expert Report, the witness shall confirm it. The Parties may agree or the Arbitral Tribunal may order that

the Witness Statement or Expert Report shall serve as that witness's direct testimony.
4. Subject to the provisions of Article 9.2, the Arbitral Tribunal may request any person to give oral or written evidence on any issue that the Arbitral Tribunal considers to be relevant and material. Any witness called and questioned by the Arbitral Tribunal may also be questioned by the Parties.

Article 9 – Admissibility and Assessment of Evidence

1. The Arbitral Tribunal shall determine the admissibility, relevance, materiality and weight of evidence.
2. The Arbitral Tribunal shall, at the request of a Party or on its own motion, exclude from evidence or production any document, statement, oral testimony or inspection for any of the following reasons:
   (a) lack of sufficient relevance or materiality;
   (b) legal impediment or privilege under the legal or ethical rules determined by the Arbitral Tribunal to be applicable;
   (c) unreasonable burden to produce the requested evidence;
   (d) loss or destruction of the document that has been reasonably shown to have occurred;
   (e) grounds of commercial or technical confidentiality that the Arbitral Tribunal determines to be compelling;
   (f) grounds of special political or institutional sensitivity (including evidence that has been classified as secret by a government or a public international institution) that the Arbitral Tribunal determines to be compelling; or
   (g) considerations of fairness or equality of the Parties that the Arbitral Tribunal determines to be compelling.
3. The Arbitral Tribunal may, where appropriate, make necessary arrangements to permit evidence to be considered subject to suitable confidentiality protection.
4. If a Party fails without satisfactory explanation to produce any document requested in a Request to Produce to which it has not objected in due time or fails to produce any document ordered to be produced by the Arbitral Tribunal, the Arbitral Tribunal may infer that such document would be adverse to the interests of that Party.
5. If a Party fails without satisfactory explanation to make available any other relevant evidence, including testimony, sought by one Party to which the Party to whom the request was addressed has not objected in due time or fails to make available any evidence, including testimony, ordered by the Arbitral Tribunal to be produced, the Arbitral Tribunal may infer that such evidence would be adverse to the interests of that Party.

International Bar Association
271 Regent Street
London W1R 7PA
England

Telephone: +44 171 629 1206
Fax: +44 171 409 0456
Website: www.ibanet.org

List of Cases

A, B and C v. D, Schweizerisches Bundesgericht, 17 December 2002, 4P.196/2002 — 180
Arab Bank plc v. John D Wood (Commercial) Ltd [1998] EGCS 34 — 208
AT&T Corp and another v. Saudi Cable Co [2000] 2 All ER (Comm) 625 — 150
Austin Hall Building Ltd v. Buckland Securities Ltd [2001] BLR 272 — 109
Baese Pty Ltd v. R A Bracken Building Pty Ltd [1989] 52 BLR 130 — 45
Balfour Beatty Construction Limited v. Lambeth London Borough Council [2002] BLR 288 — 108, 125
Balfour Beatty v. Docklands Light Railway Limited [1996] 78 BLR 42 — 33
Bank Mellat v. Helleniki Techniki SA [1983] 3 All ER 428; [1984] QB 291 — 250, 296
Bank Saint Petersburg PLC v. ATA Insaat Sanayi ve Ticaret Ltd 2 March 2001, ASA 3/2001 531 — 177
Beaufort Developments (N.I.) Ltd v. Gilbert-Ash N.I. Ltd [1998] 2 All ER 778 — 34
Biotronik Mess-und Therapiegeraete GmbH & Co. v. Medford Medical Instrument Co. 415 F. Sup. 133, 140 (DNJ 1976) — 290
Birse Construction Limited v. St David Ltd [1999] BLR 194 — 101
Bramall & Ogden Ltd v. Sheffield City Council [1983] 29 BLR 73 — 39
Bremer Oeltransport GmbH v. Drewry [1933] 1 KB 753 — 285
British Airways Pension Trustees Ltd v. Sir Robert McAlpine & Sons [1995] 72 BLR 26 — 199, 201
British Sugar plc v. NEI Power Projects Limited [1997] 87 BLR 42 — 45
Bulgarian Foreign Trade Bank Ltd v. Al Trade Finance Inc, Swedish Supreme Court, 27 October 2000, Case No. T 1881-99 — 163
Cable & Wireless plc v. IBM United Kingdom Ltd [2003] BLR 89 — 57, 129

Cala Homes (South) Ltd v. Alfred McAlpine Homes East Ltd [1995]
FSR 818   208
Cellulose Acetate Silk Co Ltd v. Widnes Foundry (1925) Ltd
[1933] AC 20   45
Channel Tunnel Group Ltd v. Balfour Beatty Construction Ltd
[1993] AC 334   57, 85, 87, 255
Chen Hang Chu v. China Treasure Enterprise Ltd [2000] 2 HKC
814   304
Chromalloy Aeroservices v. The Arab Republic of Egypt 939 F.
Supp. 907 (DDC 1996)   284, 294
Corfu Channel Case (1948) ICJ Reps 15   206
Corporacion Transnacional de Inversiones SA de CV v. STET
International SpA [2000] 49 O.R. (3d) 414, O.J. No. 3408   306
Courtney & Fairbairn Ltd. v. Tolaini Brothers (Hotels) Ltd [1975]
1 All ER 716   59
Dallal v. Bank Mellat [1986] 1 All ER 239   296
Discain Project Services Ltd v. Opecprime Development Ltd [2001]
BLR 285   108, 109, 125
Edinburgh Magistrates v. Lownie [1903] 5 F (Ct of Sess) 71   151
Egmatra A.G. v. Marco Trading Corporation [1999] 1 Lloyd's
Rep 862   179
Elanay Contracts Ltd v. The Vestry [2001] BLR 33   109
Esso/BHP v. Plowman [1995] 128 ALR 391   78
Fidelitas Shipping Co Ltd v. V/O Exportchleb [1965] 2 All ER 4   296
George Fischer Holding Ltd (formerly George Fischer (Great
Britain) Ltd) v. Multi Design Consultants Ltd (Roofdec Ltd
and others, third parties) [1998] 61 Con LR 85   203
Gleeson (M J) (Contractors) Ltd v. London Borough of Hillingdon
[1970] 215 Estates Gazette 165   39
Glencot Development and Design Co Ltd v. Ben Barrett & Son
(Contractors) Ltd [2001] BLR 207   108
Great Eastern Hotel Co Ltd v. John Laing Construction Ltd [2005]
EWHC 181   209, 211
Guardcliffe Properties Ltd v. City & St James Property Holdings
[2003] EWHC 215 (Ch)   178
Hadley v. Baxendale [1843-60] All ER Rep 461   44
Hassneh Insurance Co of Israel v. Mew [1993] 2 Lloyd's Rep 243   78
Henderson v. Henderson (1843) 3 Hare 100   281
Henry Boot Construction (UK) Ltd v. Malmaison Hotel
(Manchester) Ltd [1999] 70 Con LR 32   233
HIM Portland, LLC v. Devito Builders, Inc., 211 F.Supp.2d 230
(1st Cir. 2002)   57
Hiscox Underwriting Ltd v. Dickson Manchester & Co Ltd [2004]
EWHC 479   174

List of Cases 417

Hitachi Limited et al. v. Mitsui & Company Deutschland and Rupali Polyester et al [1998] Supreme Court Monthly Review 1618-1687 — 287

Holland (John) Construction & Engineering Pty Ltd v. Kvaerner RJ Brown Pty Ltd [1996] 82 BLR 81 — 199

International Bechtel Co Ltd v. Department of Civil Aviation of the Government of Dubai 300 F. Supp. 2d 112 (DDC. 2004) — 284

Investment Pty Ltd v. Klockner East Asia Ltd. [1993] 2 HKLR 39 (Hong Kong) — 303

Iran Aircraft Industries v. Avco Corp., 980 F 2d 141 (2nd Cir. 1992) — 303

Jones v. Sherwood Computer Services plc [1992] 2 All ER 170 — 124

Kalmneft JSC v. Glencore International AG [2002] 1 All ER 76 — 182

Kemiron Atlantic, Inc. v. Aguakem International, Inc., 290 F. 3d 1287, 1290 (11th Cir. 2002) — 57, 87

London Borough of Hounslow v. Twickenham Garden Developments Ltd [1970] 3 All ER 326 — 108

London Borough of Merton v. Stanley Hugh Leach Ltd [1985] 32 BLR 51 — 201

London Underground Ltd v. Kenchington Ford plc (Harris & Sutherland, third party) [1998] All ER (D) 555 — 208

Longley (James) & Co Ltd v. South West Thames Regional Health Authority [1983] 25 BLR 56 — 207, 210

Macob Civil Engineering Ltd v. Morrison Construction Ltd [1999] BLR 93 — 108

Mertens v. Home Freeholds Co [1921] 2 KB 526 — 202

Mitsubishi Motors Corp. v. Soler Chrysler-Plymouth Inc. 473 US 614, 628 (1985) — 292

Munkenbeck & Marshall v. The Kensington Hotel Ltd [2000] All ER (D) 561 — 208

Myron (Owners) v. Tradax Export S A Panama City R P [1969] 1 Lloyd's Rep 411 — 147

National Justice Cia Naviera SA v. Prudential Assurance Co Ltd, The Ikarian Reefer [1995] 1 Lloyd's Rep 455 — 211

National Thermal Power Corporation v. The Singer Co 3 Supreme Court Cases (1992) 551-573 — 287

Northern Regional Health Authority v. Derek Crouch Construction Co. Ltd [1984] 26 BLR 1 — 34

Omnium de Traitement et de Valorisation v. Hilmarton, Cour de Cassation, 10 June 1997, YBCA XXII (1997), 696–698 — 294

Oxford Shipping Co Ltd v. Nippon Yusen Kaisha, The Eastern Saga [1984] 3 All ER 835 — 154

Pacol Ltd v. Joint Stock Co Rossakhar [1999] 2 All ER (Comm) 778 — 178

Pando Compania Naviera SA v. Filmo SAS [1975] 2 All ER 515 — 144

*Parsons & Whittemore Overseas Co. v. Societe Generale de
L'Industrie du Papier*, RAKTA 508 F 2d 969 (2d Cir 1974)   289, 301
Peak Construction (Liverpool) Ltd v. McKinney Foundations Ltd
[1970] 1 BLR 114   40
Porter v. Magill, Weeks v. Magill [2001] UKHL 67; [2002] 1 All
ER 465   150
Pozzolanic Lytag Ltd v. Bryan Hobson Associates [1999] BLR 267   210
R v .Gough [1993] 2 All ER 724   150
*Rapid Building Group Ltd v. Ealing Family Housing Association
Ltd* [1984] 29 BLR 5   40
RC Pillar & Sons v. Edwards and another [2001] All ER (D) 232   181
Royal Brompton Hospital NHS Trust v. Hammond (No 7) [2001]
EWCA Civ 206; 76 Con LR 148   207, 208
RSL (South West) Ltd v. Stansell Ltd [2003] EWHC 1390   125
*Rugby Landscapes Ltd (Bernhard's) v. Stockley Park Consortium
Ltd* [1997] 82 BLR 39   200
Ruxley Electronics and Construction Ltd v. Forsyth [1996] AC 344   203
SA Coppée Lavalin NV v. Ken-Ren Chemicals and Fertilizers Ltd
[1995] 1 AC 38   250
Saipem SpA & Ors v. Rafidain Bank & Ors [1994] CLC 252   50
*Sanghi Polyesters Ltd (India) v. International Investor KCSC
(Kuwait)* [2000] All ER (D) 93   178
Sellar v. Highland Railway Company [1919] SC (HL) 19   151
Shell UK Ltd v. Enterprise Oil plc [1999] 2 Lloyd's Rep 456   124
*Skanska Construction UK Ltd (Formerly Kvaerner Construction
Ltd) v. Egger (Barony) Ltd* [2004] EWHC 1748   211
Smith (Paul) Ltd v. H & S International Holding Inc [1991] 2
Lloyd's Rep 127   59
Société BKMI et Siemens c/ Société Dutco, Cour de Cassation,
7 January 1992, reported in Yearbook Commercial Arbitration
XVIII (1993), pp. 140-142   86, 165
Soleimany v. Soleimany [1999] 3 All ER 847   167, 306
Sonatrach Petroleum Corp v. Ferrell International Ltd [2002] 1 All
ER (Comm) 627   163
Stanor Electric Ltd v. R Mansell Ltd [1988] CILL 399   39
*Starrett Housing Corp. v. The Government of the Islamic Republic
of Iran*, Award No. 314-21-1...   206
*Surrey Heath Borough Council v. Lovell Construction Ltd and
Haden Young Ltd (third party)* [1990] 48 BLR 108   45
T Mackley & Co Ltd v. Gosport Marina Ltd [2002] BLR 367   96
Taieb Haddad and Hans Barett v. Société d'Invesstissement Kal,
Tunisian Cour de Cassation, 10 November 1993   299
Tate & Lyle Food and Distribution Ltd v. Greater London Council
[1982] 1 WLR 149   196

Temloc Ltd v. Errill Properties Ltd [1987] 39 BLR 30	45
Texaco Pananma Inc. v. Duke Petroleum Transport Corp., 3 September 1996 95 Civ. 3761 (LMM)	299
Unichips v. Gesnouin, Paris, 12 February 1993, [1993] Rev Arb 255	284
Virgilio De Agostini and Loris and Enrico Germani v. Milliol SpA, Pia and Gabriella Germani and Andrea De Agostini Corte di Appello di Milano, 24 March 1998	303
Walford v. Miles [1992] 2 AC 128	59
Wena Hotels Limited v. Arab Republic of Egypt (ICSID Case No. ARB/98/4)	286
Whalley v. Roberts & Roberts [1990] 1 EGLR 164	208
Wharf Properties Ltd v. Eric Cumine Associates (no 2) [1991] 53 BLR 1	199
Woods Hardwick Ltd v. Chiltern Air Conditioning [2001] BLR 23	125
Wraight Ltd v. PH & T Holdings [1980] 13 BLR 26	195
Zornow (Bruno) (Builders) Ltd v. Beechcroft Developments Ltd [1989] 51 BLR 16	39

Index

A
AAA. *See* American Arbitration Association (AAA)
Acceleration, 43, 197
Accepted industry practice, 204
Access to information, rights to, 96–97
Accounting expert, 183, 194, 204
Ad hoc arbitration
    appointing arbitrator, 148
    institutional versus, 81–82, 143–144
    number of arbitrators, 145–146
Adjudication
    advantages and disadvantages, 68–69
    dispute review boards, 64–68
    drafting provisions, factors to consider when, 70–72
    European Convention of Human Rights, 108–109
    panels of experts, 64, 67
    statutory, 72–76, 108–109
    timetable, 108
    United Kingdom experience, 72–76
Administration, 32–34
    claims management. *See* Claims administration

    construction contracts, 10
Admissibility, 171, 172, 173, 191
ADR. *See* Alternative dispute resolution (ADR)
Advance payment bonds, 47–48
AIA. *See* American Institute of Architects (AIA)
AIA Standard Forms, 16–17
    Standard Form of Agreement Between Owner and Design/Builder, 17
        turnkey contract, A191
        DB-1996, 23
    General Conditions of the Contract for Construction, A201-1997, 17
Alliancing. *See* Partnering and alliancing
Alternative dispute resolution (ADR), 57, 60, 127–142
    arbitration, interface with, 139–142
    conciliation. See Mediation
    early neutral evaluation, 62–63, 137–138
    early settlement as goal, 130
    mediation. See Mediation

mini-trial/executive tribunal, 62, 136–137
response to, 309–311
rules of civil procedure requirements, 128
selecting technique, 138–139
timing, 139–141
types, 129–138
Woolf Report, 127
American Arbitration Association (AAA), 63
    Construction Industry Disputes Review Board Procedures, 66
    Construction Industry Mediation Rules, 57
    Dispute Resolution Board Guide Specifications, 112–113, 337–348
    Dispute Review Boards, 112–113
    documents/evidence, 171
    hearings, 257
    roster of experienced persons, 112
    sample model arbitration clause, 318
    Three-Party Agreement, 112
    tribunal's duty to act expeditiously, 180–182
American Institute of Architects (AIA), 13
    Standard Forms. See AIA Standard Forms
Amman Convention, 308
Appointing arbitrator, methods of, 148–149
Arbitral awards, 12, 279–280
    bilateral conventions, 306–307
    challenge. See Challenging the award
    correcting errors, 286
    domestic, 297
    enforcement, 11, 76, 130, 295–298
    domestic, 297
    foreign, 297–298
    methods, 295–296
    New York Convention. See New York Convention
    recognition and, 296
    final, 179–180, 286, 295
    foreign, 297–298
    interim, 179, 286
    monetary, 280, 284
    multilateral conventions, 307–308
    notification, 283
    options for losing party, 283–284
    partial, 165, 279–280
    remitting, 293
    res judicata effect, 281–283
    setting aside, 308
    types, 280
    varying, 293
Arbitral tribunal, 144–152
    delay, claims based on, 218
    duties of arbitrators, 149
    duty to act expeditiously, 180–182
    duty to act fairly, 176–180
    evidence, 169–173
    experts, 173–174
    hearings, 257–258
    interim measures, 251–255
    language determined by, 146
    powers, source of, 161–163
    selection of, 143–152
    site visits, 255
    tribunal-appointed experts, 205–207
    ultra vires, 308
    witness examination, role in, 269–270
    written submissions, 168–169
Arbitration
    ad hoc. See Ad hoc arbitration
    ADR, interface with, 139–142

Index 423

advantages and disadvantages, 79–80, 315
benefit, 143
consolidation. See Consolidation
decisions, 9
expert determination, compared, 107
expert, role of, 205–209
    appointing, 205–206
    party-appointed expert, 205, 206–207
    tribunal-appointed expert, 205–207
institutional. See Institutional arbitration
joinder. See Joinder
litigation, or, 76–80
mediation, or, 61–62
party autonomy, 163–164
procedural rules. See Procedural rules
request for, 157–158
    form of, 12
timing of, 139–141
tribunal, selection of, 143–152
Arbitration clauses, sample model, 317–318
Arbitration provisions, drafting, 81–88
    continuing performance, 84
    institutional versus ad hoc arbitration, 81–82
    interim and injunctive relief, 85
    language, 83
    limitation periods, 86–87
    multiple parties or contracts, 85–86
    number of arbitrators, 83
    right to appeal, 83–84
    seat of arbitration, 83
Arbitrator(s)
    appointment, methods of, 148–149
    commercial background, 79
    duties, 11, 149–152
    ex parte orders, 253–254
    impartiality, 151
    independence, 151
    legal background, 147
    Med-Arb, 141–142
    mediator, as, 141–142
    multi-lingual, 146
    neutrality, 151
    number of, 145–146
    odd number, 145
    previous contact with party, 151
    sole, 145
    statement of independence, 150
Articles of agreement, 20
Assumptions, list of, 211, 213
Attendance at hearings, 272–273
Attorney. See Lawyer
Audit rights, 97–98
Authenticity of documents, 260–261
Award of costs, 183
Awards. See Arbitral awards

B
Bespoke contract, 8, 56
    Channel Tunnel, construction of, 56
    specific drafting, including, 109
Bilateral conventions, 306–307
Bilateral disputes, 152
Bilateral investment treaties (BITs), 88–89
Bill of quantities, 20–21
BITs. See Bilateral investment treaties (BITs)
Bonds, performance, 48–49
Briefing party-appointed experts, 209–211
Build only contract, 21–22, 23
Building and Construction Industry Security of Payment Act of 2004, 9
Building contract disputes, 196
Bundles, 259–260

C

Case management
  acting expeditiously, 180–182
  procedural fairness, 176–180
CCSJC. *See* Conditions of Contract Standing Joint Committee (CCSJC)
CECA. *See* Civil Engineering Contractors Association (CECA)
CEDR. *See* Centre for Effective Dispute Resolution (CEDR)
Centre for Effective Dispute Resolution (CEDR), 63
Certificates
  challenging, 34
  completion, 33, 38
  conclusive final certificate, 46–47
  final, 33
  handover, 33
  interim, 31
  interim payment, 33
  making good defects, of, 33
  non-completion, of, 33
  role of, 33–34
Challenging the award
  challenge defined, 284
  in court, 286–287
  prior exhaustion of other options, 285–286
  purpose, 284–285
  remedies available after successful challenge, 293–294
  UNCITRAL Model Law, 288–293
Change orders. *See* Variation orders
Channel Tunnel project
  adjudication, 72
  bespoke tiered dispute resolution provision, 56
  panel of experts, 65
  pricing methods, 30
Channel Tunnel Rail Link project
  panel of experts, 65

Chartered Institute of Arbitrators (CIA), 63, 147
China International Economic and Trade Arbitration Commission, 146, 148–149, 150
Chronology, 184–186
CIA. *See* Chartered Institute of Arbitrators (CIA)
CIETAC, *See* China International Economic and Trade Arbitration Commission
Civil Engineering Contractors Association (CECA)
  CCSJC, 16
Civil law arbitrators
  evidence, 170
Civil law codes
  reasonableness, 96
Civil law jurisdictions, 12
  arbitral tribunal, powers in, 162
  challenge, 284
  cross-examination, 269
  disclosure, 104
    documents, 244–246
    extent of, 241–242
  engineer, role of, 32
  evidence, preparation of, 242–243
  expert, 247
  fusion with civil law, 243–249
  hearings, 248
  hearings, tribunal, 175
  mediation, 11
  procedural approaches, 240–243
  recourse, 284
  remission, 294
  res judicata effect of award, 281
  submissions, exchange of, 248
  tribunal-appointed experts, 205
  witness statements, exchange of, 246–247
  written statements of case, exchange of, 244
  written submissions, 242, 263

Index

Claims administration, 10, 93
    audit rights, 97–98
    communication/reporting lines, provision for, 98–99
    conditions precedent to claims, 95–96
    early warning provisions, 94
    effective contract management, 102–104
    keep working provisions, 98
    partnering and alliancing, 99–101
    provision of particulars, requirement for, 94–95
    rights to access information, 96–97
Claims for additional time and cost, 193
Claims management. *See* Claims administration
Client. *See* Employer
Closing submissions
    fusion between civil and common law, 248–249
Collateral warranties, 21
Combined Dispute Boards, 66, 113
Commencement of arbitration, 143–159
Commercial agreements
    construction contracts differentiated, 7–9
Common law arbitrators
    evidence, 169
Common law jurisdictions, 12
    ADR requirement, 128
    ADR, enforceability, 128–129
    arbitral awards, quality control over, 282
    arbitral tribunal, powers in, 162
    arbitration models, 144
    challenge, 284
    closing submissions, 248–249
    cross-examination, 269
    delay, liability for, 46
    disclosure
        documents, 187, 244–246
        extent of, 241–242
    evidence, 170, 242–243
    expert reports, exchange of, 247
    extensions of time, 40
    fusion with civil law, 243–249
    hearings, 175, 248
    limitations of liability, 44
    liquidated damages for delay, 39, 40
    oath of witness, 268
    oral submissions, length of, 262
    order production of documents by third parties, 162
    party-appointed experts, 205
    procedural approaches, 240–243
    remitting award to tribunal for review, 293
    res judicata effect of award, 281
    role of tribunal in examination, 269
    security for costs orders, 249
    submissions, exchange of, 248
    sue for breach of contract, 46
    techniques, ADR, 11
    tribunal administering oaths, 162
    tribunal-appointed experts, 205
    witness statements, exchange of, 246–247
    written statements of case, exchange of, 244
    written submissions, 242
Communication/reporting lines, provision for, 98–99
Communications, improved, 311–312
Completion, 37–39
    certificates, 38
    date for, 38, 216–222
    defects liability period, 38
    handover certificates, or, 33
    liquidated damages for delay, 39
    practical, 37
    role of, 37

sectional, 38
staged, 38
substantial, 37
tests, 39
Conciliation. *See* Mediation
Concurrent delays, 42, 230–233
Conditions of Contract Standing Joint Committee (CCSJC), 15–16
Conditions precedent, 9, 10, 33, 37, 41, 46, 51, 93, 103, 249
    claims, to, 95–96
Confidentiality
    disclosure to mediator, 133
    documents, 102, 104
    drafting adjudication provisions, 70, 86
    hearings, 78
    lawyers/client, 242, 246
    mediation, 133
    trade secrets, 171
    video conferencing testimony, 272
Conflict of interest, 11
Consent
    consolidation by, 153–154
    drafting adjudication provisions, 70, 74
    environmental, 90
    to joinder, 152, 154
    licensing and administrative, 90
    to waive mediator confidentiality, 133
    to modify institutional rules, 82
    to period of extension, 123
    planning, 90
    public award, 76
    third party, 135
Conservation orders, 250–255
Consolidation
    advantages and disadvantages, 155
    consent of parties, 153–154
    court order, 154
    umbrella agreement, 154

Construction claims
    acceleration, 197
    additional cost, for, 193
    additional time, for, 193
    delay or disruption, 194–197
    disruption, 197
    evidence required, 193–203
    global claims, 193, 199–201
    head office overheads, for, 195–196
    increased costs in executing remaining works following delay, 197
    net lost profits, for, 196–197
    overheads, for, 195–196
    primary document, 193–194
    repudiation, contractors' claim for, 198–199
    site overheads and preliminaries, for, 195
    specific events, arising out of, 193
Construction contracts
    acceleration, 43
    administration, 10, 32–34
    build only, 21–22, 23
    certificates. See Certificates
    commercial agreements, differentiated, 7–9
    completion. See Completion
    completion date, 216
    concurrent delay, 42, 230–233
    construction management model, 24–25
    constructive acceleration, 43
    cost plus, 27
    critical path analysis, 41–42
    design and build, 23–24
    design team, 19
    design-bid-build, 21–22
    documents, 20–21
    extension of time. See Extension of time
    forms of, 21–27

Index

guaranteed maximum price (GMP), 28–29
independent engineer as administrator, 32–33
key players, 19
liability. See Liability
liquidated damages for delay, 39, 40
lump sum, 28
management contracting, 25–26
measured works, 29–30
milestone payments, 31
partnering and alliancing, 26–27, 99–101
payment arrangements, 31
performance failures, liquidated damages for, 43–44
pricing methodologies, 27–30
programme, role of, 41
progress payments, 31
project financed construction projects. See Project financed construction projects
project security. See Project security
provisional lump sum, 28
target cost, 27–28
turnkey, 23–24
unit price, 29–30
variation orders. See Variation orders
Construction disputes
construction contracts differentiated from commercial agreements, 7–9
dispute resolution generally, 9–12
extension of time claims, 36
variations, 36–37
Construction Industry Mediation Rules, AAA, 57
Construction management model, 24–25
Construction operations, defined, 73

Constructive acceleration, 43
Contemporaneous records, 8, 218, 227
Contra proferentem, 308
Contract management, 10
confidentiality of documents, 102
Contractor, 19
concurrent delay, 230–233
delay or disruption, claims for, 194–197
disruption, claims for, 197
increased costs in executing remaining works following delay, claims for, 197
net lost profits, claims for, 196–197
overheads, claims for, 195–196
ownership of float, 233–235
pay-when-paid provisions, 156
programme, 218–222, 226
project financed construction projects, 50–51
repudiation, claim for, 198–199
turnkey model, 23–24
Control of arbitration. *See* Arbitration
Convention on the Recognition and Enforcement of Foreign Arbitral Awards of 1958. *See* New York Convention
Cost plus contracts, 27
Construction costs
claims for additional time and cost, 193
construction claims
additional cost, for, 193
increased costs in executing remaining works following delay, 197
construction contracts
cost plus, 27
target cost, 27–28
increased, 101, 102, 197, 199, 202

increased costs in executing remaining works following delay, claims for, 197
inspection, 203
pricing methodologies
    cost plus, 27
    target cost, 27–28
prime cost, 29
prolongation cost items, 46
target cost contracts, 27–28
variation orders, 36
Credibility
    expert witness, of, 137, 207–208, 243, 246, 269
Critical path analysis, 41–42, 222–224
Cross examination of witness, 267
Cultural contributions, 312–313

D
Daywork, 29
Declaratory relief, 280
Defective works, 203
Defects liability period, 38
Delay, 215
    analysis, 222–223
    concurrent, 42, 230–233
    construction claims for, 193, 194–197
    extensions of time. See Extensions of time
    increased costs in executing remaining works following delay, 197
    liability for expense due to, 46
    liquidated damages for, 39
    programme. See Programme
    work-around measures, 42–43
Depreciation, 195
Design and build model, 23–24
Design build, conditions of contract for
    FIDIC Orange Book, 14
Design team, 19
Design-bid-build contract, 21–22

Directing tribunal-appointed experts, 212
Disclosure, 244–246
    arbitrators' statement of independence, 150
    contrasting civil and common law procedural approaches, 241–242
    Dispute Review Boards, 122
    document review and, 186–189
    fusion between civil and common law, 244–246
    scope, 187
Dispute Adjudication Boards, 66, 67. See also Dispute Review Boards
    comparison of rules and procedures, 114–121
    FIDIC, 111–112
    ICC, 113
Dispute avoidance, 9–10, 55–58
    advocating in reports, 68
Dispute Boards. See Dispute Adjudication Boards, Dispute Review Boards
Dispute resolution clauses
    drafting considerations, 319–322
Dispute resolution provisions, 9–12, 55, 57
    adjudication. See Adjudication
    governing law, 90
    jurisdiction and service abroad, 91–92
    management discussions, 57, 59–60
    state immunity, 90–91
Dispute Review Boards, 7, 10, 67, 105, 111
    AAA, 112–113
    adjudication, 64–68
    combined, 66, 113
    comparison of rules and procedures, 114–121
    decision, 123–125
        enforcing, 124–125

Index

jurisdiction, 124–125
lack of fairness, 125
status, 123–124
disclosure of documents, 122
enforceability of decisions, 11
enforcing decision, 124–125
FIDIC, 111–112
hearings, 122–123
ICC, 113
jurisdiction, 124–125
natural justice, rules of, 107–109
notice to refer, 106
powers, 106–107
preparation of submissions, 122
proceedings, 11
recommendation, 110, 113
reference, commencing the, 105–106
World Bank, 109–111
written submissions, 106
Disputes Review Expert, 110
Disruption
contractors' claim for, 194–197
liability for expense due to, 46
Document management
document review and disclosure, 186–189
electronic, 187–189
hearings, 190–191
paper based, 188, 191, 312
Scott Schedules, 191–193
Document review and disclosure, 186–189
Documents
admissibility, 191
authenticity, 260–261
bundles, 259–260
construction contract, in 20–21
presentation of documentary evidence, 261
primary, 193, 194
Domestic awards, 297
Draft mediation agreement, 133

Drawings, 20
Due process provisions
dispute review boards, 107–109
Due process rights, 176–180
equal treatment, 176
impartial and independent arbitrator, 176
party autonomy, 165
right to be heard, 176

E
Early neutral evaluation, 62–63, 137–138
Early warning provisions, 94
ECC. *See* ICE; Engineering and Construction Contract (ECC)
EDMS. *See* Electronic document management systems (EDMS)
Eichleay formula, 196
Electronic document management, 187–189
Electronic document management systems (EDMS), 187–189
market trends, 189
objective data fields, 188
software and web-based packages, 189
subjective data fields, 188
Emden formula, 196
Employer
construction contract, player in, 19
construction management model, 24–25
defective works, claims for, 203
delay claims, 201–203
design-bid-build contract, 21–22
engineer's liability to contractor, 8
independent engineer, appointment of, 32–33
loss of profits, claims for, 201–202
management contracting, 25–26

power of forfeiture, 8
representative as administrator, 33
repudiation by contractor, claims for, 202–203
turnkey contract, 23–24
wasted expenditure, claims for, 202
Employer's power of forfeiture, 8
ENAA. *See* ENAA Model Forms
ENAA Model Forms
   Model Form International Contract for Power Plant Construction, 15
      early warning provisions, 94
      mutual consultation, resolution by, 56
   Model Form International Contract for Process Plant Construction, 15
      turnkey contract, 23
ENE. *See* Early neutral evaluation
Energy Charter Treaty, 88, 89
Enforcement
   ADR, 128–129
   arbitral awards. *See* Arbitral awards
   challenging, 166
   dispute review board decisions, 124–125
   expert panels, 11
   New York Convention. *See* New York Convention
   recognition and, 296
   settlement agreement in mediation, 130
   tiered dispute resolution procedures, 57
Engineer
   Association of Consulting Engineers, 16
   decision maker in disputes, 64

independent engineer as administrator, 32–33
liability to the contractor, 9
project financed construction projects, 51
Engineering Advancement Association of Japan (ENAA), 13
   Model Forms. *See* ENAA Model Forms
English Arbitration Act, 1996
   arbitral tribunal, powers of, 162
   duty to act expeditiously, 180–182
   number of arbitrators, 145
   party autonomy, 163
Equal treatment, right to, 176–180
Ertan Project in Sichuan Province, China
   panel of experts, 65
European Convention of Human Rights
   restrictions on party autonomy, 166
   statutory adjudication, 108–109
Evidence, 12, 169–173
   admissibility, 171, 172, 173, 191
   chronology, 184–186
   civil and common law procedural approaches, 242–243
   collection, 183–213
   construction claims. *See* Construction claims
   delay or disruption, claims for, 194–197
   document management. *See* Document management
   documentary, 122, 173, 184, 190, 194, 199, 243, 260, 261
   employers' delay claims, 201–203
   expert. *See* Expert evidence
   factual, 204, 212, 213, 269, 271
   global claims, 199–201

Index

materiality, 171–172
new, 157, 176
oral evidence, 207, 213, 265, 267, 271
preparation, 183–213
primary documents, 193–194
relevance, 171–173
repudiation, claims for, 198–199
truthful, 268
weight, 171–173
Evidence in chief, 186, 263, 269
Ex parte hearings, 276–277
Ex parte interim relief, 252–253
Executive tribunal, 62
Exhibits to witness statements, 190
Expert conclaves, 213, 247
Expert determination, 2, 105, 107, 129
compared to arbitration, 107
Expert evidence, 204–213
appointing experts, 205–207
arbitration, role in, 205–209
briefing an expert, 209–212
form, 212–213
oral evidence, 213
party-appointed experts, 205, 206–207, 209–211
reports, 212–213
selecting an expert, 207–209
tribunal-appointed experts, 205–207
Expert panels
enforceability of decisions, 11
proceedings, 11
Expert reports, exchange of
fusion between civil and common law, 247
Expert testimony, 206
Expert witnesses, 173–174, 271–272
accounting, 183, 194, 204
appointing experts, 205–207
arbitration, role in, 205–209
briefing an expert, 209–212
credibility, 137, 207–209, 243, 246, 269

dispute review expert, 110
evidence. See Expert evidence
finding candidates, 207
party-appointed experts, 205, 206–207, 209–211
professional, 107, 207, 208
qualifications and expertise, 207–208
quantum, 183
recommendation, 110
selecting, 207–209
timing of appointment, 209
tribunal-appointed experts, 205–207, 212
Extended preliminaries calculation, 195, 196
Extensions of time, 40–41, 216–217
acceleration, 43
claims and disputes, 43
concurrent delay, 41
constructive acceleration, 43
critical path analysis, 41–42
programme, role, 41
work-around measures, 42–43

F
Factual evidence, 204, 212, 213, 269, 271
Factual matrix, 213, 217
Fast-track adjudication, 9
UK experience, 72–75
Fast-track procedures, 10
mandatory, 13
turnkey model, 23
Fédération Internationale des Ingénieurs-Conseils (FIDIC)
Conditions of Contract. See FIDIC Conditions of Contract
dispute adjudication boards, 111–112
FIDIC Conditions of Contract, 13–15
adjudication procedures, 66
Green Book, 14–15
Orange Book, 14

dispute boards, 111
Red Book, 13–14, 56
   access to information,
     rights to, 97
   adjudication, 66
   African Development Bank
     (AfDB), 14
   Asian Development Bank
     (ADB), 14
   build-only contract, 13–14
   Caribbean Development
     Bank (CDB), 14
   Commission of the
     European Communities
     (CEC), 14
   conditions precedent to
     claims, 95
   dispute review boards, 110
   European Bank for
     Reconstruction and
     Development (EBRD), 14
   European Investment Bank
     (EIB), 14
   Inter-American
     Development Bank (IDB),
     14
   International Bank for
     Reconstruction and
     Development (IBRD), 14
   United Nations
     Development Programme
     (UNDP), 14
Silver Book, 14, 56
   access to information,
     rights to, 96–97
   adjudication, 66
   conditions precedent to
     claims, 95
   early warning provisions,
     94
   notice of force majeure, 95
Yellow Book, 14, 56
   access to information,
     rights to, 97

   adjudication, 66
   conditions precedent to
     claims, 95
   turnkey contract, 23
Final awards, 179–180, 286, 295
Final certificate, 33
   conclusive, 46–47
Finance charges, 195
Finance of project, 10
   international projects, 8
Float, ownership of, 233–235
Foreign awards, 297–298

G
Geographical neutrality, 83
Global claims, 193, 199–201
   mini global claim, 195
Governing law, 90
Green Book. *See* FIDIC Conditions of
   Contract
Guaranteed maximum price (GMP)
   contracts, 28–29
Guarantees, 48–49
   parent company, 49
   payment, 49–50

H
Head office overheads, claims for,
   195–196
Hearings, 12, 175–176
   attendance, 272–273
   authenticity of documents,
     260–261
   bundles, 259–260
   cancellation penalties, 273–274
   conduct of, 257–277
   Dispute Review Boards,
     122–123
   document management,
     190–191
   documents, 259–261
   duration, 273
   ex parte, 276–277

Index

fusion between civil and common law, 248
hearing room, arrangement of, 275
logistics, 274–276
mediation, 134–135
oral, 133, 175
post-hearing matters, 176
PowerPoint presentations, 264
practicalities, 272–277
presentation of documentary evidence, 261
scheduling, 273–274
stenographers, 275
submissions. See Submissions
transcribers, 275
translators, 275–276
venue, 274–275
video conferencing of witness, 272
visual aids, 264
who organizes, 274
witnesses. See Witnesses
HKIAC. See Hong Kong International Arbitration Centre (HKIAC)
Hong Kong Airport project
panel of experts, 65
Hong Kong International Arbitration Centre (HKIAC), 63
Hostile local environment, 9
Hudson formula, 196

I
IBA Guidelines on the Conflicts of Interest in International Arbitration 2004, 151
IBA Rules on the Taking of Evidence in International Commercial Arbitration 1999 (IBA Rules), 172, 405–414
ICC. See International Chamber of Commerce (ICC)
ICE. See Institute of Civil Engineers (ICE)
ICSID. See International Centre for the Settlement of Investment Disputes (ICSID)
Impartiality of arbitrator, 11, 151, 176–180
Increased costs, 101, 102, 197, 199, 202
Independence of arbitrator, 151, 176–180
Independent engineer
construction contracts, administrator of, 32–33
Injunctive relief
arbitration provisions, drafting, 85
Institute of Civil Engineers (ICE), 13
Standard Forms
CCSJC, 16
Conditions of Contract, 15–16
Design & Construction Conditions of Contract (ICE D&C), 16
conditions precedent to claims, 95
turnkey contract, 23
Engineering and Construction Contract (ECC), 16
early warning provision, 94
New Engineering Contract (NEC), 16
Institutional arbitration, 11, 143
ad hoc versus, 81–82, 143–144
appointing arbitrator, 148
number of arbitrators, 145–146
tribunal, appointment of, 11
Instructions, 99, 102, 103, 107, 190, 211, 212
Insurance premiums, 203
Insurers
reporting requirements, 8
Interests based solutions, 142

Interim measures, 174–175, 250–255. *See* also Interim relief and Injunctive relief
    arbitral tribunal, 250–252
    national courts, 252
Interim payment certificates, 33
Interim relief, 251
    arbitration provisions, drafting, 85
    ex parte, 252–253
International arbitrators
    evidence, 170
International Centre for the Settlement of Investment Disputes (ICSID), 89, 273, 283, 286, 295, 307
International Chamber of Commerce (ICC), 63
    administrative expenses and fees, 82
    ADR Rules, 63, 64, 145, 148–149
    appealing arbitral award, 84
    appointment of arbitrator, 148
    awards, summaries or extracts from, 9
    Combined Disputes Board, 113
    confidentiality, 78
    Construction Arbitration Report, 185–186, 187, 190, 193, 240
    Dispute Adjudication Boards, 67, 113
    Dispute Board Clauses (ICC DB Clauses), 113, 349–369
    Dispute Board Rules (ICC DB Rules), 66, 67, 68, 349–369
    Dispute Boards, 113
    Dispute Review Board, 113
    evidence, 170
    hearings, 122, 175
    independence, 151
    interim relief, 251
    International Centre for Expertise, 207
    International Court of Arbitration Bulletin, 9
    joinder or consolidation, 85
    language, 146
    Model Dispute Board Member Agreement (DBMA), 63, 349–369
    modification of rules by parties, 82
    party autonomy, 163
    Report on Construction Industry Arbitrations, 7
    request for arbitration, 157
    sample model arbitration clause, 317
    Scott Schedules, 192–193
    service, 91
    sole arbitrator, 145
    statement of case, 122
    tribunal's duty to act expeditiously, 180–182
    written submissions, 168–169
International commercial arbitrations, 147, 172, 191, 250, 261, 266, 268, 269
International commercial disputes, 170
International construction arbitration
    documentary evidence, 184
    electronic document management, 187
    expert evidence, 204
    experts, 207
    preliminary issues, 249
    response to ADR, 309
International construction disputes
    complicating factors, 7–9, 152–153
    construction contracts differentiated from commercial agreements, 7–9
    parties from different jurisdictions, legal systems, cultures, 8

Index 435

International construction projects
    investment treaties, 10
International standard forms. *See*
  Standard Forms
Internationalization of arbitration, 12, 13
Investment treaties, 10

J
Joinder, 8, 11
    advantages and disadvantages, 155
    court order, 154
    lack of, 155–156
    late joinder of parties, 155
Joint report, 213, 271
Jurisdiction, 91–92

K
Keep working provisions, 98

L
Language of arbitration, selecting, 146
Lawyer
    arbitral tribunal, used in, 158
    arbitrator, 147
    lawyer/client confidentiality, 242, 246
LCIA. *See* London Court of
  International Arbitration (LCIA)
Legal costs
    award for, 183
    mediator, 132
    security for costs orders, 249
Lenders, 10
    interests of, 8
    project financed construction projects
        engineer, 51
        step-in rights, 51–52
    reporting requirements, 8
    security to, 9
Liability
    conclusive final certificate, 46–47
    defects, regard to, 45–46
    delay, 45
    employer's engineer to the contractor, 9
    expense due to delay and disruption, 46
    limit in monetary terms, 45
    limitations, 44
    loss, entitlement to, 46
    performance, 45
    prolongation cost items, 46
Limitation periods, 81, 86–87
Liquidated damages for delay, 39, 40
    performance failures, 43–44
Litigation
    advantages and disadvantages, 315
    arbitration, or, 76–80
    mediation advantages and disadvantages over, 61–62
    support agencies, 189
Local law, 13
London Court of International
  Arbitration (LCIA), 12, 63
    appealing arbitral award, 84
    appointment of arbitrator, 146
    evidence, 171
    independence and impartiality, 151
    interim relief, 251
    joinder or consolidation, 85
    jurisdictional approaches, 158
    language, 146
    modification of rules by parties, 82
    number of arbitrators, 145
    party autonomy, 163
    request for arbitration, 157
    sample model arbitration clause, 317
    selecting an arbitrator, 148–149
    site visits, 256
    sole arbitrator, 145

tribunal's duty to act
expeditiously, 180–182
written submissions, 168–169
Long-term concession
dispute resolution provisions, 57
Loss
consequential loss, 45
profits, claims for loss of,
201–202
Lost hire fees, 195
Lump sum contracts, 28

M
Management contracting, 25–26
Management discussions, 57
advantages and disadvantages,
59–60
Market trends
electronic document
management systems (EDMS),
189
Measured works. See Unit price
contracts
Med-Arb, 141–142
Mediation, 11, 60–62
AAA Construction Industry
Mediation Rules
57
advantages and disadvantages
over litigation and arbitration,
61–62
conducting procedure through
institution, 63
confidentiality, 133
draft mediation agreement, 133
early settlement as goal, 131
hearing, 134–135
settlement agreement
documenting, 135–136
enforceability, 130
timing, 139–141
Mediator, 61, 63
arbitrator as, 141–142
authority to settle, 133

choosing, 131–132
confidentiality, 133
cost, 132
position papers presented to, 134
prejudices, 132
pupil, 132
skills needed, 132
Meta-data, 188
Milestone payments, 31
Mini global claim, 195
Mini-trial/executive tribunal, 62,
136–137
MITs. See Multilateral investment
treaties (MITs)
Model Form International Contract
for Power Plant Construction. See
ENAA Model Forms
Model Form International Contract
for Process Plant Construction. See
ENAA Model Forms
Multi-lingual arbitrator, 146
Multi-tiered dispute resolution
provisions. See Tiered dispute
resolution provisions
Multilateral conventions, 307–308
1965 Washington Convention,
307
1975 Panama Convention, 307
Amman Convention, 308
Multilateral investment treaties
(MITs), 88–89
Multiple parties
arbitration agreement, 152
complications in construction
projects, 7–8
tribunal, selecting, 145

N
National arbitration laws
written submissions, 169
National courts, 80, 90, 91, 159
challenge of award, 285, 287
checks and balances, 254
fair hearing, 289

Index 437

interim measures, 252
interim relief, 85
public policy, 292
recourse, 144
security for costs, 250
support orders of arbitrators, 245
tribunals, 252
Natural justice, rules of
dispute review boards, 107–109
NEC. *See* Institute of Civil Engineers (ICE), Standard Forms
Net lost profits, claims for, 196–197
Netherlands Arbitration Institute
appointing an arbitrator, 149
Neutral
arbitrator, 151
assessors, 205
forum, 76
third party. See Mediator
New Engineering Contract. *See* Institute of Civil Engineers (ICE), Standard Forms
New York Convention, 371–376
challenge of award, 166
double-exequatur, 298
enforcement of arbitration awards, 76, 123, 130, 298, 299
formalities, 300
recognition and, 299
refusal, 301–306
resisting, 300–301
party autonomy, 164
reciprocity reservation, 299
recognition of awards, 299
refusal, 301–306
scope of application, 298–299
seat of arbitration, 78, 79, 83
Non-payment, right to stop work for, 9
North American Free Trade Agreement, 88
Notice to refer, 106
Notification of award, 283

O
Oral communications/instructions, 103, 254
Oral evidence, 207, 213, 265, 267, 271
Oral hearings, 133, 175
Orange Book. *See* FIDIC Conditions of Contract
Overheads, claims for, 195–196
Ownership of float, 233–235

P
Panama Convention of 1975. *See* Multilateral Conventions
Panel of experts
decision makers in disputes, 64–68
reference, commencing, 105–106
Paper based document management, 188, 191, 312
Parent company guarantees, 49
Partial awards, 165, 279–280
Parties' representative
oral evidence, 213
selection of, 158
Partnering and alliancing, 26–27, 99–101
Party autonomy, 163–164
restrictions, 165–167
Party-appointed experts, 205, 206–207
briefing, 209–211
Pay-when-paid provisions, 9, 156
Payment guarantees, 49–50
Performance bonds and guarantees, 48–49
Pleadings, 62, 122, 178, 181, 183, 184, 185, 186, 190, 191, 192, 199, 200, 244, 259, 263, 267, 308
Position papers
mediation, function, 134
mini-trial, 136
Possession of works, 218, 220
Post-hearing matters, 176
PowerPoint presentations, 264

Practical completion, 37, 38
Practice and precedents, 196
Pre-hearing conference, 205, 210
Precedent, 281, 282
    practice and, 196
Prejudice, mediator, 132
Preliminary issues, 249
Preliminary meeting, 168
Preliminary steps, 168
Preparation of witnesses
    contrasting civil and common law procedural approaches, 243
Preservation orders, 250–255. *See also* Interim measures
Pricing methodologies, 27–30
    cost plus, 27
    guaranteed maximum price (GMP), 28–29
    lump sum, 28
    measured works, 29–30
    provisional lump sum, 28
    standard methods of measurement, 29
    target cost, 27–28
    unit price, 29–30
Prime cost, 29
Principal. *See* Client
Procedural rules
    civil and common law approaches, contrasting, 240–243
    closing submissions, 248–249
    derivation, 238–240
    disclosure
        documents, 244–246
        extent of, 241–242
    evidence, preparation of, 242–243
    exchange of written statements of case, 244
    expert reports, exchange of, 247
    fusion between civil and common law, 243–249
    hearings, 248
    interim measures, 250–255
    objective, 237–238
    preliminary issues, 249
    security for costs, 249–250
    site visits, 255–256
    submissions, exchange of, 248
    witness statements, exchange of, 246–247
    written submissions, 242
Procedure, 12
Professional expert witnesses, 107, 207, 208
Programme
    accuracy, 220
    as-built, but for, 226–227, 228–229
    as-planned vs. as-built, 224–225, 228–229
    as-planned, impacted, 225–226, 228–229
    choosing method of analysis, 228–229
    concurrent delay, 230–233
    contractor, 218–222, 226
    critical path analysis, 222–224
    date for completion, 216–222
    delay analysis, 222–223
    detailed, 219–222
    e-mail traffic, 218–219
    original, 222, 223, 224, 225
    ownership of float, 233–235
    role of, 41, 216–222
    time impact, 227–229
Progress payments, 31
Project financed construction projects, 10, 50–51
    contractor, 50
    lenders' engineer, 51
    lenders' step-in rights, 51–52
    project security, 52–53
    shareholders, 50
    variation orders, 52
Project management
    programme, 216–222

Project security, 50
    advance payment bonds, 47–48
    letters of credit, 50
    parent company guarantees, 49
    payment guarantees, 49–50
    performance bonds and
        guarantees, 48–49
    project financed construction
        projects, 52–53
    retention funds, 48
    trust funds, 50
    types, 47
Prolongation cost items, 46
Property value, diminution of, 203
Provision of particulars, requirement
    for, 94–95
Provisional lump sum, 28, 29
Public policy, 165, 166, 167, 180, 281,
    291, 292, 305, 306, 308
Punitive damages, 280
Pupil mediator, 132

Q
Quantum experts, 183
Quantum meruit, 30, 193

R
Reasonableness, 95–96
Recognition and enforcement, 296
    New York Convention, 299,
        301–306
Recommendation, 110, 113
Red Book. *See* FIDIC Conditions of
    Contract
Report on Construction Industry
    Arbitrations (ICC), 7
Reporting requirements
    insurers, 8
    lenders, 8
Repudiation
    contractor, by, 202–203
    contractors' claim for, 198–199
Request for arbitration, 157–158,
    168–169

Res judicata effect of award, 281–283
Retention funds, 48
Rights based analysis, 142
Rights to access to information, 96–97
Rolled-up claims, 103, 199

S
Sample model arbitration clauses,
    317–318
SBDW. *See* Standard Bidding
    Documents–Procurement of Works
    (SBDW)
Scott schedules, 12, 191
Seat of arbitration, 13, 78, 79, 83, 158,
    240, 272
Sectional completion, 38
Security
    costs, for, 249–250
    lenders, 9
    project. *See* Project security
Service abroad, 91–92
Settlement agreement
    documenting, 135–136
    mediation, 130
Shareholders
    project financed construction
        projects, 50–51
Silver Book. *See* FIDIC Conditions of
    Contract
Site access, 97
Site diary, 185
Site overheads and preliminaries,
    claims for, 195
Site visits, 255–256
Society of Construction Law's Delay
    and Disruption Protocol, 228
Sole arbitrator, 145
Specification, 20
Staged completion, 38
Standard Bidding Documents–
    Procurement of Works (SBDW), 66,
    110–111
Standard form contracts, 13
    adjudication provisions, 66

multi-tiered dispute resolution provisions, 10
practical completion, 216
proceedings before dispute review boards or expert panels, 11
Standard Forms
   AIA. See AIA Standard Forms
   FIDIC. See FIDIC Conditions of Contract
   ICE. See Institute of Civil Engineers (ICE), Standard Forms
Standing panel. See Panel of experts
Stare decisis, 281
State immunity, 90–91
Statement of independence, 150
Statutory adjudication, 72–76, 108–109
Stenographers, 275
Submissions
   approach to, 263
   closing, 248–249
   Dispute Review Boards, 122
   fusion between civil and common law, 248
   length of oral submission, 262–263
   post-hearing, 263
   visual aids, 264
   who may appear, 261–262
   written, 106, 263
Substantial completion. See Practical completion
Swift solutions, 10

T
Target cost contracts, 27–28
Technology, introduction of new, 311–312
Terms of reference, 169, 173, 184, 205, 206, 239, 244, 290
Test of reasonableness, 95–96
Tests on completion, 39

Third parties, 11
   award to, 78
   binding, 167, 251, 252
   claim against, 78
   disputes, 62
   impartial, 60
   independent expert, 137
   interest in outcome, 133
   joinder, 152
   lack of fairness, 125
   Med-Arb, 141
   neutral, 63
   provision of particulars, 94
   state court judgments, 80
   video conferencing, 272
   witnesses, 162
Third party security, 47, 50
Three-Party Agreement, 112
Tiered dispute resolution procedures, 10
   adjudication. See Adjudication
   enforceability, 57
   mandatory discussions, 58–60
Time limits, 86, 96, 103, 123, 145, 180, 181, 249, 261, 262, 287
Time line, 184
Timing of appointment, 209
Transcribers, 275
Translators, 275–276
Tribunal
   arbitral. See Arbitral tribunal
   mini-trial/executive tribunal, 62
Tribunal-appointed experts, 205–207
   directing, 212
Truthful evidence, 268
Tunnel projects
   dispute review boards, 109
   panel of experts, 65
Turnkey, 8, 23–24
   build only and, 22
   contract programme, 41
   contract, ENAA, 15
   FIDIC orange Book, 14
   lump sum contracts, dispute in, 28

Index

U

UK Housing Grants Construction and Regeneration Act of 1996, 9, 72–76
    adjudication timetable, 108
Umbrella agreement, 154
UNCITRAL. *See* United Nations International Trade Law branch (UNCITRAL)
Unfair Contract Terms Act (UCTA) of 1977
    test of reasonableness, 95–96
Unforeseen physical conditions, 9
Unit price contracts, 29–30
United Nations International Trade Law branch (UNCITRAL)
    Arbitration Rules, 89, 148–149, 150
        ad hoc arbitration, 81, 82, 148
        appointment of arbitrator, 149
        evidence, 171
        hearings, 257
        impartiality, arbitrators', 150
        interim relief, 251
        language, 146
        number of arbitrators, 146
        tribunal, powers of, 162
        written submissions, 169
    Conciliation Rules, 63
    Model Law on International Commercial Arbitration, 77–78, 130, 142, 146, 377–403
        challenging award, 288–293, 301–306
        evidence, 171–172
        interim measures, 252–255
        party autonomy, 163
    Notes on Organizing Arbitral Proceedings, 164, 172, 239

V

Variation orders, 35–37
    controlling order and costs, 36
    disputes, 36–37
    project financed construction projects, 52
    requirement, 35–36
Video conferencing at hearing, 272
Visual aids in submissions, 264

W

Washington Convention of 1965. *See* Multilateral Conventions
Wasted expenditure, claims for, 202
Web-based packages, 189
Witness statements
    exhibits to, 190
    fusion between civil and common law, 246–247
Witnesses
    conferencing, 270–271
    contrasting civil and common law procedural approaches, 243
    cross examination, 267, 269
    direct examination, 267, 269
    examination, 267–270
    expert. See Expert witnesses
    oath, 268
    preparation, 265–267
    re-examination, 267
    scope of examination, 270
    statements, 264–265
    truthful evidence, 268
    video conferencing, 272
Woolf Report, 127
Working files, 190
World Bank
    dispute review boards, 109–111
    Standard Bidding Documents for the Procurement of Works, 66, 110–111, 323–335
Written statements of case, 244

Written submissions, 168–169, 263
   civil and common law
      procedural approaches, 242

Y
Yellow Book. *See* FIDIC Conditions
   of Contract